Categorical data analysis for
geographers and environmental scientists

Categorical data analysis for geographers and environmental scientists

Neil Wrigley
University of Bristol

Longman
London and New York

Longman Group Limited
Longman House, Burnt Mill, Harlow
Essex CM20 2JE, England
Associated companies throughout the world

**Published in the United States of America
by Longman Inc., New York**

© Longman Group Limited 1985

First published 1985

British Library Cataloguing in Publication Data

Wrigley, N.
  Categorical data analysis for geographers
  and environmental scientists.
  1. Mathematical statistics
  I. Title
  519.5   QA276

0-582-30127-0

Library of Congress Cataloging in Publication Data

Wrigley, Neil.
  Categorical data analysis for geographers and environmental scientists.

  Bibliography: p.
  Includes index.
  1. Geography—Statistical methods. S. Environmental
protection—Statistical methods. I. Title.
G70.3.W74 1984   519.5'3'02491   83–18771
ISBN 0–582–30127–0

set in 10/11 pt. Linotron 202 Times
Printed in Great Britain at
The Pitman Press, Bath

For

Susan and Hannah

CONTENTS

PREFACE

In 1971, as a postgraduate student at the University of Cambridge, I was faced with the problem of using a multiple regression model in which my response/dependent variable took only a limited number of values. Whilst searching through the available literature on the issue, I came, almost simultaneously, across three remarkable items published in the previous eighteen months: Henri Theil's paper 'On the estimation of relationships involving qualitative variables' published in the *American Journal of Sociology*, David Cox's invaluable monograph on *The Analysis of Binary Data*, and James Grizzle, Frank Starmer and Gary Koch's paper on the 'Analysis of categorical data by linear models' published in *Biometrics*. It was immediately clear to me that something exceptional was beginning to take place in the statistical analysis of categorical data which, at the very least, required me to completely rethink the arbitrary divisions between statistical methods for low and high order data which I had been taught as an undergraduate. I rapidly became hooked on the concept of an intrinsic unity between conventional continuous linear models and their categorical data analogues, and this conviction was strengthened as I delved deeper into the subject and began to read the influential work of Leo Goodman, and the earlier papers of such scholars as Berkson, Birch, and Mantel. I began to use the new methods in my own Ph.D. research and, as no appropriate computer software was available, wrote my own multiple response category logistic/logit model program using numerical optimization routines from the Harwell subroutine library.

By the time I moved into university lecturing at the University of Southampton in 1973 and began to inflict logistic/logit, log-linear models, etc. on my students and colleagues, the literature on categorical data analysis was beginning to mushroom at a truly extraordinary rate. A younger generation of statisticians and econometricians which included Fienberg, Bishop, Holland, Koch, Haberman, and McFadden were beginning to produce a series of exceptional papers and books, and Nelder and Wedderburn had, in 1972, spelt out the unity of the family of 'generalized linear models'. Scarcely an issue of any major statistical journal passed without some further article advancing the new approaches, and I was convinced of the immense potential of these methods for social and environmental scientists. As such, I resolved to hang on to the coat tails of this rapid advance in statistical methodology for as long as possible, and to continue to teach the methods to my undergraduates and postgraduates. During the intervening ten years this is essentially what I have done and, at various stages, the editors of several geography journals have allowed me the opportunity to summarize my impressions of these developments for a wider audience.

This book has developed directly out of those earlier summaries, particularly the encouraging response to my paper in *Progress in Human Geography* in 1979. It has taken a long time to write and has grown to a monumental size. Nevertheless, I make no apologies for its length for I believe that there are considerable pedagogic advantages in attempting to survey all the major features of categorical data analysis within a unified framework. As such, the book sets out to be much more comprehensive in scope than existing texts in the field. In addition it is designed specifically to meet the needs of geographers and environmental scientists. It, therefore, contains extensive discussion of more than forty empirical illustrations which are designed to form an integrated part of the text, and includes a major section on the use of categorical data models in the research field which has become known as 'discrete choice modelling'. Furthermore, it attempts to be sensitive to the likely statistical backgrounds of geographers and environmental scientists and, as such, devotes considerable space to clarifying statistical and notational concepts which are simply assumed by those textbooks written primarily for statisticians.

The structure of the book and the material presented within it reflect more than ten years of teaching such topics within quantitative geography courses. To Peter Haggett and my colleagues at Bristol I owe a considerable debt for providing the intellectual environment in which such courses are feasible. I am also indebted to R.G. Golledge and T.R. Lakshmanan for giving me the opportunity to hone the structure of the book within courses I have taught at the University of California, Santa Barbara and the University of Boston, and to the ESRC for sponsoring the Bristol research training courses on 'Data Collection and Analysis for Geographical Research' at which I have presented much of this material. I am also grateful to David Hensher for keeping me in touch with many developments in discrete choice modelling, to John Davis of the Kansas Geological Survey for providing the oil well data used in Examples 2.1, 2.4, 2.5 and 2.7, to several British statisticians to whom I have turned for advice, and to a group of my own postgraduate students and research assistants: Larry O'Brien, Paul Longley, Leigh Griffin, Kelvyn Jones, Richard Dunn, Mark Uncles, Steven Reader and Bill Halperin who have contributed to the development of the book in many ways. Finally, I owe thanks to Longman for taking on and overcoming the major problems which a book of this nature and size creates for a publishing house.

The book is dedicated to my wife for her assistance and support, and to our daughter who was born as I began to write it and has gracefully accepted it as part of her life.

Neil Wrigley
Bristol, Spring, 1983

ACKNOWLEDGEMENTS

The Author and Publishers wish to express their thanks to the following for permission to reproduce material from copyright works:

The Biometric Society for Fig. 8.3 from Fig. 3 p. 491 (D. Pregibon, 1982); The Econometric Society for Table 4.8, Fig. 4.1 from Table 1, Fig. 1B pp. 1092, 1084 (M. M. Li, 1977); Geological Survey of Canada for Table 2.5, Fig. 2.9 from Table 1, Fig. 2 p. 6 (C. F. Chung, 1978); Institute of British Geographers for Figs. 8.6, 8.7 from Figs. 3, 4a pp. 136–7 (N. Wrigley, 1977); International Economic Review for Table 3.1 from Table 10, p. 480 (P. Schmidt & R. P. Strauss, 1975a); Ohio State University Press for Figs. 8.8a,b from Figs. 2, 4 pp. 333–4 (J. Odland & R. Barff, 1982); Pergamon Press Ltd for Fig. 2.11 from Fig. 1, p. 316 (F. Southworth, 1981); Pion Ltd for Fig. 1.1 from Fig. 3.1, p. 53 (P. Haggett, 1981); Pion Ltd & the author, S. R. Lerman, for Table 10.4 adapted from Tables 8, 9 pp. 742–3 (E. J. Miller & S. R. Lerman, 1981); Royal Dutch Geographical Society for Fig. 8.9 from the article 'Scaling residential preferences: a methodological note' by P. A. Longley & N. Wrigley in *Tijdschrift voor Economische en Sociale Geographie* 1984; John Wiley & Sons Inc. for Fig. 2.6 from Fig. 7.3, p. 196 (J. W. Harbaugh, J. H. Doveton & J. C. Davis, 1977).

PART 1

Some essential preliminaries

CHAPTER 1

Introduction

Pick up any statistics or quantitative social science journal of the late 1970s or early 1980s and it will not be long before you encounter the terms and acronyms: logistic model; logit model; probit model; log-linear model; multidimensional contingency table; the GSK approach; iterative proportional fitting; discrete choice modelling; product multinomial sampling schemes; GLIM; ECTA; GENCAT; QUAIL; BLOGIT; and so on. These are the visible signs of a revolution which has swept through an area of statistical methodology and which has transformed the practice of data analysis in an important area of social science research. This book is an account of this revolution and those models.

Such is the importance of these new methods that the need for introductory accounts, oriented to the needs and mathematical back-grounds of social scientists, has been recognized for a number of years. In response have come books such as those by Fienberg (1977, 1980), Everitt (1977), Reynolds (1977), Upton (1978a) and Haberman (1978, 1979). This book differs from those in four important ways.

1. It is designed specifically to meet the needs of the geographer. Not only does this imply examples which stress the geographer's concern with location, place, environment, region, distance, destination, spatial structure, spatial relationships, spatial interaction, and so on, but also the need to take account of geography's traditional bridging role between the social and environmental sciences. A glance at the listing of examples in the contents pages will indicate to the reader that this concern has played an important role in the design of the book, and the book is offered as a text for geographers *and* environmental scientists.

2. The book must take account of geography's well-established quantitative tradition stretching back to the so-called 'quantitative revolution' of the 1960s (see Wrigley and Bennett 1981 for a review) and the strong links in that tradition with econometric modelling (e.g. Cliff and Ord 1973, 1981; Hepple 1974; Bennett 1979; Wrigley 1979b; Haining 1980). The result of this is that most geographers will come to this book having completed a multivariate statistics course oriented to the 'general linear model' (e.g. Johnston 1978; Mather 1976). As such, the book is designed as a logical sequel to such courses and texts and this is reflected in the structure of presentation adopted.

3. It is a consequence of geography's bridging role, both between the social and environmental sciences and within the social sciences, that the quantitative methods used by geographers in their research often cover a much broader range than those in, say, sociology, political science, geology or even economics. As such, this book aims to be much more comprehensive in scope than existing texts in the field. Such texts often concentrate upon one particular aspect of the developments in categorical data analysis (e.g. log-linear modelling) in order to meet the needs of certain social science disciplines with, traditionally, a much narrower focus in their use of quantitative methods.

4. Finally, for many geographers, the developments in statistical methodology which underlie the new approaches to categorical data analysis cannot be divorced from a linkage of these developments with associated developments in micro-economic theory. This linkage has produced a new research field which is of considerable importance in transportation science, economics, and human (particularly urban) geography, and which is referred to as 'discrete choice modelling' (see Hensher and Johnson 1981a; Manski and McFadden 1981; Williams 1981a; Wrigley 1982). Many geographers have made important contributions in this area of research, and their students are likely to have first encountered the elements of the new approaches to categorical data analysis (logit models, probit models, etc.) via this research. In addition, they will have been exposed to the specialized terminology which has developed in this area. Existing textbooks in the field of discrete choice modelling (e.g. Hensher and Johnson 1981a) are not oriented to the task of providing an overview of the categorical data models which they utilize, neither do existing categorical data textbooks pay more than passing attention to this major area of empirical application. Consequently, it falls to a quantitative geographer and to this book to provide this linkage. As such, the book can be regarded, in part, as providing the statistical background to further, more specialized, courses in discrete choice modelling.

1.1 Classifications, categorical data and a methodological transformation

Classifications are a familiar part of everyday life. Individuals are classified by sex, marital status, nationality, occupation, political affiliation, etc. Places are classified by country, region, locality, etc. Our opinions and attitudes are sought by political pollsters, market research firms, pressure groups, etc. and we are classified as pro-Government or anti-Government, against nuclear weapons or in favour of nuclear weapons, extrovert or introvert, and so on. Some of these classifications (e.g. sex of the individual) define just two exhaustive and mutually exclusive categories (male/female). Others (e.g. nationality, occupation) used singly, or simultaneously in the form of a cross-classification, define multiple categories.

Data which consist of *counts* of the number of individuals/households/places in particular categories, or of simple indentifications of which category each individual/household/place belongs to, are known collectively as *categorical* data. If the categories are recognized as standing in some kind of relationship to each other then a ranking or ordering of the categories is possible (e.g. low/average/high socio-economic status) and the data can be said to be measured at the ordinal scale. If, on the other hand, the categories are unordered then the data can be said to be measured at the nominal scale. Categorical data are, therefore, data measured at the low level, nominal or ordinal scales. Sometimes the terms qualitative, non-metric or discrete are used to describe such data, but whatever term is adopted it is used essentially to distinguish the low-level qualitative measurement characteristic of the social sciences (and to a lesser extent environmental sciences) from the high-level quantitative measurement characteristic of the physical sciences.

Traditionally, the analysis of categorical data is an area where statistical methodology has been weak. Despite the fact that much of the data collected and analysed in the social, environmental and medical sciences is of a categorical type, it was not until the late 1960s that an integrated approach to the analysis of such data began to emerge. At that time several influential statements on the analysis of categorical data were published in quick succession in Britain and the United States by David Cox (1970), Henry Theil (1969, 1970), Leo Goodman (1968, 1970; 1971a, b, 1972a, b), James Grizzle, Frank Starmer and Gary Koch (1969). These succeeded in reaching and influencing a wide audience, and alerted that audience to the important contributions of scholars such as Berkson, Birch, Good, Kastenbaum, Lancaster, Mantel and Plackett. At the same time, doctoral dissertations or early research papers on the topic were completed at Harvard, Berkeley and Chicago by Stephen Fienberg (1968), Yvonne Bishop (1967), Dan McFadden (1968) and Shelby Haberman (1970) and this group was to go on to produce equally influential books and papers on the topic in the mid-1970s (Bishop, Fienberg and Holland 1975; Fienberg 1977, 1980; McFadden 1974; Domencich and McFadden 1975; Haberman 1974a, 1978, 1979). Suddenly, the analysis of categorical data became one of the most rapidly developing research frontiers in statistical analysis, and in the 1970s scarcely an issue of any major statistical journal passed without some further article advancing the new approaches.

In retrospect, it is now clear that the major advances achieved since the late 1960s have succeeded in creating a long-awaited unified approach to the analysis of categorical data, and that this unified approach has significant advantages over traditional methods. Amongst the most important of these are that:

(a) it links the analysis of categorical data to the general linear model;
(b) it provides a comprehensive and unified scheme for the analysis of multidimensional contingency tables;
(c) it allows the development of computer programs comparable in generality to those available for the analysis of linear models;
(d) it demonstrates how many statistical procedures seen and taught as distinct entities can be viewed as part of an integrated and powerful general methodology;
(e) it provides the basis of the significant progress which has been made in developing traditional micro-economic consumer theory to encompass choice among discrete alternatives.

In the light of these advantages, many recent commentators on the advances in categorical data analysis take the view that these new methods do not merely supplement traditional methods, but make obsolete and replace many of the older methods. For example Duncan (1974) (see also Maxwell in Everitt 1977, p. vii) has stated:

these methods solve the problems in a definitive way. That is, it is possible to recognise that the solutions actually supersede and do not merely compete with previous procedures, recipes and rules of thumb.

In this book, we accept and build upon the spirit of such statements.

Given the characteristic low level of measurement of much social science data, particularly social survey data, the unified approach to categorical data analysis has proved to be particularly appropriate to the needs of the social scientist and recognition of its importance has increased steadily throughout the 1970s and early 1980s. In geography, however, the new methods did not begin to play a major role until the late 1970s. One reason for this is that spatial-time series and spatial process modelling continued to be the dominant focus of statistical modelling in geography throughout most of the 1970s (such work is best summarized in Haggett et al. 1977; Bennett 1979; Cliff and Ord 1981). Such work, with its links to econometrics, control engineering and stochastic process theory, was far removed from the practical pressures to analyse large quantities of social survey data and/or to examine multidimensional contingency tables which was leading to the rapid acceptance of the new categorical data methods in some of the other social science disciplines. In addition, much spatial analysis in geography continued to assume the availability of data measured at high levels, and many spatial analysts were, at best, somewhat distrustful of low-order data, the analysis of which they associated with non-parametric methods; methods which they regarded with some suspicion and which they saw as lacking an integrated structure.

Despite this, an overall view of the 1970s and early 1980s shows a gradual and, in recent years, increasingly rapid shift in quantitative geography towards an increased concern with what has variously been termed low-order, non-metric, categorical or soft data, and most geographers now acknowledge, as Nijkamp (1982) has noted, that: 'In practical research, soft information is very often the rule rather than the exception.' The trend can be summarized using a reformulation of the classical geographical data cube suggested by Haggett (1981). In this cube shown in Fig. 1.1 we can point to a gradual shift which has taken place in the centre of gravity of quantitative geography over the past ten years, away from the cell labelled $S_4 T_4 D_4$ in which space, time and data are all measured in continuous terms at the metric level, and towards the cell labelled $S_1 T_1 D_1$ in which space, time and data are all measured in categorical terms at the nominal scale. There are many aspects of this shift: the increasing use of multidimensional scaling (see Gatrell 1981a for a review); the introduction of Q-analysis (see Chapman 1981; Gould 1980; Gatrell 1981b; Johnson and Wanmali 1981); the development of non-parametric matrix comparison and evaluation procedures (see Hubert and Golledge 1981, 1982; Halperin et al. 1984), and so on (see Wrigley 1981b for further comments). For our purposes, however, the most important aspect of this shift is the increasingly wide use in geography of the integrated family of statistical

Figure 1.1
STD cube showing the interrelationships of space, time and data at each of four measurement levels (after Haggett 1981)

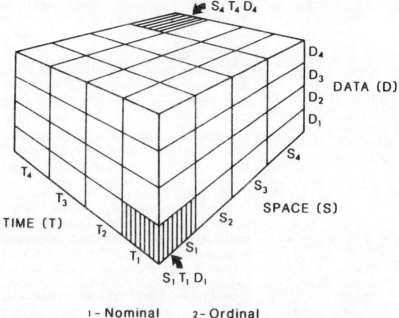

1 – Nominal 2 – Ordinal
3 – Interval 4 – Ratio

models for categorical data; the log-linear, logistic, logit, probit and related models which will be the concern of subsequent chapters. It now appears certain that the use of these models and associated computer software will increase rapidly in geographical research in the 1980s and the book is offered as a guide to those geographers and environmental scientists who wish to follow this path in their own research and/or to evaluate the work of others who have adopted these methods.

1.2 A framework for discussion

A useful framework for understanding the new approaches to categorical data analysis is that shown in Table 1.1 (see Wrigley 1979a, 1981a). What it shows is a classification of statistical problems on the basis of the type of response (or dependent) and explanatory (or independent) variables involved.

Continuous variables are those measured at the high level, interval or

Table 1.1
Classes of statistical problems

		Explanatory variables			
		Continuous	*Mixed*	*Categorical*	
Response variables	Continuous	(a)	(b)	(c)	
	Categorical	(d)	(e)	(f)	(g)

ratio scales. Categorical variables are those measured at the low-level, nominal or ordinal scales, and mixed variables are a mixture of continuous and categorical. Categorical variables can be of three different types:

(i)   dichotomous (e.g. presence or absence, yes or no);
(ii) unordered polytomous† (e.g. car, bus, train, cycle);
(iii) ordered polytomous (e.g. high income, middle income, low income).

Table 1.1 is organized in such a way that as we move from cell (a) to cells (f) and (g) so the categorical data problem becomes more pervasive and the level of expertise of most geographers and environmental scientists declines. This sequence also serves the useful purpose of linking together the methods discussed in existing statistics textbooks for geographers and environmental scientists and the new methods for analysing categorical data. The methods discussed in existing statistics textbooks for geographers and environmental scientists are confined largely to handling the problems in the first row, cells (a) to (c). In other words, existing textbooks (e.g. Johnston 1978; Mather 1976; Taylor 1977; Davis 1973; Ferguson 1977; Yeates 1974) discuss conventional regression models which are appropriate for the problems of cell (a), regression models with 'dummy' explanatory variables which are appropriate for the problems of cell (b), and occasionally regression models appropriate for the problems of cell (c) in which all the explanatory variables are categorical and which can be shown to be equivalent to traditional analysis of variance models. The new unified approach to categorical data analysis, on the other hand, provides us with methods for handling the problems in cells (d) to (f) and, in addition, allows us to handle the problems of cell (g) in which all variables are categorical but in which there is no division of the variables into response and explanatory (i.e. all variables in cell (g) are considered to be response variables).

The framework outlined in Table 1.1 is central to the organization of this book. In Part 2 we will work through cells (d) to (g) one by one and discuss the form of an integrated family of statistical models appropriate for such problems. Subsequently, in all other parts of the book, the reader will be referred to Table 1.1 so that the statistical problem under discussion can be located within the general typology of statistical problems.

1.3   Some important tables

Before launching into a discussion of the statistical models appropriate for the categorical data problems of cells (d) to (g) of Table 1.1, it is appropriate first to establish the forms in which categorical data are typically summarized and displayed. This involves a consideration of

---

† Although polytomous rather than 'polychotomous' is the correct Greek, the term 'polychotomous' has often been used in the literature (including in papers by this author), e.g. Gurland et al. 1960; Mantel 1966; Aitchison and Bennett 1970.

what are termed *contingency tables*, and we will see that the form of such tables changes as we move from cells (g) to (d) in Table 1.1 and vice versa.

The simplest form of contingency table is that shown in Table 1.2. In this case we have two categorical variables, A and B. Variable A is divided into I categories and variable B into J categories and thus the table is referred to as an $I \times J$ two-dimensional contingency table. If we make the assumption that the variables A and B are both response variables (i.e. are not divided into response and explanatory) we have the simplest form of data display which will be encountered in cell (g) of Table 1.1.

Table 1.2
An $I \times J$ two-dimensional contingency table

		Variable B			
	$j = 1$	2	... J	*Total*
$i = 1$	n_{11}	n_{12}	... n_{1J}	n_{1+}
$i = 2$	n_{21}	n_{22}	n_{2J}	n_{2+}
.	.	.	.	.
.	.	.	.	.
.	.	.	.	.
$i = I$	n_{I1}	n_{I2}	... n_{IJ}	n_{I+}
Total	n_{+1}	n_{+2}	... n_{+J}	$n_{++} = N$

(Variable A labels the rows)

Within the contingency table the data displayed consist of counts of the number of individuals/households/places/etc. in each category (or cell) of the cross-classification of the two variables A and B. These counts are termed frequencies, and we represent the observed frequency or count in the typical *ij*th cell of the table by the term n_{ij}. Summing these frequencies for a category of a single variable produces what is termed a marginal frequency or marginal total. For example, the total count in the *i*th category of variable A is denoted by n_{i+}, where the + sign denotes summation over a variable. That is to say:

$$n_{i+} = n_{i1} + n_{i2} + \ldots + n_{iJ}$$

$$= \sum_{j=1}^{J} n_{ij} \qquad \text{[1.1]}$$

Likewise, the marginal total for the *j*th category of variable B is denoted by n_{+j}, where

$$n_{+j} = n_{1j} + n_{2j} + \ldots + n_{Ij}$$

$$= \sum_{i=1}^{I} n_{ij} \qquad \text{[1.2]}$$

In a similar fashion $n_{++} = N$ represents the overall total

$$N = n_{++} = \sum_{i=1}^{I} \sum_{j=1}^{J} n_{ij} \qquad\qquad [1.3]$$

Having seen the form of the simple two-dimensional contingency table, we must next consider the form of a three-dimensional table. Table 1.3 shows a typical example of such a table in which the opinions of a sample of people taking part in a planning participation exercise in a single British county have been cross-classified by the location of their homes (urban/rural), their sex (male/female), and by their attitudes to the scheme proposed by the planners (in favour/against/undecided). In this case we have three variables which are divided into 3, 2 and 2 categories and thus the table is referred to as a $3\times2\times2$ three-dimensional contingency table. Once again, we make the assumption that all the variables are response variables and that we are, therefore, considering the form of data display which will be encountered in cell (g) of Table 1.1

Table 1.3
Data on attitudes to planning proposal, sex and location of respondents

	Rural		Urban		
	Female	Male	Female	Male	Total
In favour	265	350	325	411	1,351
Against	334	307	322	258	1,221
Undecided	143	86	96	73	398
Total	742	743	743	742	2,970

Table 1.3 can then be generalized to the form shown in Table 1.4. The notation used is a simple extension of that employed in Table 1.2. The observed frequency in the typical ijkth cell is denoted n_{ijk}, and these frequencies can be summed as in Table 1.2 to produce marginal totals. If we sum over two variables we obtain what are termed 'single variable', 'one-way', or 'one-dimensional' marginal totals. For example:

$$n_{i++} = \sum_{j=1}^{J} \sum_{k=1}^{K} n_{ijk}$$

$$n_{+j+} = \sum_{i=1}^{I} \sum_{k=1}^{K} n_{ijk} \qquad\qquad [1.4]$$

$$n_{++k} = \sum_{i=1}^{I} \sum_{j=1}^{J} n_{ijk}$$

To illustrate this, consider Table 1.3 where:

$$
\begin{aligned}
n_{2++} &= 334 + 307 + 322 + 258 = 1{,}221 \\
n_{+2+} &= 350 + 307 + 86 + 411 + 258 + 73 = 1{,}485 \qquad [1.5] \\
n_{++2} &= 325 + 322 + 96 + 411 + 258 + 73 = 1{,}485
\end{aligned}
$$

Table 1.4
A $3 \times 2 \times 2$ three-dimensional contingency table

		Variable C				
		k = 1		k = 2		
		Variable B		Variable B		
		j = 1	j = 2	j = 1	j = 2	*Total*
Variable A	i = 1	n_{111}	n_{121}	n_{112}	n_{122}	n_{1++}
	i = 2	n_{211}	n_{221}	n_{212}	n_{222}	n_{2++}
	i = 3	n_{311}	n_{321}	n_{312}	n_{322}	n_{3++}
	Total	n_{+11}	n_{+21}	n_{+12}	n_{+22}	$n_{+++} = N$

If, on the other hand, we sum over just one variable, we obtain what are termed 'two variable', 'two-way', or 'two-dimensional' marginal totals:

$$n_{ij+} = \sum_{k=1}^{K} n_{ijk}$$

$$n_{i+k} = \sum_{j=1}^{J} n_{ijk} \tag{1.6}$$

$$n_{+jk} = \sum_{i=1}^{I} n_{ijk}$$

In terms of the data of Table 1.3 this implies, for example,

$$n_{21+} = 334 + 322 = 656$$
$$n_{2+2} = 322 + 258 = 580 \tag{1.7}$$
$$n_{+12} = 325 + 322 + 96 = 743$$

Moving from a two-dimensional to a three-dimensional table involves, therefore, only a very simple extension of the notation established in Table 1.2. This notation can then readily be generalized to higher-dimensional tables. For example, the typical *ijkl*th cell of a four-dimensional table is denoted n_{ijkl}, and a typical 'three-way' or 'three-dimensional' marginal total would be

$$n_{ijk+} = \sum_{l=1}^{L} n_{ijkl} \tag{1.8}$$

So far we have considered tables in which all variables are assumed to be response variables. Let us now consider the situation in which the variables are divided into response and explanatory. As an example we will consider Table 1.3 and assume that attitude to the planning proposal is a response variable and sex and location of respondents are explanatory variables. In this case we can reorganize Table 1.3 into the form shown in Table 1.5. This is the form of data display which will be encountered in cell (f) of Table 1.1.

Table 1.5 can then be generalized and made applicable to the case in which there are any number of categories of the response variable and any number of explanatory variables. As Table 1.6 shows, we do this by

Explanatory variables		Variable A = *Response variable*			
B	**C**	*In favour*	*Against*	*Undecided*	*Total*
Female	Rural	265	334	143	742
Male	Rural	350	307	86	743
Female	Urban	325	322	96	743
Male	Urban	411	258	73	742
					2,970

indexing the response categories $r = 1, \ldots R$ and indexing the groups formed by the cross-classification of the explanatory variables $g = 1, \ldots G$ (e.g. in Table 1.5 there are four groups formed from the cross-classification of the explanatory variables, $g = 1$: female rural, $g = 2$: male rural, $g = 3$: female urban, $g = 4$: male urban). We call the groups formed from the cross-classification of the explanatory variables: sub-populations. The typical element in Table 1.6, $n_{r|g}$, then denotes the number of choices of the rth response category conditional upon or given that the individuals/households/places/etc. belong to the gth sub-population. As an example of this, consider the element $n_{2|3}$ in Table 1.5. This is the number of individuals who choose the second response (against) given that they belong to the third sub-population, i.e. given that they are individuals who are of the $j = 1$ category of variable B (female) and $k = 2$ category of variable C (urban). As can be seen in Table 1.6, dividing the observed frequencies by the marginal totals (e.g. $n_{r|g}/n_{+|g}$) produces the observed proportions ($f_{r|g}$) of each sub-population who choose each response category and these observed proportions sum to one for each sub-population.

Sub-populations	Response categories				Total
	$r = 1$	2	$\ldots R$		
$g = 1$	$n_{1|1}\ (f_{1|1})$	$n_{2|1}\ (f_{2|1})$	$\ldots$	$n_{R|1}\ (f_{R|1})$	$n_{+|1}\ (1.0)$
$g = 2$	$n_{1|2}\ (f_{1|2})$	$n_{2|2}\ (f_{2|2})$	$\ldots$	$n_{R|2}\ (f_{R|2})$	$n_{+|2}\ (1.0)$
.	.	.		.	.
.	.	.		.	.
$g = G$	$n_{1|G}\ (f_{1|G})$	$n_{2|G}\ (f_{2|G})$	$\ldots$	$n_{R|G}\ (f_{R|G})$	$n_{+|G}\ (1.0)$
					$n_{+|+} = N$

Note: $f_{r|g} = n_{r|g}/n_{+|g}$ = observed proportion; $\sum_{r=1}^{R} f_{r|g} = 1.0$

Tables 1.5 and 1.6 represent the form of data display which will be encountered in cell (f) of Table 1.1 All contingency tables in which there is a division of variables into response and explanatory can be reorganized from the form of Table 1.4, and its higher-dimensional equivalents, into the form of Table 1.6.

So far we have seen the form of data display encountered in cells (g) and (f) of Table 1.1. We next must consider what form of display will be encountered in cells (d) and (e) of Table 1.1. To do this we must imagine that the number of sub-populations in Table 1.6 is increased substantially, whilst N, the number of individuals (observations), remains fixed. In this case the number of individuals in each sub-population will fall consistently until, ultimately, each sub-population is composed of only one individual. In this case each row total in Table 1.6 will have the value 1, each $n_{r|g}$ in the body of the table will take either the value 1 or 0, and G the number of sub-populations will equal N the number of individuals (observations) in the sample. In this case, instead of indexing the sub-populations $g = 1, \ldots G$ we will index[†] them $i = 1, \ldots N$ as each sub-population is now a single individual (case) i. The result is as shown in Table 1.7 and this is the form of data display which will be encountered in cell (d) of Table 1.1 where all explanatory variables are continuous. It is equivalent to listing the individual cases one by one, with for each case the values of the continuous explanatory variables given first, and then the chosen category of the response variable. Table 1.7 is also the most usual form of data display encountered in cell (e) of Table 1.1 where the explanatory variables are a mixture of continuous and categorical. In this case, however, if the proportion of categorical explanatory variables in the mixture is large, and the proportion of continuous variables is small, then it becomes increasingly possible to approximate the data display of Table 1.6 rather than that of Table 1.7.

Table 1.7
Form of Table 1.6 when explanatory variables are continuous or mixed

Individual cases	Response categories				Total
	1	2	. . .	R	
$i = 1$	$n_{1|1}=1$ or 0	$n_{2|1}=1$ or 0	. . .	$n_{R|1}=1$ or 0	1
$i = 2$	$n_{1|2}=1$ or 0	$n_{2|2}=1$ or 0	. . .	$n_{R|2}=1$ or 0	1
.	.	.		.	.
.	.	.		.	.
$i = N$	$n_{1|N}=1$ or 0	$n_{2|N}=1$ or 0	. . .	$n_{R|N}=1$ or 0	1
					N

Note: remember that the index i for the individual cases acts as a summary of the values of the (continuous or mixed) explanatory variables for that case, just as g in Table 1.6 summarizes the values of the (categorical) explanatory variables

1.4 Sampling schemes

Having described the forms of some of the tables used to display categorical data, the question then arises of how data of the type

---

[†] In this context it should be noted that we have already used i to help index a typical element in contingency tables of the type shown in Tables 1.2 and 1.4. This should not be confused with its use in Table 1.7 to index an individual case. The subscript i is conventionally used for both these purposes, and we will adhere to this practice and allow the context to clarify how i is being used. Experience suggests that this is likely to create no serious problems for the reader.

displayed in Tables 1.2–1.7 are generated. More specifically, what types of sampling schemes lead to the types of data tables we have just discussed? At this stage in the presentation it is sufficient to note that there are a number of possibilities and to introduce these in a non-mathematical fashion.

1.4.1 Poisson

One possible sampling scheme which would lead to the type of contingency table (e.g. Table 1.4) encountered in cell (g) of Table 1.1 involves the imposition of no restriction on the total sample size and the observation of a set of Poisson processes, one for each cell in the contingency table, for a fixed period of time. Each Poisson process yields a count or frequency for that particular cell, and the observed cell frequencies (e.g. n_{ijk}) are viewed as having independent Poisson distributions with expected frequencies (e.g. $m_{ijk} = E(n_{ijk})$) as their means. This situation will occur only rarely in the survey sampling used by geographers and environmental scientists, but it is important to note it as the Poisson is a probability distribution of fundamental importance in the analysis of categorical data. Moreover, we will see later that the Poisson scheme often gives rise to the same desirable results as those produced by the more probable sampling schemes discussed below.

1.4.2 Multinomial

A second possible sampling scheme which would lead to the type of contingency table (e.g. Table 1.4) encountered in cell (g) of Table 1.1 involves taking a random sample of *fixed* size N and then the allocation of each member of the sample to a cell of the table on the basis of which categories of the cross-classified variables the sample member falls into. The complete table of observed frequencies then has a multinomial distribution. In the survey sampling used by geographers and environmental scientists, this scheme is much more likely than the previous Poisson scheme. However, it should be noted that the two schemes are closely linked by what is termed the *conditional Poisson distribution*. This states that if the observed cell frequencies (e.g. n_{ijk}) have been generated according to a Poisson sampling scheme, then the conditional distribution of the cell frequencies given the fixed sample size N is multinomial. This close relationship can often be usefully exploited in the analysis of categorical data.

1.4.3 Product-multinomial

A third possible sampling scheme applies when, as in cells (d) to (f) of Table 1.1, we can divide our variables into explanatory and response. In this case (e.g. Table 1.6) we take a stratified random sample in which the sample size for each stratum is fixed and the strata are the groups (or sub-populations) formed from the cross-classification of the explanatory variables. The distribution of responses within each sub-population is then governed by a multinomial distribution, and thus each row of response frequencies in the table has an independent multinomial distribution. This scheme will occur frequently in the survey sampling used by geographers and environmental scientists and, moreover, it is often possible to argue that much secondary-source data could conceivably have arisen from an underlying product-multinomial scheme. In addition, it should be noted that, just as the Poisson and multinomial schemes are linked by a conditional probability distribution, so are the multinomial and product-multinomial. If the observed cell frequencies have been generated according to a multinomial scheme, then the conditional distribution of the cell frequencies, given the fixed sub-population (row) totals, is product-multinomial.

1.5 Integration and transition

The major themes of the previous sections have been integration and transition. Table 1.1 displays a framework which integrates categorical data problems into the class of statistical problems which can be handled using the general linear model, and in this way it links together the new methods for analysing categorical data and the methods discussed in existing statistical methods textbooks for geographers and environmental scientists. Similarly, the introduction to this chapter committed us to a consideration of how the integration of these new methods for analysing categorical data with developments in micro-economic theory has resulted in the development of discrete choice modelling as an important research field.

Within this overall theme of integration, smooth transition from one type of problem to the next is the theme. We have seen how data tables are progressively transformed as we move from cell (g) to (d) in Table 1.1, and as we recognize the division of variables into explanatory and response and progressively increase the number of variables and sub-populations. Similarly, we have seen how the sampling schemes which lead to these data tables are transformed as the number of sample size constraints increases and how these schemes are linked by conditional probability distribution arguments. In terms of statistical models similar transitions occur. Many geographers and environmental scientists will be familiar with the transition in the form of regression models as we move from cells (a) to (c) in Table 1.1 and the proportion of categorical explanatory variables increases. In Part 2 of the book a similar transition in the form of statistical models will be shown to occur as we move from cells (d) to (g) in Table 1.1 and as models of the logistic, linear logit, and probit types merge into log-linear models and vice versa. Similarly, within the class of discrete choice models we will see in Chapter 11 that we can conceive of a transition along a continuum from restrictive but computationally tractable models to general non-restrictive but computationally and conceptually complex models.

1.6 How to use the book and what it assumes

In their undergraduate curriculum most geographers and environmental scientists can be expected to complete two courses in statistical methods and data analysis. The first course will introduce the essential components of descriptive and inferential statistics and geographers and environmental scientists will find such a course supported by textbooks written by Gregory (1978), Norcliffe (1977), Till (1974), Hammond and McCullagh (1978), Silk (1979), Ebdon (1977) and Matthews (1981). The second course will be a multivariate statistics course oriented to the 'general linear model'. The aim of this will be to develop the skills of model building and model fitting, and to enable the student to appreciate a research literature in which multivariate statistical procedures often play a crucial role. Geographers will use textbooks by Johnston (1978) and Mather (1976) (and to a lesser extent by Taylor 1977; Yeates 1974; King 1969) for their area of study, and environmental scientists can turn to well-established textbooks such as Davis (1973) and to older classics such as Krumbein and Graybill (1965) and Sokal and Rohlf (1969). This book is designed as the logical sequel to such a multivariate statistics course and such textbooks, and it can be thought of, in some senses, as a thrid course in statistical methods. As such, it assumes that the reader is familiar with the classical regression model and its associated least squares estimation methods, with some of the most important extensions of the classical regression model, and with problems encountered in the use of regression models, e.g. heteroscedasticity, multicollinearity, autocorrelation, etc. It also assumes familiarity with concepts which are normally taught in an introductory statistics course, for example: principles of hypothesis testing; error types; critical regions; statistical distributions; parameters and sample estimates; certain statistical tests (including the chi-square), etc. Finally, it assumes (particularly in Chapters 4 and 8) some familiarity with matrix expressions and the principles of matrix algebra of the type provided for geographers and environmental scientists by Davis (1973), Mather (1976), Wilson and Kirkby (1980) and Rogers (1971). To derive the maximum advantage from the book the reader should, if necessary, revise such topics and he will find directions to appropriate sources in the footnotes which accompany the text.

Assuming this background knowledge, the structure of the book has a straightforward logic. In Part 2 we begin with the regression models of the reader's second course in statistical methods. We show how these can be extended to handle categorical response variables and how the extension can be integrated with other members of the family of generalized linear models to provide a unified family of models appropriate for the categorical data problems of cells (d) to (g) in Table 1.1. It is in this part of the book that the reader is introduced to the building blocks of categorical data analysis, and it represents an essential prerequisite for all but the most advanced and/or specialized readers. Part 3 then considers some useful extensions of the basic models introduced in Part 2, and Part 4 links the statistical models for categorical data analysis to associated developments in micro-economic theory and shows how the statistical models of the earlier parts of the book serve to operationalize the research field which has come to be known as 'discrete choice modelling'. Finally, in Part 5 the book concludes with a brief glimpse at an alternative framework for discussing the unified family of categorical data models and linking that family to the general linear models considered in the student's second course in statistical methods.

In designing the book considerable attention has been paid to four important pedagogic features.

1. Every effort has been made to ensure that concepts, models, measures, procedures, symbols, etc. are introduced in a successive and *developmental* manner, building upon and out from the reader's assumed background in the general linear model. In particular, the chapters in Parts 1 and 2 are designed to be read *in sequence*, and concepts introduced in one chapter are carried over into subsequent chapters. Although individual chapters in Parts 1 and 2 can be read as separate entities, the reader will gain maximum benefit from considering each chapter in the context of the other chapters and from a strict sequential reading. Similarly within a chapter, few sections or subsections truly stand alone, and the reader is, once again, advised to read them in sequence. Parts 3 and 4 of the book are, however, an exception to such statements. Some readers who are using the book to provide the statistical background to courses in discrete choice modelling can move directly from Part 2 to Part 4 without great loss, whilst others may choose (at least on a first reading) to ignore Part 4.

2. Considerable attention has been paid to the choice of examples. Once again, these examples are introduced in a developmental manner and they should not be regarded as stand-alone items divorced from the main text. Instead, they are designed to form an *integrated* part of the text, illustrating previously discussed concepts and measures, and providing essential reinforcement to the technical sections. Many of the examples have been cut and shaped to meet the exact needs of the particular section of the book in which they are to be found, and others have been developed specially to meet such needs. The reader is advised to resist any temptation he might feel to move directly from one technical section to the next without considering the examples. Maximum benefit is to be gained from regarding them as part of the main text and reading them in sequence.

3. As a result of the fact that it is a relatively new and rapidly developing subject, categorical data analysis suffers from a profusion of notation systems. This forms a real, and often underestimated, burden for the student who is encountering the literature for the first time. Elsewhere (Wrigley 1980c) the author has drawn attention to such problems and has argued for a more unified notation system, which at the very least might be adopted in the use of these methods by geographers and environmental scientists. Considerable efforts have been made in the design of this book to develop a logical, internally consistent, notation system; one which, moreover, is maximally consistent with the many notation systems currently used in the categorical data analysis and discrete choice modelling literature. Clearly, the reader will still encounter notational difficulties as he turns to the papers and books in the accompanying reference list, but it is hoped that he might find the notation system adopted in this book to be sufficiently logical, consistent and portable to use it in his own research.

4. Every effort has been made to facilitate the flow of the text and to increase its overall readability. To this end, certain necessary technical

elaborations, etc. have been taken out of the main text and set as footnotes immediately beneath the point in the text to which they refer. Readers with different backgrounds in statistics and with different requirements from the book must choose for themselves on which reading of the book it is appropriate to consult these footnotes.

Although the book is conceived of as providing a complete third course in statistical methods for geographers and environmental scientists and/or as providing the statistical background to more specialized courses in discrete choice modelling, it clearly can be used in other ways by both students and teachers. For example: suitable elements of Parts 1 and 2 can be integrated into existing basic multivariate statistics courses; also elements of Chapter 5 can usefully enrich the introduction to contingency tables and the teaching of the chi-square test which geographers and environmental scientists normally receive in their first course in statistical methods. In addition, the book can be used, together with a matching course and text on social survey methods, to provide higher-level instruction on the collection and analysis of social survey data. Finally, for some courses, perhaps those in urban, behavioural or transport geography, the examples contained in the book can be utilized and extended to provide the required illustration of lecture material. As a number of geographers have now suggested, categorical data methods are certain to become major research tools of the geographer and environmental scientist in the next decade. It is hoped that this book will prove to be sufficiently comprehensive and flexible to serve these various needs.

PART 2

The basic family of statistical models

At the heart of the unified approach to categorical data analysis lies an integrated family of statistical models. In Part 2 of the book, the basic features of this family of models will be discussed and illustrated. As such, this part is an essential prerequisite for all but the most advanced and/or specialized readers, for the later parts of the book will build upon the foundations established here.

The organization of Part 2 is straightforward. We will simply examine, cell by cell, the types of statistical problems which fall within the lower row of Table 1.1. To aid the reader, Table 1.1 is reproduced again and it is recommended that this table should be consulted before beginning each new chapter in Part 2.

Table 1.1
Classes of statistical problems

		Explanatory variables			
		Continuous	*Mixed*	*Categorical*	
Response variables	Continuous	(a)	(b)	(c)	
	Categorical	(d)	(e)	(f)	(g)

As noted in Part 1, this book assumes that the reader is familiar with the classical regression model and some of its most important extensions, and will have come to the book having completed the type of course on multivariate statistical methods which most geographers and environmental scientists are likely to be exposed to during their undergraduate careers. As such, this part of the book begins logically in Chapter 2 with a consideration of the statistical problems encountered in cell (d) of Table 1.1, for it will be seen that these problems can be handled using extensions of the classical regression models appropriate for cells (a) to (c). Subsequent chapters then build upon these extensions and move progressively towards the problems encountered in cells (f) and (g) where all variables are categorical.

CHAPTER 2

Categorical response variable, continuous explanatory variables

It is now well known that when we attempt to extend the conventional regression models adopted in cells (a), (b) and (c) of Table 1.1 to the problems of cell (d) a number of difficulties are encountered which necessitate an alternative form of model. In this chapter we will consider what these difficulties are and discuss feasible alternative models. For simplicity, we will focus initially upon the case in which we have a dichotomous (2–category, 2–alternative) response variable. We will then generalize our results and consider the case in which we have a polytomous (multiple–category, multiple–alternative) response variable.

2A TWO RESPONSE CATEGORIES: THE DICHOTOMOUS CASE

2.1 Introduction

To aid our discussion it is useful to think in terms of a specific example. We will assume that a local government authority wishes to build a large new mental hospital. As part of the planning process a social survey of opinions about the scheme and its proposed siting is conducted, and at the simplest level each survey respondent is classified as being 'in favour' or 'against' the scheme. In addition the survey provides information on a range of other variables including the location of the respondent's residence. From this the distance of the home of each respondent from the proposed site of the mental hospital can be calculated, and we can build a simple statistical model relating the likelihood that a respondent will be 'in favour' of the mental hospital scheme to his distance from the proposed site. Clearly, this will be a very naïve model, as many other variables are likely to be important determinants of a respondent's attitude to the mental hospital proposal and should therefore be included in the model. However, for our purposes, the basic modelling principles can be illustrated using just the single explanatory variable: distance from the proposed site. Moreover, it should be noted that several geographers (e.g. Smith 1980, 395–7; Dear and Wittman 1980) have stressed the importance which people attach to distance as a means of coping with what is often regarded as a 'noxious' public facility.

2.2   A conventional regression model approach

The simplest way to model the relationship between attitude to the mental hospital scheme and distance from the proposed site would be to attempt a simple extension of the regression models used for the problems of cells (a) to (c) in Table 1.1 and to consider a regression model of the form.[†]

$$Y_i = \alpha + \beta X_i + \varepsilon_i \qquad\qquad [2.1]$$

where

X_i = distance from the home of the respondent i to the proposed site of the mental hospital

Y_i = 1 if respondent i is 'in favour' of the scheme
  = 0 if respondent i is 'against' the scheme

ε_i = an independently distributed random error term with zero mean.

The two possible outcomes of the dichotomous response variable have been coded 1 and 0 and this dichotomous variable has been substituted in place of the continuous response variable familiar in the regression models encountered in cell (a) of Table 1.1.

Unfortunately, although this regression model has a conventional appearance, it can easily be demonstrated that it suffers from a number of problems which seriously reduce its value.

The first problem concerns the estimated or predicted values which may be generated using this regression model, and it arises because of the logical interpretation of the regression model as a probability model. To see this, we first take the expected value of both sides of the regression model.[‡]

$$\begin{aligned} E(Y_i) &= E(\alpha + \beta X_i + \varepsilon_i) \\ &= \alpha + \beta X_i + E(\varepsilon_i) \\ &= \alpha + \beta X_i \end{aligned} \qquad\qquad [2.2]$$

[†] A simple regression model can also be written as $Y_i = \beta_1 + \beta_2 X_i + \varepsilon_i$ where α, the intercept or constant term of [2.1], is now denoted by β_1. It is conventional, however, in most standard textbooks to use α in simple regression models and then to change to the use of β_1 for the constant term in multiple regression models (e.g. see Kmenta 1971, 347; Pindyck and Rubinfeld 1976, 55). This is done to simplify matrix expression of multiple regression models. We will adopt a similar convention in this book.
  [‡] This result follows from the classical regression model assumptions that X is a fixed value and that $E(\varepsilon_i) = 0$.

We then note that since Y_i can only assume two different values, 1 and 0, its expected value is simply a weighted average of the two possible values of Y_i with weight given by the respective probabilities of occurrence of the two possible values. By letting these probabilities be[†] $P_{1|i} = \text{Prob}\,(Y_i = 1)$ and $P_{2|i} = 1 - P_{1|i} = \text{Prob}\,(Y_i = 0)$, the expected value of Y_i is therefore:

$$E(Y_i) = 1(P_{1|i}) + 0(P_{2|i})$$
$$= P_{1|i} \qquad\qquad\qquad \text{[2.3]}$$

Combining [2.2] and [2.3] we then have:

$$P_{1|i} = E(Y_i) = \alpha + \beta X_i \qquad\qquad \text{[2.4]}$$

Thus the regression model can be interpreted[‡] as describing the probability that an individual will select the first response category given information about X_i the level of the explanatory variable. In terms of our example, therefore, it is useful and reasonable to interpret the regression model as describing the probability that the respondent will be 'in favour' of the mental hospital scheme given information about the distance of his home from the proposed site of the mental hospital.

The problem with this interpretation, however, concerns the estimated or predicted values,

$$\hat{Y}_i = \hat{\alpha} + \hat{\beta} X_i \qquad\qquad\qquad \text{[2.5]}$$

generated from the regression model [2.1]. In view of the probability interpretation of $E\,(Y_i)$ these predicted values are interpreted as predicted probabilities, i.e. $\hat{Y}_i = \hat{P}_{1|i}$. However, whereas probability is defined to lie between 0 and 1 these predictions generated from the regression model [2.1] are unbounded and may take values from $-\infty$ to $+\infty$. Consequently they may lie outside the meaningful range of probability and thus be inconsistent with the probability interpretation of the model.

The second problem concerns the violation of one of the standard assumptions of the classical linear regression model; the assumption of constant error variance or homoscedasticity (see any standard econometric text, e.g. Pindyck and Rubinfeld 1976, 17, 95; Huang 1970, 25; Theil 1971, 160; Johnston 1972, 214; also Chatterjee and Price 1977, 102). This follows from the fact that the error term in the regression model

$$\varepsilon_i = Y_i - (\alpha + \beta X_i) \qquad\qquad \text{[2.6]}$$

[†] Note that this notation follows the convention established in Table 1.7. We could alternatively have written: $P_{1|i} = P_i$ and $P_{2|i} = 1 - P_i$.

[‡] It should now be clear that $P_{1|i}$ is actually a shorthand way of writing the probability that response category one will be selected by respondent i conditional upon X_i (the value of the explanatory variable relating to respondent i) and the values of the parameters α and β. In other words $P_{1|i}$ could be written more fully and more formally as:

$$P_{1|i} = P_{(1|X_i,\alpha,\beta)}$$

However, as this expression (and its extensions in the multiple explanatory variable, multiple response category cases) is rather cumbersome to use, we will adopt the shorthand expression $P_{1|i}$ from this point onwards, unless explicitly stated otherwise. The reader should be aware, however, of what is implied by this shorthand expression.

can only have two possible values,

$$\varepsilon_i = \begin{array}{ll} 1 - (\alpha + \beta X_i) & \text{if } Y_i = 1 \\ -(\alpha + \beta X_i) & \text{if } Y_i = 0 \end{array} \qquad [2.7]$$

These two possible values of ε_i must occur with probabilities $P_{1|i}$ and $(1 - P_{1|i}) = P_{2|i}$ respectively. Thus the standard assumption that the error term in the regression model has zero mean (i.e. $E(\varepsilon_i) = 0$) implies,

$$E(\varepsilon_i) = P_{1|i}\,(1 - \alpha - \beta X_i) + (1 - P_{1|i})(-\alpha - \beta X_i) = 0 \qquad [2.8]$$

Solving for $P_{1|i}$, or from [2.4] directly, we have that

$$P_{1|i} = \alpha + \beta X_i \qquad [2.9]$$

$$P_{2|i} = (1 - P_{1|i}) = 1 - \alpha - \beta X_i \qquad [2.10]$$

The error variance can then be calculated as:

$$\begin{aligned} \text{Var }(\varepsilon_i) = E(\varepsilon_i^2) &= P_{1|i}(1 - \alpha - \beta X_i) + (1 - P_{1|i})(-\alpha - \beta X_i) \\ &= P_{1|i}(1 - P_{1|i})^2 + (1 - P_{1|i})(-P_{1|i})^2 \\ &= P_{1|i}(1 - P_{1|i}) = P_{1|i}P_{2|i} \\ &= (\alpha + \beta X_i)\,(1 - \alpha - \beta X_i) \end{aligned} \qquad [2.11]$$

Clearly, the error variance is not constant but depends upon the values of X_i, the explanatory variable. Observations for which $P_{1|i}$ is close to 0 or close to 1 will have relatively low variances while observations for which $P_{1|i}$ is close to 0.5 will have higher variances. When the constant error variance assumption is violated, the problem of heteroscedasticity is said to be present. The presence of heteroscedasticity implies that if ordinary least squares (OLS) estimation of the unknown regression parameters (α and β) is used, then the OLS estimators ($\hat{\alpha}$ and $\hat{\beta}$) of the parameters will remain unbiased and consistent but will no longer be the most efficient estimators. In addition, the estimated variances of $\hat{\alpha}$ and $\hat{\beta}$ will be biased estimators of the true variances of $\hat{\alpha}$ and $\hat{\beta}$, and if these are used then the statistical tests commonly used in regression modelling will be incorrect.

2.3 Alternative solutions

As a means of modelling the relationship between a dichotomous response variable and a continuous explanatory variable, the simple extension [2.1] of the conventional regression model has been shown to be seriously deficient. We will now consider alternative models which attempt to overcome these problems, and we will see that a characteristic of all these alternatives is that they attempt to conform with the probability model interpretation discussed above.

2.3.1 *The linear probability model*

This is merely a slight extension of [2.1]. It can be written (compare it with [2.4]) in the form:

$$P_{1|i} = E(Y_i) = \begin{array}{ll} 0 & \text{if } \alpha + \beta X_i \leq 0 \\ \alpha + \beta X_i & \text{if } 0 < \alpha + \beta X_i < 1 \\ 1 & \text{if } \alpha + \beta X_i \geq 1 \end{array} \qquad [2.12]$$

As can be seen in Fig. 2.1, the difference between models [2.1], [2.4] and [2.12] concerns the manner in which they treat extreme values. Model [2.12] conforms with the probability interpretation by assuming a discontinous function at the boundaries 0 and 1.

Conventionally, the parameters of the linear probability model are estimated by assuming that all observations lie in the range where the probabilities are between zero and one, and by employing the OLS estimation procedure (despite the fact that the error variance in the model is not constant and the problem of heteroscedasticity is present). This procedure, however, can give rise to a number of problems. The first of these concerns the predicted values which may be generated from the model. Even when the specification of the model is correct it is possible that, as a result of normal sampling effects, a given sample of observations obtained from the true model will result in OLS parameter estimates which produce a *fitted* linear probability function, $\hat{P}_{1|i} = \hat{\alpha} + \hat{\beta} X_i$, like that shown in Fig. 2.2(a). This will give rise to predicted values which lie outside the 0–1 interval for values of the explanatory variable at the extremes of the observed range. In this case, we can correct the problem by setting extreme predictions to 1 if $\hat{\alpha} + \hat{\beta} X_i \geq 1$ and to 0 if $\hat{\alpha} + \hat{\beta} X_i \leq 0$ (as shown in Fig. 2.2b). However, this may not be very satisfactory because it may have the result that for some values of the explanatory variable we are required to predict that a response category/alternative will be chosen with a probability of 1 when, in fact, we may actually observe that it is sometimes not chosen (or vice versa for a probability of 0). Thus, although the OLS estimation procedure might yield unbiased parameter estimates, the predicted probabilities obtained from the estimation process may be rather badly biased.

Figure 2.1
A diagrammatic representation of the contrast between models [2.1], [2.4] and [2.12]

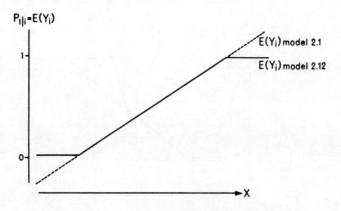

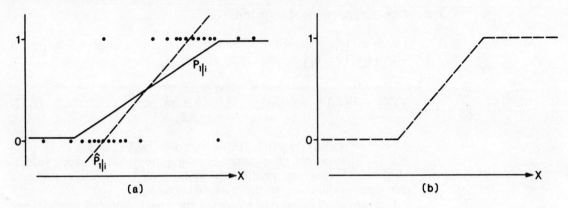

Figure 2.2
OLS estimation of model
[2.12]

A second and related problem occurs when observations in a given sample are bunched together. Figure 2.3 shows such a sample which has been drawn excessively from the ranges of the explanatory variable where the probabilities take the extreme values. In this case, the value of $\hat{\beta}$, the OLS slope coefficient of the fitted linear probability function, substantially underestimates β, the true slope parameter. Conversely, if the bunching of the sample occurs in the rather different manner shown in Fig. 2.4, the fitted linear function may overestimate the true slope. It can be seen, therefore, that the particular sample of observations can have a disconcertingly powerful effect on the OLS parameter estimates.

Figure 2.3
OLS estimation when sample drawn from range of explanatory variable where probabilities take extreme values

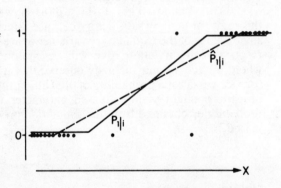

Figure 2.4
Converse situation to that shown in Fig.2.3

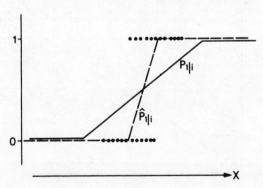

Conventional OLS estimation[†] of the linear probability model can be seen, therefore, to have a number of drawbacks but, in addition, there is a further more general problem associated with the model. This concerns the functional form of the model. Experience suggests the functional form of the linear probability model is unlikely to be a reasonable approximation to the true model. Empirical observation by geographers of phenomena such as the progressive adoption of innovations over time and space suggests that cumulative probability functions are often likely to display a general S-shape. That is to say, the S-shaped (two-tailed ogive) curve shown in Fig. 2.5 is likely to be a more reasonable functional form than the discontinuous form of the linear probability model. Such S-shaped curves imply that a given marginal change in the probability is more difficult to obtain when the probability is close to 0 or 1. For example, changing the probability from 0.96 to 0.97 requires a larger change in X than changing from 0.49 to 0.50. Furthermore, if the S-shaped curve in Fig. 2.5 corresponds to the true functional form, it can be seen that the fitted linear probability function is likely to deviate in a manner which varies systematically with X. In the case of Fig. 2.5 it can be seen that the linear probability function will underestimate the effect of a change in the explanatory variable in the intermediate probability range, will overestimate the effect of a change in the explanatory variable when the probabilities are near to 0 or 1, and will predict no effect when 0 and 1 are reached and the linear probability function is truncated.

Figure 2.5
A more reasonable form for the probability model

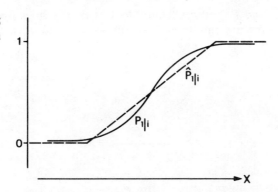

The limitations of the linear probability model point to the need for more suitable model specifications and, as we have seen, these are likely to be associated with S-shaped cumulative probability functions. One

---

[†] An alternative to the OLS procedure is to estimate the α and β parameters of the linear probability model by least squares, subject to the inequality constraint

$$0 \leqslant \hat{\alpha} + \hat{\beta} X_i \leqslant 1$$

This a nonlinear estimation problem which can be solved using a mathematical programming routine. Domencich and McFadden (1975, 105) suggest that the imposition of the inequality constraint leads to parameter estimates with lower variances, but that these are no longer unbiased. Given that this inequality-constrained procedure is more costly, more sensitive to specification error, and does not eliminate the problems discussed above, Domencich and MacFadden suggest that it is preferable to use the OLS version of the linear probability model.

way of considering these alternatives is to first write the general form of our model as:

$$P_{1|i} = F(\alpha + \beta X_i) \qquad \qquad \text{[2.13]}$$

where F denotes a cumulative probability function. The linear probability model can then be seen to be produced from [2.13] when F is a cumulative uniform probability function (see Bulmer 1967, 37–8; Domencich and McFadden 1975, 59), whereas more promising models can be produced from [2.13] when F is a S-shaped, two-tailed ogive such as the cumulative *normal* probability function, the cumulative *logistic* probability function, or cumulative *Cauchy* probability function (Domencich and McFadden 1975, 56). We will now consider the models which result from two of the most important of these functions; the cumulative logistic and the cumulative normal probability functions.

2.3.2 *The logit model*

This model is produced from [2.13] when F is the cumulative logistic probability function. In this case, [2.13] takes the form†:

$$P_{1|i} = F(\alpha + \beta X_i) = \frac{e^{\alpha + \beta X_i}}{1 + e^{\alpha + \beta X_i}} \qquad \qquad \text{[2.14]}$$

and $P_{1|i}$ ranges from 0 to 1 as $\alpha + \beta X_i$ goes from $-\infty$ to $+\infty$. We then rewrite this model as follows to produce a linear expression:

$$P_{1|i}\left(1 + e^{\alpha + \beta X_i}\right) = e^{\alpha + \beta X_i}$$

$$P_{1|i} = \left(1 - P_{1|i}\right)e^{\alpha + \beta X_i}$$

$$\frac{P_{1|i}}{1 - P_{1|i}} = e^{\alpha + \beta X_i}$$

$$\log_e \frac{P_{1|i}}{P_{2|i}} = \log_e \frac{P_{1|i}}{1 - P_{1|i}} = \alpha + \beta X_i \qquad \qquad \text{[2.15]}$$

The left-hand side of this linear model is known as the log odds or the *logit* transformation (we can abbreviate it as $L_{12|i}$) and the model is known as the *linear logit model*. The important point to note about this logit transformation is that it increases from $-\infty$ to $+\infty$ as $P_{1|i}$ increases from 0 to 1. Thus, whilst the probabilities are bounded, the logits are unbounded with respect to the values of X. The predicted logit values

$$\hat{L}_{12|i} = \log_e \frac{\hat{P}_{1|i}}{\hat{P}_{2|i}} = \hat{\alpha} + \hat{\beta} X_i \qquad \qquad \text{[2.16]}$$

---

† Readers familiar with seeing the logistic function written as $P_{1|i} = 1/1 + e^{-(\alpha + \beta X_i)}$ should note that this is equivalent to [2.14]. In [2.14] we follow Cox (1970).

are likewise unbounded, but the predicted probabilities which can be found by substituting $\hat{\alpha}$ and $\hat{\beta}$ into [2.14] are confined to the range 0 to 1.

The nonlinear model [2.14] from which the linear logit model was derived can be referred to as the *logistic* model. As $P_{1|i}$ represents the probability that individual i selects the first response category,

$$1 - P_{1|i} = P_{2|i} = \frac{1}{1+e^{\alpha+\beta X_i}} \qquad [2.17]$$

represents the probability that individual i selects the second response category.

2.3.3 *The probit model*

This model (see Finney 1971; Stopher and Meyburg 1979, 299–318) is produced from [2.13] when F is the cumulative normal probability function. In this case [2.13] takes the form:

$$P_{1|i} = F(\alpha + \beta X_i) = \frac{1}{\sqrt{2\pi}} \int_{-\infty}^{\alpha+\beta X_i} e^{-u^2/2} du \qquad [2.18]$$

where u is a standard normal deviate (i.e. it has a mean of zero and variance of one), and it states that $P_{1|i}$ (the probability that the first response category will be selected by the ith individual) is the area under the standard normal curve between $-\infty$ and $(\alpha+\beta X_i)$. The greater the value of $(\alpha+\beta X_i)$ the more likely that the ith individual will select the first response category.

Just as the linear logit model was derived from [2.14], so in the case of [2.18] we can derive a linear model known as the *probit* model[†] by applying the inverse[‡] F^{-1}, of the cumulative normal probability function to [2.18]

$$F^{-1}(P_{1|i}) = \alpha + \beta X_i \qquad [2.19]$$

Like the logits on the left-hand side of [2.15], the probits on the left-hand side of [2.19] increase in value from $-\infty$ to $+\infty$ as $P_{1|i}$ increases from 0 to 1. Moreover, logits and probits change in value in virtually the same fashion, except close to extreme probability values of 1 and 0. This reflects the close similarity of the S-shaped cumulative logistic and cumulative normal probability functions, both of which imply that a unit change in the explanatory variable X has a larger effect on the probability close to the centre of the distribution and a progressively smaller effect close to the limits of 0 and 1. As a result, the choice between logit and probit models as alternatives to the linear probability model, and as suitable models of the relationship between a dichotomous response variable and a continuous explanatory variable,

---

[†] This model is also referred to as a *normit* model. Historically, in probit analysis, a constant (with a value of 5) was added to $(\alpha+\beta X_i)$ so that calculation of the parameter estimates involved only positive numbers. With modern computers, this is no longer necessary, and the distinction between normit and probit models is no longer important.

[‡] This is referred to as the integrated normal transformation (see Bishop, Fienberg and Holland 1975, 367).

is essentially a matter of computational convenience. As the probit model is slightly less convenient in computational terms, substantially more difficult to comprehend for those readers meeting such models for the first time, and as it produces very similar results to the logit model in the dichotomous response variable case, we will set it aside for the moment and concentrate upon the logit model. However, the probit model will reappear in Chapter 11. We will see in that chapter that, in its more general polytomous form, it has useful advantages over the logit model for a certain class of problems, though it is computationally much more taxing.

2.4 Estimating the parameters of a logistic/logit model

Having selected logistic/logit models as models which are capable of handling dichotomous response variables, how do we then estimate the parameters of these models? Basically, there are two possible strategies depending on whether our sample data can be treated as falling into distinct groups or if each data unit must be treated as a separate observation.

2.4.1 Grouped data and weighted least squares

Let us imagine in our mental hospital example that: our survey provides information on 1,000 respondents; these respondents have been selected using a locationally stratified sample; and at each of forty distances from the proposed site of the mental hospital, information from twenty-five respondents has been collected. Recalling that our explanatory variable, X_i, is the distance from the home of respondent i to the proposed site of the mental hospital, we therefore have a sample divided into forty groups; each group providing twenty-five repetitions of the same explanatory variable value. That is to say, we have $n_1 = 25$ respondents having explanatory variables value X_1, $n_2 = 25$ respondents having explanatory variable value X_2, and so on. For each of these groups we know the numbers falling into each of the response categories, i.e. the numbers who are 'in favour' or 'against' the mental

Table 2.1
Form of contingency table
for grouped data

Sub-populations	Response categories		Total
	$r = 1$	$r = 2$	
$g = 1$	$n_{1\|1}\ (f_{1\|1})$	$n_{2\|1}\ (f_{2\|1})$	$n_{+\|1} = 25\ (1.0)$
$g = 2$	$n_{1\|2}\ (f_{1\|2})$	$n_{2\|2}\ (f_{2\|2})$	$n_{+\|2} = 25\ (1.0)$
.	.	.	.
.	.	.	.
.	.	.	.
$g = 40$	$n_{1\|40}\ (f_{2\|40})$	$n_{2\|40}\ (f_{2\|40})$	$n_{+\|40} = 25\ (1.0)$

hospital scheme, and our sample data can, therefore, be displayed in the form of Table 2.1 (i.e. in the manner of Table 1.6) where $f_{r|g}$ is used to denote the observed proportion of group (sub-population) g who fall into each response category ($r = 1 = $ 'in favour', $r = 2 = $ 'against').

The observed proportions $f_{1|g}$ and $f_{2|g}$ can then be used to approximate the unknown probability values[†] in the left-hand side of the linear logit model [2.15], and the linear logit model can be written as[‡]

$$\bar{L}_{12|g} = \log_e \frac{f_{1|g}}{f_{2|g}} = \alpha + \beta X_g + (\bar{L}_{12|g} - L_{12|g}) \qquad [2.20]$$

where the term $(\bar{L}_{12|g} - L_{12|g})$ is introduced to take account of the fact that the left-hand side now contains only an estimate $(\bar{L}_{12|g})$ of the true logit; an estimate based upon the observed proportions. Under the assumption that the observed proportions $f_{1|g}$ and $f_{2|g}$ are the results of independent random drawings from a binomial population with probabilities $P_{1|g}$ and $P_{2|g}$, it follows (Theil 1970, 137–8) that $\bar{L}_{12|g}$ is asymptotically normally distributed with mean and variance.[§]

$$E_A(\bar{L}_{12|g}) = L_{12|g} \qquad \mathrm{Var}_A(\bar{L}_{12|g}) = 1/[n_{+|g}P_{1|g}(1-P_{1|g})] \qquad [2.21]$$

and the error term $(\bar{L}_{12|g} - L_{12|g})$ has asymptotic properties

$$E_A(\bar{L}_{12|g} - L_{12|g}) = 0$$

$$\mathrm{Var}_A(\bar{L}_{12|g} - L_{12|g}) = 1/[n_{+|g}P_{1|g}(1 - P_{1|g})] \qquad [2.22]$$

Furthermore, it can be shown that the asymptotic properties of $\mathrm{Var}_A(\bar{L}_{12|g} - L_{12|g})$ are not affected when we replace $P_{1|g}$ by $f_{1|g}$.

From [2.22] it can be seen that the error term $(\bar{L}_{12|g} - L_{12|g})$ in the linear logit model is heteroscedastic. That is to say, the error variance is not constant but depends upon the probabilities of occurrence of each response ($P_{1|g}$ and $P_{2|g} = (1 - P_{1|g})$) and on the sample size of each sub-population. Because of this heteroscedasticity we do not use the conventional ordinary least squares (OLS) procedure to estimate the values of the parameters α and β. Instead, we adopt a procedure known as *weighted least squares* (WLS) which does not require a constant error variance (homoscedasticity) assumption.

The weighted least squares procedure we use is a very straightforward extension of the ordinary least squares procedure. In the OLS procedure, parameter estimates are obtained by minimizing the sum of squared residuals, whereas in the WLS procedure parameter estimates are obtained by minimizing the weighted sum of squared residuals. In

---

[†] This is particularly useful, for in [2.15] $P_{1|i}/P_{2|i}$ will equal zero or infinity and $\log_e (P_{1|i}/P_{2|i})$ will be undefined when $P_{1|i}$ equals 0 or 1. Thus, if we simply replace $P_{1|i}$ in [2.15] by 1 or 0 to denote the observed response category of individual i, the logit will be undefined. We will only avoid this problem by observing the responses of a group of individuals and working with observed proportions.

[‡] Clearly, the left-hand side of [2.20] can be written in terms of observed frequencies as $\log_e(n_{1|g}/\{n_{+|g} - n_{1|g}\})$, and the reader will often see the logit written in this way. He will also see a modified version of this (see Cox 1970, 33) which takes account of the fact that it is undefined when $n_{1|g} = 0$ or $n_{+|g}$. This modified version takes the form $\log_e(\{n_{1|g}+\frac{1}{2}\}/\{n_{+|g}-n_{1|g}+\frac{1}{2}\})$.

[§] E_A means asymptotic expectation (or mean) and Var_A means asymptotic variance.

the case of a simple regression model, for example, this implies that the OLS estimates are obtained by minimizing

$$\sum_i (Y_i - \hat{Y}_i)^2 = \sum_i (Y_i - \hat{\alpha} - \hat{\beta} X_i)^2 \qquad [2.23]$$

whereas the WLS estimates are obtained by minimizing

$$\sum_i w_i (Y_i - \hat{Y}_i)^2 = \sum_i w_i (Y_i - \hat{\alpha} - \hat{\beta} X_i)^2 \qquad [2.24]$$

where $w_i = 1/\mathrm{Var}(\varepsilon_i)$. In the case of our linear logit model [2.20] it follows, therefore, that we obtain WLS parameter estimates by minimizing

$$\sum_g w_g (\bar{L}_{12|g} - \hat{L}_{12|g})^2 = \sum_g w_g (\bar{L}_{12|g} - \hat{\alpha} - \hat{\beta} X_g)^2 \qquad [2.25]$$

where the weights are defined as

$$w_g = n_{+|g} f_{1|g} (1 - f_{1|g}) \qquad [2.26]$$

Weights of this form imply that, given the value of $f_{1|g}$, more weight is given to those sub-populations which have larger sample sizes $(n_{+|g})$. On the other hand, given $n_{+|g}$, as $f_{1|g}$ approaches 0 or 1 less weight is allocated, for in these situations the logit $\bar{L}_{12|g}$ takes large negative or positive values and is highly sensitive to small changes in $f_{1|g}$. In the case where $f_{1|g}$ is either 0 or 1 the weight is zero for in this case the logit becomes infinitely large and cannot be handled in the computations. The weights thus effectively exclude a sub-population in which the observed proportion choosing a particular response category is 0 or 1. Berkson (1953, 1955) has suggested that such an exclusion represents an unwarranted waste of information. He advocates that we use instead replacement working values of the form[†]

$$1/2n_{+|g} \text{ instead of } 0 \text{ when } f_{1|g} = 0$$
$$1 - 1/2n_{+|g} \text{ instead of } 1 \text{ when } f_{1|g} = 1 \qquad [2.27]$$

As in the case of OLS estimation of a simple regression model (see Pindyck and Rubinfeld 1976, 7), we can find the values of $\hat{\alpha}$ and $\hat{\beta}$ which minimize the expression [2.25] by using elementary calculus. We do this by taking the first partial derivatives with respect to $\hat{\alpha}$ and $\hat{\beta}$, setting each of these to 0, and solving the resulting pair of simultaneous equations. That is to say:

$$\frac{\partial}{\partial \hat{\alpha}} \sum_g w_g (\bar{L}_{12|g} - \hat{\alpha} - \hat{\beta} X_g)^2 = -2 \sum_g w_g (\bar{L}_{12|g} - \hat{\alpha} - \hat{\beta} X_g) = 0 \qquad [2.28]$$

$$\frac{\partial}{\partial \hat{\beta}} \sum_g w_g (\bar{L}_{12|g} - \hat{\alpha} - \hat{\beta} X_g)^2 = -2 \sum_g w_g X_g (\bar{L}_{12|g} - \hat{\alpha} - \hat{\beta} X_g) = 0 \qquad [2.29]$$

[†] Recalling that $f_{1|g} = n_{1|g}/n_{+|g}$ these replacement working values can be derived automatically by replacing any zero value for $n_{1|g}$ or $n_{2|g}$ in Table 2.1 by ½.

Dividing by -2 and rewriting [2.28] and [2.29], we obtain the pair of simultaneous equations known as the 'normal equations'

$$\sum_g w_g \bar{L}_{12|g} = \hat{\alpha}\sum_g w_g + \hat{\beta}\sum_g w_g X_g \qquad [2.30]$$

$$\sum_g w_g X_g \bar{L}_{12|g} = \hat{\alpha}\sum_g w_g X_g + \hat{\beta}\sum_g w_g X_g^2 \qquad [2.31]$$

We then solve[†] these two simultaneous equations for the two unknown terms $\hat{\alpha}$ and $\hat{\beta}$, and this produces (see Berkson 1953, 566) the required weighted least squares estimators

$$\hat{\beta} = \frac{\sum_g w_g(L_{12|g} - L^*)(X_g - X^*)}{\sum_g w_g(X_g - X^*)^2} \qquad [2.32]$$

$$\hat{\alpha} = L^* - \hat{\beta}X^* \qquad [2.33]$$

where

$$L^* = (\sum_g w_g \bar{L}_{12|g})/\sum_g w_g \qquad \text{and} \qquad X^* = (\sum_g w_g X_g)/\sum_g w_g$$

The asymptotic variances of $\hat{\alpha}$ and $\hat{\beta}$ can be estimated from the expressions

$$\text{Var}(\hat{\beta}) = 1/\sum_g w_g(X - X^*)^2 \qquad [2.34]$$

$$\text{Var}(\hat{\alpha}) = (1/\sum_g w_g) + (X^*)^2\text{Var}(\hat{\beta}) \qquad [2.35]$$

To a significant extent, the weighted least squares method which we have just discussed depends for its justification on there being a large number of repetitions of each explanatory variable value. That is to say, $n_{+|g}$, the size of each sub-population in Table 2.1 must be sufficiently large to justify the approximation of the unknown probability values in the linear logit model by the observed proportions $f_{1|g}$ and $f_{2|g}$ and also to justify the estimation of $\text{Var}_A(\bar{L}_{12|g} - L_{12|g})$ by $1/[n_{+|g}f_{1|g}(1-f_{1|g})]$. (Pindyck and Rubinfeld 1976, 250; Domencich and McFadden 1975 110; Koch *et al.* 1977, 157 discuss the issue of what represents 'sufficiently large'.) When there is only one explanatory variable, a large overall sample, and a sample design similar to that which underlies Table 2.1, then the number of repetitions of each explanatory variable value is likely to be sufficient. However, when the number of explanatory variables increases and/or the sample design does not result in exact repetitions of values of the explanatory variables, then serious difficulties can be encountered.

[†] Divide through both sides of [2.30] by $\sum_g w_g$ and solve for $\hat{\alpha}$. Substitute this result into [2.31] and rearrange to produce

$$\hat{\beta} = \frac{\sum_g w_g X_g \bar{L}_{12|g} - \dfrac{\sum_g w_g \bar{L}_{12|g}\sum_g w_g X_g}{\sum_g w_g}}{\sum_g w_g X_g^2 - \dfrac{\left(\sum_g w_g X_g\right)^2}{\sum_g w_g}}$$

This expression can be simplified (see Berkson 1953, 566) to the form [2.32].

When the number of explanatory variables increases, the sub-populations in Table 2.1 are defined on the basis of distinct combinations of values of the explanatory variables. Thus, in the case in our mental hospital example, if we postulate a model which contains three explanatory variables, and there are fifteen distinct values of the first explanatory variable in the sample, ten values of the second explanatory variable, and six values of the third explanatory variable, there will be $15 \times 10 \times 6 = 900$ sub-populations which can be defined on the basis of distinct combinations of values of the explanatory variables. Given our assumption that the total sample size in our example is 1,000, this number of sub-populations will result in there being insufficient repetitions of each combination of values of the explanatory variables to justify use of the WLS estimation procedure. Clearly, for a fixed total sample size, as the number of explanatory variables increases, we will rapidly reach the position in which there is at most only one observation of each distinct combination of values of the explanatory variables (i.e. a situation like that displayed in Table 1.7). In this case the number of sub-populations equals the total sample size; $f_{1|g}$ equals 0 or 1; it is not possible to calculate $\log_e f_{1|g}/f_{2|g}$; and the weighted least squares procedure breaks down. The only exception to this statement occurs in the case in which all the explanatory variables in the model are categorical and each variable is divided into only a small number of categories (see Table 1.5). In this case the number of sub-populations will remain reasonably small as the number of explanatory variables increases.

The reader should recall, however, that in this chapter we are concerned with the problems of cell (d) of Table 1.1 in which all the explanatory variables are continuous. Here, not only will the explanatory variables not be divided into a small number of natural categories, but also it is unlikely that there will be many, if any, exact repetitions of values of the explanatory variables. Instead there will be approximate repetitions. For example, instead of there being only forty or fifteen distinct values of the first explanatory variable amongst our sample of 1,000, there may be 500, 600 or more distinct values, some of which differ from each other in terms of only the minor decimal places (e.g. 1.250, 1.262, 1.253, etc.). In order to employ the weighted least squares procedure in this case, it is necessary, therefore, to group together these approximate repetitions. This is done by partitioning each of the continuous explanatory variables into a limited number of categories and by defining the sub-populations in a table such as Table 2.1 on the basis of the cross-classifications of the partitioned variables (see Wrigley 1976, 16–18 for an example). Unfortunately, not only is this partitioning and grouping procedure laborious, particularly when a series of models is to be fitted, but it also becomes increasingly difficult to apply when there is more than a small number of explanatory variables. In addition, the procedure can result in biased parameter estimates due to the linked problems of an 'errors-in-explanatory variables' effect (see Domencich and McFadden 1975, 110), and the violation of the implicit assumption of the linear logit model that each individual in a sub-population has the same unknown probability of choosing a particular response category.[†]

[†] In order to assume that $f_{1|g}$ has a binomial distribution with mean $P_{1|g}$ and thus that $\hat{L}_{12|g}$ asymptotically has mean $L_{12|g}$, it is necessary to assume that each individual in a sub-population g has the

For these reasons, the WLS estimation procedure is rarely used when all the explanatory variables in a logistic/logit model are continuous.

2.4.2   Individual data and maximum likelihood

The difficulties associated with the weighted least squares approach when all the explanatory variables are continuous point to the need for an estimation procedure which does not require the partitioning of the continuous explanatory variables and the grouping of observations and which, instead, works directly with the individual observations. Such a procedure is in fact available. It is known as the method of *maximum likelihood* and it enables us to deal directly with sample data sets like that displayed in Table 1.7 where there is only one observation of each distinct combination of values of the explanatory variables. Using this procedure, we deal directly with individual observations i rather than groups of observations g.

The logic of the maximum likelihood[‡] procedure is straightforward. First, we derive an expression for the likelihood of observing the pattern of response category choices in our data set. Second, we note that the value of this expression depends upon a number of unknown parameters (α, β, etc.) which we wish to estimate. Third, we take as our estimates those values of the parameters which maximize the likelihood of the observed pattern of responses.

Given that the response category choices are viewed as independent drawings from a binomial distribution, the likelihood of obtaining the particular sample of response category choices which we observe in our data set is simply the joint probability of occurrence of the observed choices. If we arbitrarily order the choices in our data set so that the N_1 choices of the first response category (each of which occur with probability $P_{1|i}$) come prior to the $N - N_1$ choices of the second response category (each of which occur with probability $P_{2|i} = 1 - P_{1|i}$), the likelihood of the particular sample of response category choices in our data set is

$$\Lambda = \prod_{i=1}^{N_1} P_{1|i} \prod_{i=N_1+1}^{N} P_{2|i} \qquad\qquad [2.36]$$

Assuming, as in our original mental hospital scheme example, that we have only one explanatory variable, we can then substitute into [2.36] the expressions for $P_{1|i}$ and $P_{2|i}$ from [2.14] and [2.17]. This gives:

$$\Lambda = \prod_{i=1}^{N_1} \frac{e^{\alpha + \beta X_i}}{1 + e^{\alpha + \beta X_i}} \prod_{i=N_1+1}^{N} \frac{1}{1 + e^{\alpha + \beta X_i}} \qquad\qquad [2.37]$$

---

same unknown probability $P_{1|g}$ of choosing a particular response category. In the linear logit model [2.20] it is assumed that $f_{1|g}$ is not equal to $P_{1|g}$ and that $\hat{L}_{12|g}$ is only an estimate of the true logit, simply because only $n_{+|g}$ individuals are sampled, not because some of the individuals might have a different probability $P_{1|g}$. If, on the other hand, individuals in a sub-population *do* have different probabilities, then the error term in the model [2.20] will not be independent of the explanatory variables (s) and this will result in biased parameter estimates. This latter situation is likely to occur when continuous explanatory variables are partitioned into a limited number of categories and when approximate repetitions of explanatory variable values are grouped together into such categories (see Hanushek and Jackson 1977, 199–200).

[†] The method of maximum likelihood was developed by R.A. Fisher and arose out of his investigations into the efficiency and sufficiency of estimators (see Edwards 1972 for a general review of the method, and Kmenta 1971, 174–82 for a useful brief introduction).

It can be seen that the value of the likelihood [2.37] depends upon the values of the unknown parameters α and β and we, therefore, must estimate these parameters. We do this by taking as parameter estimates those values which maximize the overall value of the likelihood. In other words, the maximum likelihood method involves choosing those parameter values which would be most likely to have produced the observed sample of response category choices in our data set.

There are several possible methods by which we can determine which values of the parameters maximize the likelihood. One way is simply to compute Λ directly by hand for different values of α and β and to search for the maximum by trial and error. This method is clearly inefficient, particularly in more complex examples where there are a greater number of parameters to be estimated. As a result, we normally resort to the use of calculus. This involves taking partial derivatives, and since taking partial derivatives of the products in the likelihood [2.37] is difficult, it proves to be much more convenient to transform the likelihood and to maximize the logarithm of the likelihood instead of the likelihood itself. The log likelihood in this case is

$$\log_e \Lambda = \sum_{i=1}^{N_1} (\alpha + \beta X_i) - \sum_{i=1}^{N} \log_e (1 + e^{\alpha + \beta X_i}) \qquad [2.38]$$

which is a simple sum of the logs of the arguments and is much easier to differentiate than [2.37]. Since the logarithmic transformation is a monotonic transformation, the parameter values which maximize Λ also maximize $\log_e \Lambda$.

To find the parameter values that maximize the log likelihood, we take the first partial derivatives of $\log_e \Lambda$ with respect to α and β and set these equal to zero (these are known as the first-order conditions for a maximum[†])

$$\frac{\partial \log_e \Lambda}{\partial \alpha} = N_1 - \sum_{i=1}^{N} \frac{e^{\alpha + \beta X_i}}{1 + e^{\alpha + \beta X_i}} = 0 \qquad [2.39]$$

$$\frac{\partial \log_e \Lambda}{\partial \beta} = \sum_{i=1}^{N_1} X_i - \sum_{i=1}^{N} X_i \frac{e^{\alpha + \beta X_i}}{1 + e^{\alpha + \beta X_i}} = 0 \qquad [2.40]$$

Evaluating these derivatives at $\hat{\alpha}$ and $\hat{\beta}$, the solutions of these simultaneous equations are the maximum likelihood parameter estimators $\hat{\alpha}$ and $\hat{\beta}$. It can be seen, however, that these simultaneous equations are non-linear and are difficult to solve. As a result, we must solve them by resorting to a computer which can be programmed to find a numerical solution by means of some sort of iterative procedure. This, in fact, turns out to be much less complex than it sounds, and there are many package programs (see Ch. 7) which will perform the operation automatically.

To find the asymptotic variances of the maximum likelihood estimators $\hat{\alpha}$ and $\hat{\beta}$ we first take the second partial derivatives of $\log_e \Lambda$

[†] The second-order conditions for a maximum are that the second partial derivatives of $\log_e \Lambda$ with respect to α and β should both be negative.

with respect to α and β:

$$\frac{\partial^2 \log_e \Lambda}{\partial \alpha^2} = -\sum_{i=1}^{N} \frac{e^{\alpha + \beta X_i}}{\left(1 + e^{\alpha + \beta X_i}\right)^2} = -\sum_{i=1}^{N} P_{1|i}P_{2|i} \qquad \text{[2.41]}$$

$$\frac{\partial^2 \log_e \Lambda}{\partial \beta^2} = -\sum_{i=1}^{N} X_i^2 \frac{e^{\alpha + \beta X_i}}{\left(1 + e^{\alpha + \beta X_i}\right)^2} = -\sum_{i=1}^{N} X_i^2 P_{1|i}P_{2|i} \qquad \text{[2.42]}$$

$$\frac{\partial^2 \log_e \Lambda}{\partial \alpha \partial \beta} = -\sum_{i=1}^{N} X_i \frac{e^{\alpha + \beta X_i}}{\left(1 + e^{\alpha + \beta X_i}\right)^2} = -\sum_{i=1}^{N} X_i P_{1|i}P_{2|i} \qquad \text{[2.43]}$$

The matrix formed by the negative of the expected values of these second partial derivatives is called the *information matrix* (see Kmenta 1971, 160, 182, 216) and it has the form

$$\begin{bmatrix} -E\left(\dfrac{\partial \log_e \Lambda}{\partial \alpha^2}\right) & -E\left(\dfrac{\partial \log_e \Lambda}{\partial \alpha \partial \beta}\right) \\ -E\left(\dfrac{\partial \log_e \Lambda}{\partial \alpha \partial \beta}\right) & -E\left(\dfrac{\partial \log_e \Lambda}{\partial \beta^2}\right) \end{bmatrix} \qquad \text{[2.44]}$$

If we then evaluate these partial derivatives at $\hat{\alpha}$ and $\hat{\beta}$ and take the diagonal elements of the *inverse* of the information matrix we obtain consistent estimators of the asymptotic variances of $\hat{\alpha}$ and $\hat{\beta}$:

$$\text{Var}(\hat{\beta}) = 1/\sum_i P_{1|i}P_{2|i}(X_i - \bar{\bar{X}})^2 \qquad \text{[2.45]}$$

$$\text{Var}(\hat{\alpha}) = (1/\sum_i P_{1|i}P_{2|i}) + \bar{\bar{X}}^2 \text{Var}(\hat{\beta}) \qquad \text{[2.46]}$$

where

$$\bar{\bar{X}} = \sum_i P_{1|i}P_{2|i}X_i / \sum_i P_{1|i}P_{2|i}$$

These expressions are usually computed in the iterative search procedure employed to find the maximum likelihood parameter estimates and are produced automatically by the package programs.

The maximum likelihood estimation procedure has intuitive appeal for it allows us to use logit models in which all the explanatory variables are continuous, and in the process to deal with individual observations directly rather than via some partitioning and grouping procedure. In addition, it can be shown (McFadden 1974) that, except when the explanatory variables are multicollinear, $\log_e \Lambda$ is a well-behaved, strictly concave function which has a unique maximum (provided one exists). This property facilitates the numerical solution of the maximization problem by allowing the use of rapid iterative search procedures which are guaranteed to converge to the maximum. Finally, it can be shown that the maximum likelihood procedure yields estimators ($\hat{\alpha}$, $\hat{\beta}$, etc.) which have very desirable properties. First, it can be shown that the maximum likelihood estimators (MLEs) $\hat{\alpha}$ and $\hat{\beta}$ are *sufficient* estimators of α and β (if such estimators exist). Second, in large samples, these MLEs become approximately normally distributed with expected values which approach the true parameters (α, β, etc.) and with variances which approach the minimum possible variances of any unbiased estimators. In colloquial terms we can say that the MLEs will produce the most likely values of the parameters given the available sample data, and that, as sample size increases, the most likely values become ever closer to the true values of the parameters.

2.5 Some simple examples

Now that we have selected logistic/logit models as suitable models for analysing dichotomous response variables, and seen that for problems in cell (d) of Table 1.1 direct maximum likelihood estimation is a suitable method of estimating the model parameters, it is appropriate to get a flavour of the typical uses of these models by considering some simple empirical applications. In contrast to our previous hypothetical mental hospital example, however, most real-world applications of these models involve models with more than one explanatory variable. Before proceeding, therefore, we must first note how our models extend to the multiple explanatory variable case.

When we have several explanatory variables, the expressions [2.14] and [2.15] for the logistic and linear logit models can be extended and written in the general forms:

$$P_{1|i} = \frac{e^{\beta_1 + \beta_2 X_{i2} + \beta_3 X_{i3} + \ldots + \beta_K X_{iK}}}{1 + e^{\beta_1 + \beta_2 X_{i2} + \beta_3 X_{i3} + \ldots + \beta_K X_{iK}}} \qquad [2.47]$$

and

$$\log_e \frac{P_{1|i}}{P_{2|i}} = \beta_1 + \beta_2 X_{i2} + \beta_3 X_{i3} + \ldots + \beta_K X_{iK} \qquad [2.48]$$

where K represents any number, and there are $K-1$ explanatory variables. These models are natural extensions of our previous single explanatory variable models, and when $K = 2$, they reduce to the equations [2.14] and [2.15]. It should be noted, however, (see footnote page 22) that when we have more than one explanatory variable we adopt the usual multiple regression convention of denoting the intercept or constant term by β_1 rather than α. This makes no difference to the models but has the great virtue that it simplifies the adoption of matrix and vector notation, which in turn allows us to write our models in a much more compact form. To see this, note that [2.48] can be expressed in vector notation as,

$$\log_e \frac{P_{1|i}}{P_{2|i}} = [1, X_{i2}, X_{i3}, \ldots, X_{iK}] \begin{bmatrix} \beta_1 \\ \beta_2 \\ \cdot \\ \cdot \\ \cdot \\ \beta_K \end{bmatrix} = x_i'\boldsymbol{\beta} \qquad [2.49]$$

where x_i' is known as a row vector of explanatory variable values, and $\boldsymbol{\beta}$ a column vector of parameters.

To derive the appropriate expression for the likelihood and log-likelihood in the multiple explanatory variable case, we simply substitute $x_i'\beta$ in place of $\alpha + \beta X_i$ in [2.37] and [2.38]. The maximum likelihood procedure then remains exactly the same. Maximum likelihood parameter estimates and estimated asymptotic variances of the parameter estimates will normally be found using a rapid iterative search procedure (with estimated asymptotic variances of the parameter estimates being obtained as before from the diagonal elements of the inverse of the information matrix) and the MLEs will have the properties discussed in the previous section.

EXAMPLE 2.1   *Oil and gas exploration in south-central Kansas*

Figure 2.6 shows the location of 124 wells drilled between 1948 and 1963 into the 'B' division of the Mississippian Osage Series in a 13 by 13 mile area of Stafford Country, south-central Kansas (Doveton 1973; Harbaugh, Doveton and Davis 1977, 192–219). The main oil and gas fields discovered in the period are indicated and exploration experience suggests that the location of these fields reflects both stratigraphic and structural controls. Using, as explanatory variables, three sub-surface

Figure 2.6
Location of oil and gas fields and 124 wells drilled into Mississippian B; Stafford County, south-central Kansas (after Harbaugh, Doveton and Davis 1977)

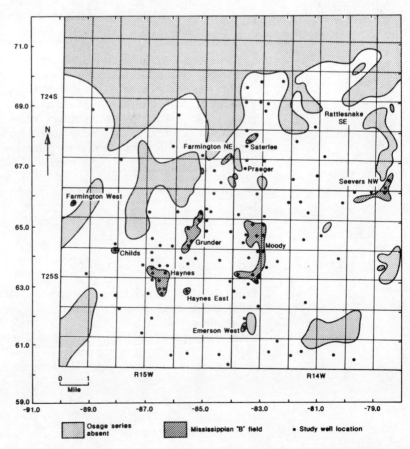

variables chosen to reflect both structural and stratigraphic influences, a logistic/logit model [2.47] was fitted by the author in an attempt to account for the probability that a well penetrating the Mississippian B at locality i is a 'producing' rather than a dry well. The parameters of this model were estimated using the maximum likelihood procedure, and the fitted model had the form:

$$\hat{P}_{1|i} = \frac{e^{x'_i\boldsymbol{\beta}}}{1 + e^{x'_i\boldsymbol{\beta}}} \quad \text{or} \quad \log_e \frac{\hat{P}_{1|i}}{\hat{P}_{2|i}} = x'_i\hat{\boldsymbol{\beta}} \qquad [2.50]$$

where

$$\begin{aligned} x'_i\hat{\boldsymbol{\beta}} &= \hat{\beta}_1 + \hat{\beta}_2 X_{i2} + \hat{\beta}_3 X_{i3} + \hat{\beta}_4 X_{i4} \\ &= -1.1265 + 0.0614\,X_{i2} + 0.0115\,X_{i3} - 8.1628\,X_{i4} \qquad [2.51] \\ &\quad\;(0.9678)\quad\;(0.0279)\qquad\;\;(0.0061)\qquad\;\;(3.0601) \end{aligned}$$

$\hat{P}_{1|i}$ = predicted probability of the well at locality i being a 'producing' well.

X_{i2} = thickness of the Mississippian B at i.

X_{i3} = local geological structure at i (as measured by the residual from a first-order trend surface of the top of the Mississippian B).

X_{i4} = the shale content of the Mississippian B at i (an indicator of permeability, measured by the mean digitized gamma ray log reading).

Estimated asymptotic standard errors of the maximum likelihood parameter estimates are shown in parentheses below the estimates.

This simple model indicates that Mississipian B 'producing' wells in this area of south-central Kansas are most likely in localities where the B zone is thick, where the shale content is low, and where there are local positive structures. It should be noted, however, that this is a very simple specification for illustration purposes only. Other explanatory variables (e.g. indicating lateral variations in the sub-surface characteristics in the areas surrounding the wells), are likely to be important determinants of whether a well is a 'producing' rather than dry well. Also, we make no attempt at this stage to assess the significance of the parameter estimates in the fitted model.

EXAMPLE 2.2   *Work trip mode choice in Sydney*

Using the maximum likelihood procedure, Hensher and McLeod (1977) estimated a large number of work trip mode choice models using information collected in journey to work surveys in the northern suburban area of Sydney, Australia. Their models were aimed at explaining and predicting the choice of car rather than train as the main mode of travel on work trips. The simplest model they fitted had the form:

$$\hat{P}_{1|i} = \frac{e^{x'_i\boldsymbol{\beta}}}{1 + e^{x'_i\boldsymbol{\beta}}} \quad \text{or} \quad \log_e \frac{\hat{P}_{1|i}}{\hat{P}_{2|i}} = x'_i\hat{\boldsymbol{\beta}} \qquad [2.52]$$

where

$$x_i'\hat{\boldsymbol{\beta}} = 0.110 - 0.0476\ X_{i2} - 0.0176\ X_{i3} \qquad\qquad [2.53]$$
$$\phantom{x_i'\hat{\boldsymbol{\beta}} = 0.110 - }(0.016)\phantom{\ X_{i2} - }(0.007)$$

$\hat{P}_{i|i}$ = predicted probability that individual i will choose to use a car.

X_{i2} = $(T_i^{car} - T_i^{train})$ = the difference between the two modes in terms of total door-to-door travel times.

X_{i3} = $(C_i^{car} - C_i^{train})$ = the difference between the two modes in terms of travel cost.

Both parameter estimates have the expected negative signs, indicating that as the time and cost of travelling by car increase relative to travel by train so the odds (recall that $P_{1|i}/P_{2|i}$ is known as the odds) of choosing to use a car as the main mode of travel to work decrease.

An important point to notice about this model is that its explanatory variables represent attributes of the choice alternatives or response categories (i.e. attributes of the car and train modes of travel to work) and that they are entered into the model as *differences* between the attributes of choice alternatives or response categories. This contrasts with our previous example [2.51] where the explanatory variables represent attributes of the localities i at which the response categories are measured *not* attributes of the response categories ('producing' or dry) themselves, and the variables are entered into the model in the conventional (non-difference) regression model form. This contrast between the models in the nature of their explanatory variables is a very important one. It reflects a slightly different derivation of the logistic/logit model in the case of model [2.52], (see Appendix 2.1 for explanation), and is related to the fact that the objectives of a field of study which has become known as 'discrete choice modelling' (which we will consider in detail in Part 4) differ somewhat from those of the more conventional statistical modelling and data analysis with which we are concerned in Parts 1 to 3 of this book. However, unless this difference in the nature of the explanatory variables is highlighted and explained, it can cause confusion for readers meeting these models for the first time. Unfortunately this is rarely, if ever, done in the existing literature.

At this stage in our discussion, it is appropriate simply to highlight this difference in the nature of the explanatory variables, to refer the reader to Appendix 2.1, and to establish that in general the explanatory variables in our logistic/logit models can be partitioned into two groups. The first group of variables (which we will term X^ϕ variables) represent attributes of the choice alternatives/response categories, and the second group (which we will term X^ψ variables) represent attributes of the choice maker/respondent i or the locality i at which the choice is made or the response category is measured. In any single model, explanatory variables from one or both of these groups may be included. After extending our logistic/logit models to the case in which we have a polytomous (multiple-category, multiple-alternative) response variable, we will return to this issue and develop in Section 2.11 a general expression for $x_i'\boldsymbol{\beta}$ which distinguishes these two groups of explanatory variables.

<u>EXAMPLE 2.3</u>   *Shopping trip mode choice in Pittsburgh*

In the previous examples, 2.1 and 2.2, we have considered models in which the explanatory variables are drawn from only one of the two groups we have just distinguished (X^ψ variables in the case of Example 2.1, and X^Φ variables in the case of Example 2.2). We now consider an example in which the explanatory variables are drawn from both groups.

Domencich and McFadden (1975, 165) investigated the mode choice on shopping trips of a sample of 140 shoppers in Pittsburgh. The models they fitted were aimed at explaining and predicting the choice of car rather than public transit (bus and street car treated as a single mode) as the mode of travel on shopping trips. The simplest model they fitted had the form:

$$\hat{P}_{1|i} = \frac{e^{x'_i\boldsymbol{\beta}}}{1 + e^{x'_i\boldsymbol{\beta}}} \quad \text{or} \quad \log_e \frac{\hat{P}_{1|i}}{\hat{P}_{2|i}} = x'_i\hat{\boldsymbol{\beta}} \quad\quad \textbf{[2.54]}$$

where[†]

$$\begin{aligned}
x'_i\hat{\boldsymbol{\beta}} = &-6.78 + 0.374TW_i - 0.065(AIV_i - TSS_i) \\
&\;\,(1.66)\quad(0.33)\qquad\;\;(0.03) \\
&-4.11(AC_i - F_i) + 2.24(A_i/W_i) \\
&\;\;\,(1.67)\qquad\qquad(1.11)
\end{aligned} \quad\quad \textbf{[2.55]}$$

$\hat{P}_{1|i}$ = predicted probability that individual i will choose to use a car.

TW_i = the walking time for individual i to and from the public transit stop.

AIV_i = time spent in the car (auto) by individual i including parking.

TSS_i = time spent in the public transit mode trip including any waiting time and transfer time.

AC_i = the cost of using the car including parking charges.

F_i = the public transit fare.

A_i/W_i = the number of cars per worker in individual i's household.

All the parameter estimates have the expected signs. The positive sign of the TW coefficient indicates that as walking time to the public transit stop increases, so do the odds of choosing to use a car rather than public transit. The negative signs of the $(AIV - TSS)$ and $(AC - F)$ coefficients indicate that as the time and cost of travelling by car increase relative to the time and cost of travelling by public transit, so the odds of choosing to use a car decrease. Finally, the positive sign of the A/W coefficient indicates that the odds of choosing to use a car increase as car availability increases.

It is important to notice that the first three explanatory variables, TW, $(AIV - TSS)$ and $(AC - F)$, represent attributes of the choice alternatives/response categories (i.e. attributes of the car and public

---

[†] Notice that in $x'_i\hat{\boldsymbol{\beta}}$ instead of X_{i1}, X_{i2}, etc. we use the acronyms TW_i, $AIV_i - TSS_i$, etc. where $TW_i = X_{i1}$, etc. This is often more convenient and from this point onwards we will often use such acronyms.

transit modes) as experienced by individual i. That is to say, they are all group one (or X^ϕ) variables. The fact that two are entered into the model as differences will have indicated this. However, it is important that the reader does not rely on seeing the difference form in order to identify X^ϕ variables for, as in the case of the first variable (walking time), the difference form will not be apparent if for that variable one choice alternative/response category has a natural or assumed zero value. For example, the form of the walking time variable was strictly $(AW_i - TW_i)$ but walking times to and from the car on shopping trips proved to be very difficult to measure and, as a result, Domencich and McFadden considered it preferable to make the simplifying assumption that walking times to and from the car on all shopping trips were zero. When walking times to and from the car were assumed to be zero $(AW_i - TW_i)$ became $(0 - TW_i)$ or simply, for convenience, TW_i. It can be seen, therefore, that some X^ϕ variables which are implicitly in difference form will appear at first sight to the reader to be entered into the model in non-difference form. Moreover, we will see in Section 2.11 that some X^ϕ variables are never entered in difference form.

The final explanatory variable in the model, A/W, is a group two (or X^ψ) variable. It represents an attribute of the choice maker/respondent i (i.e. it is a socio-economic variable which indicates his/her likely access to a car at any given time), *not* an attribute (walking time, in-vehicle time, etc.) of a choice alternative (where in this case the choice alternatives are modes of travel).

2.6 Testing hypotheses about the logistic/logit model parameters[†]

In the previous examples we have considered only the signs of the parameter estimates and asked whether these are in accordance with our prior expectations. It is possible, however, to go beyond this and to test hypotheses about the values of the logistic/logit model parameters. We can do this by using statistical tests analogous to the separate and joint tests employed in the multiple regression context (see Huang 1970, 90–103; Kmenta 1971, 366–74).

2.6.1 Separate tests

A separate test is one in which we test a hypothesis about the value of a single parameter (for example, the parameter β_k). Most frequently, we are interested in testing the null hypothesis (H_0) that β_k is equal to zero,

$$H_0 : \beta_k = 0 \qquad \qquad \text{[2.56]}$$

[†] This discussion assumes that the reader is familiar with hypothesis testing, confidence intervals, the t, F, and chi-square distributions, and the testing of hypotheses in multiple regression models. Readers who wish to revise these topics should consult Kmenta (1971, 112–51, 235–9, 366–74).

against the alternative hypothesis (H_1) that

$$H_1 : \beta_k \neq 0 \tag{2.57}$$

In a conventional multiple regression model, such tests are conducted using the test statistic

$$t = \frac{\hat{\beta}_k - \beta_k}{\sqrt{\mathrm{Var}\,(\hat{\beta}_k)}} \tag{2.58}$$

which has a t distribution with $N - K$ degrees of freedom. The logic of this test statistic is that if the sample estimate of the parameter and the hypothesized true value of the parameter are far apart, we tend not to believe in the null hypothesis. To decide how far apart the two values must be before we decide to reject the null hypothesis, we compare this difference to a measure of the sampling distribution of the estimator of β_k; that is to say, we compare it to the standard deviation or standard error, $\sqrt{\mathrm{Var}\,(\hat{\beta}_k)}$. Under the null hypothesis, β_k equals zero, and thus the test statistic takes the form

$$t = \frac{\hat{\beta}_k - 0}{\sqrt{\mathrm{Var}\,(\hat{\beta}_k)}} = \frac{\hat{\beta}_k}{\sqrt{\mathrm{Var}\,(\hat{\beta}_k)}} \tag{2.59}$$

The boundary between the rejection and acceptance (non-rejection) regions (known as the critical value t_c) can then be determined from the table of the t distribution for any given level of significance and for any number of degrees of freedom. For example, for a two-tail test at the 5 per cent level of significance and with $N - K$ degrees of freedom, the critical value t_c equals $t_{N-K,\ 0.05/2'}$ and the acceptance (non-rejection) region is defined by

$$-t_c \leq \frac{\hat{\beta}_k}{\sqrt{\mathrm{Var}\,(\hat{\beta}_k)}} \leq + t_c \tag{2.60}$$

If the computed value of the test statistic t lies outside this (i.e. in the rejection region) we can reject the null hypothesis that $\beta_k = 0$. Since $t_c = 1.96$ for a large sample size and a 5 per cent level of significance, a rule of thumb is often applied which states that if the value of the test statistic is greater than $+2$ or less than -2 we reject the null hypothesis.

By analogy with the t test of the conventional multiple regression model, it has become the custom in applications of logistic/logit models to calculate the ratio of the parameter estimates to their estimated asymptotic standard errors, i.e. $\hat{\beta}_k / \sqrt{\mathrm{Var}\,(\hat{\beta}_k)}$, and to report these as the 't statistics' (Schmidt and Strauss 1975a; and Domencich and McFadden 1975, 159 provide typical examples). In this case, however, under the usual null hypothesis, the ratios are asymptotically distributed as the standard normal distribution. As such, the reported statistics are not strictly t statistics in the conventional multiple regression sense. In practice, however, this makes little difference as it is the custom in applications of logistic/logit models to adopt the 5 per cent significance level and to apply the rule of thumb noted above (i.e. if the value of the 't statistic' is greater than $+2$ or less than -2 the null hypothesis is rejected).

2.6.2   Joint tests

A second type of statistical test which we can apply is that known in the multiple regression context as a joint test. In this case, we test the null hypothesis that *all* the parameters associated with the explanatory variables are equal to zero

$$H_0 : \beta_2 = \beta_3 = \ldots = \beta_k = 0 \qquad\qquad \text{[2.61]}$$

against the alternative that H_0 is not true, i.e. that at least one of the parameters associated with the explanatory variables is different from zero. Notice that in this test we do not include the constant term β_1 as the test is concerned with the overall value of including the chosen set of explanatory variables in the model.

In the multiple regression context the hypothesis is evaluated using a test statistic which exploits measures calculated during the least squares estimation procedure. The test statistic is based on the ratio (known as the F ratio) of regression sum of squares to residual sum of squares and has a F distribution. In the case of logistic/logit models whose parameters are estimated by the maximum likelihood procedure, this F ratio test is clearly not appropriate and, instead, we adopt a likelihood ratio test. The simplest way to express this test statistic is to state it in terms of the maximized log likelihoods. It then takes the form[†] (McFadden 1974, 121)

$$-2[\log_e \Lambda(C) - \log_e \Lambda(\hat{\boldsymbol{\beta}})] \qquad\qquad \text{[2.62]}$$

where $\log_e \Lambda(\hat{\boldsymbol{\beta}})$ is the maximized log likelihood of the fitted model which includes all the parameters $\beta_1, \beta_2, \ldots, \beta_K$, and $\log_e\Lambda(C)$ is the maximized log likelihood of the fitted model which includes only the constant term β_1 (i.e. a model in which $\beta_2, \ldots, \beta_K$ are constrained to zero). If the null hypothesis is true, the test statistic [2.62] is distributed asymptotically as chi-square with $K-1$ degrees of freedom. A value of [2.62] greater than the calculated value of chi-square at, say, the 5 per cent level of significance with $K-1$ degrees of freedom, implies that we can reject the null hypothesis $\beta_2 = \beta_3 = \ldots = \beta_K = 0$ with a 5 per cent chance of error.

2.6.3   Partial joint tests

Closely related to the joint test is a test which in the multiple regression context is sometimes referred to as a partial joint test (Huang 1970, 99). In this case, we consider jointly not *all* the explanatory variable

---

[†] This is often reported as $-2 \log_e \lambda$, where λ is the likelihood ratio $\Lambda(C)/\Lambda(\hat{\boldsymbol{\beta}})$. See Hensher and Johnson 1981a, 50.

parameters but a subset. We do this by partitioning our parameters (and hence explanatory variables) into two subsets: that is to say,

$$\sum_{k=1}^{K} \beta_k = \sum_{k=1}^{p} \beta_k + \sum_{k=q}^{K} \beta_k \qquad\qquad [2.63]$$

In vector notation this can be written as

$$\boldsymbol{\beta} = \begin{bmatrix} \boldsymbol{\beta}_A \\ \boldsymbol{\beta}_B \end{bmatrix} \qquad\qquad [2.64]$$

where the subscripts in this case indicate the subsets. We then test the null hypothesis that all parameters in the second subset are equal to zero

$$H_0 : \boldsymbol{\beta}_B = \boldsymbol{0} \qquad\qquad [2.65]$$

against the alternative hypothesis that H_0 is not true.

In our logistic/logit model case, the appropriate test statistic is similar to [2.62] and takes the form

$$-2\left[\log_e \Lambda(\hat{\boldsymbol{\beta}}_A) - \log_e \Lambda(\hat{\boldsymbol{\beta}})\right] \qquad\qquad [2.66]$$

where $\log_e \Lambda(\hat{\boldsymbol{\beta}}_A)$ is the maximized log likelihood of the fitted model which includes only the $\boldsymbol{\beta}_A$ subset of parameters (i.e. a model in which the $\boldsymbol{\beta}_B$ subset of parameters have been constrained to zero). If the null hypothesis is true, the test statistic is distributed asymptotically as chi-square with degrees of freedom equal to the difference in the number of parameters in the full set $\boldsymbol{\beta}$ and the subset $\boldsymbol{\beta}_A$. This implies that the degrees of freedom equal the number of parameters in the $\boldsymbol{\beta}_B$ subset, i.e. the number constrained to zero. As before, if the value of [2.66] is greater than the tabulated value of chi-square at a given level of significance, and with the appropriate degrees of freedom, we reject the null hypothesis and suggest that the $\boldsymbol{\beta}_B$ subset of parameters are not all equal to zero.

This partial joint test is a very useful one. It enables us to compare different specifications of a model. That is to say, it enables us to test whether a postulated model can be simplified without a significant loss of explanatory power or, alternatively, whether adding extra variables to a model leads to a significant improvement in the explanatory power of the model.

EXAMPLE 2.4   *Oil and gas exploration in south-central Kansas (continued)*

In Example 2.1 we considered a simple model which attempted to account for the probability that a well penetrating the 'B' division of the Mississippian Osage Series will be a 'producing', rather than a dry, well. We can now add some useful additional pieces of information to those presented in Example 2.1. Specifically, we can add the maximized log likelihood of the fitted model, the maximized log likelihood of the model which includes only the constant term β_1, and the ratios of the parameter estimates to their estimated asymptotic standard errors (the '*t* statistics'). We can then use these to test hypotheses about the parameters of the model. In this case $x'_i\hat{\beta}$ now becomes:

$$x'_i\hat{\beta} = -1.1265 + 0.0614X_{i2} + 0.0115X_{i3} - 8.1628X_{i4} \qquad [2.67]$$
$$\phantom{x'_i\hat{\beta} =} (0.9678) \quad (0.0279) \qquad (0.0061) \qquad (3.0601)$$
$$\phantom{x'_i\hat{\beta} =} [-1.1640] \quad [2.2007] \qquad [1.8945] \qquad [-2.6675]$$

where the, 't statistics' are given in square brackets below the estimated asymptotic standard errors, and

$$\log_e \Lambda(C) = -71.8419 \qquad \log_e \Lambda(\hat{\beta}) = -56.2673 \qquad [2.68]$$

Using the joint test described above, we can test the null hypothesis

$$H_0 : \beta_2 = \beta_3 = \beta_4 = 0 \qquad\qquad [2.69]$$

using the test statistic [2.62]. For this model the test statistic has the value $-2[-71.8419 - (-56.2673)] = 31.1492$, whilst the tabulated value of chi-square at the 5 per cent level of significance and with $K-1$ (in our case 4–1) degrees of freedom is 7.81. Clearly, the null hypothesis can be rejected and we can suggest that at least one of the explanatory variables has an associated non-zero parameter. This is confirmed by inspecting the 't statistics'. With the exception of that for β_3, they are all greater than $+2$ or less than -2. This implies that at the conventional 5 per cent level of significance, and using the rule of thumb described above, we can reasonably reject the separate null hypotheses $\beta_2 = 0$ and $\beta_4 = 0$, and suggest that thickness of the Mississippian B zone and its shale content are significant influences on the likelihood of a well being a 'producing' well rather than a dry well.

On the basis of the separate tests, it appears that, in this particular area of Kansas, the explanatory variable (local geological structure) is not a significant determinant of whether a well is a 'producing' rather than a dry well. Nevertheless, its associated 't statistic' value (1.89) is rather close to the 'rule of thumb' critical value (i.e. 2). Moreover, it is closer in value to the 't statistic' of β_2 than the 't statistic' of β_2 is to that of β_4. Does this imply that we should draw similar inferences about both β_2 and β_3 and about the associated explanatory variables (thickness and local geological structure)? To answer the question and confirm our previous inference, we can consider using the partial joint test procedure discussed above. We do this in two ways.

1. We can use the test statistic [2.66] directly to test the null hypothesis that $\beta_3 = 0$. We do this by dividing the parameters into two subsets with β_1, β_2 and β_4 in the first subset (β_A) and β_3 in the other subset (β_B), and by fitting a new model which includes only the explanatory variables X_{i2} (thickness of Mississippian B) and X_{i4} (shale ratio) and the parameters β_1, β_2 and β_4. This new fitted model provides the required value of $\log_e \Lambda(\hat{\beta}_A)$ in [2.66]. In this case its value is -58.0695. The maximized log likelihood value (-56.2673) of the original model [2.67] provides the required value of $\log_e \Lambda(\hat{\beta})$. Substituting these maximized log likelihood values into [2.66] gives an observed test statistic value of $-2(-58.0695 + 56.2673) = 3.6044$. This observed value is *less* than the tabulated value of chi-square at the 5 per cent level of significance with $(4-3 = 1)$ degrees of freedom (3.84) and we, therefore, cannot reject the null hypothesis. $H_0 : \beta_B = 0$ (i.e. in our case $\beta_3 = 0$). This confirms the result of the separate 't test'.

2. We can use the test statistic [2.66] to test whether we should draw similar inferences about the variables: thickness of the Mississippian B, and local geological structure. In other words, because of the relative similarity of their 't statistics' compared to that of the shale ratio variable, and because of the non-significance of the local geological structure variable, can both variables be treated in the same way and regarded as non-significant influences on the likelihood of a well being a 'producing' well rather than a dry well? This question is equivalent to posing the null hypothesis $H_0 : \beta_2 = \beta_3 = 0$ or, in others words, asking whether both explanatory variables (thickness and local geological structure) can be removed from the original model [2.67] without a significant loss of explanatory power. To answer this question using the partial joint test statistic [2.66] we divide the parameters into two subsets with β_1 and β_4 in the first subset ($\boldsymbol{\beta}_A$) and β_2 and β_3 in the other subset ($\boldsymbol{\beta}_B$), and fit a new model which includes only the variable X_{i4} (shale ratio) and the parameters β_1 and β_4. The new fitted model provides the required value of $\log_e \Lambda(\hat{\boldsymbol{\beta}}_A)$ in [2.66]. In this case its value is -62.2093. The maximized log likelihood value (-56.2673) of the original model [2.67] provides the required value of $\log_e \Lambda(\hat{\boldsymbol{\beta}})$. Substituting these vaues into [2.66] gives an observed test statistic value of $-2(62.2093 + 56.2673) = 11.8840$. This observed value is *greater* than the tabulated value of chi-square at the 5 per cent level of significance with $(4-2=2)$ degrees of freedom (5.9915), and we can reasonably *reject* the null hypothesis $H_0 : \boldsymbol{\beta}_B = \boldsymbol{0}$ (i.e. $\beta_2 = \beta_3 = 0$) and suggest that it is not reasonable to simplify the original model by removing *both* explanatory variables: thickness of the Mississippian B and local geological structure. Given the previous evidence (from the separate and partial joint tests) that $\beta_3 = 0$ (i.e. that local geological structure is not a significant explanatory factor), we can conclude that $\beta_2 \neq 0$ and that the explanatory variable, thickness of the Mississippian B must be included in our model.

On the basis of these tests, it seems reasonable to draw the inference that whilst thickness of the Mississippian B zone and its shale content are significant influences on the probability of a well in this area of Kansas being a 'producing' rather than a dry well, local geological structure (at least when it is entered into the model in the simple form shown in [2.51] and [2.67]) is not a statistically significant influence. As a result, it is not necessary to retain X_{i3} in our model and we can, therefore, remove it and fit a simplified model in which $x_i'\hat{\boldsymbol{\beta}}$ becomes:

$$x_i'\hat{\boldsymbol{\beta}} = -1.1601 + 0.0615X_{i2} - 7.8369X_{i4} \qquad [2.70]$$
$$\phantom{x_i'\hat{\boldsymbol{\beta}} = } (0.9383) \quad (0.0230) \qquad\quad (3.0487)$$
$$\phantom{x_i'\hat{\boldsymbol{\beta}} = } [-1.2364] \quad [2.6750] \qquad [-2.5706]$$

$$\log_e \Lambda(C) = -71.8419 \qquad \log_e \Lambda(\hat{\boldsymbol{\beta}}) = -58.0695$$

When interpreting these results, however, the reader must be cautious. It should be borne in mind that the model ([2.51], [2.67]) was a very simple specification used for illustration purposes. Other explanatory variables (e.g. indicating lateral variations in sub-surface characteristics in the areas surrounding the wells) are likely to be important determinants of whether a well is a 'producing' rather than a dry well, and a full analysis would include such variables. In a more realistically specified model the inferences we have drawn from the simple model would not necessarily be confirmed.

2.7 Goodness-of-fit measures, residuals and predicted values

Having tested hypotheses about the values of the parameters of a model, it is normal practice in conventional regression modelling to consider the overall fit of the model and to examine discrepancies between observed values of the response variable and values estimated or predicted by the model (i.e. to examine residuals). Analogous procedures for our logistic/logit models can now be considered.

2.7.1 Goodness-of-fit measures

In the conventional multiple regression context, the overall fit of a model is measured by the R^2 (coefficient of determination) statistic which is defined as the ratio of the regression sum of squares to the total sum of squares. R^2 measures the proportion of the variability in the response variable accounted for or 'explained' by the explanatory variables, and it ranges in value from 0 to 1. The closer R^2 is to 1, the better the fit of the model, i.e. the better the performance of the explanatory variables.

In the case of our logistic/logit models, an analogous goodness-of-fit measure would be useful. However, unlike R^2, such a measure cannot be based upon a sums of squares ratio, for we have seen that for cell (d) problems a maximum likelihood rather than a least squares procedure will be used to estimate the parameters of the logistic/logit model. As a result, a pseudo R^2 measure must be defined which utilizes a ratio of maximized log likelihood values rather than a ratio of sums of squares. This pseudo R^2 (known as rho-square) is defined (see Domencich and McFadden 1975, 123; Tardiff 1976; Hensher and Johnson 1981a, p.51) as:

$$\rho^2 = 1 - \frac{\log_e \Lambda(\hat{\boldsymbol{\beta}})}{\log_e \Lambda(C)} \qquad [2.71]$$

(i.e. one minus the ratio of the maximized log likelihood values of the fitted and constant-only-term models). In the case of our logistic/logit models the maximized log likelihood values will always be negative numbers. That for the fitted model, $\log_e \Lambda(\hat{\boldsymbol{\beta}})$, will always be a smaller negative number than $\log_e \Lambda(C)$ and the ratio of the two maximized log likelihoods will lie between zero and one. The closer to zero this ratio, the larger ρ^2 (one minus the ratio) and the better the fit of the model. Although ρ^2 ranges in value from 0 to 1, it should be noted, however, that its values tend to be considerably lower than those of the R^2 index, and they should not be judged by the standards of what is normally considered a 'good fit' in conventional regression analysis. Domencich and McFadden (1975, 24) have provided a graph of the relationship between R^2 and ρ^2 and McFadden (1979, 307) has suggested that ρ^2

values of between 0.2 and 0.4 should be taken to represent a very good fit of the model[†].

A second method of assessing the overall fit of a model is to consider what is termed the *prediction success* of the model. This involves a comparison of the observed number of selections of each response category/choice alternative with the predicted selections derived from the model. To facilitate this comparison, McFadden (1979; see also Hensher and Johnson 1981a, 52–5) has proposed the prediction success table shown in Table 2.2. The observed response category selections form the vertical axis of this table, and the predicted response category selections form the horizontal axis. The typical entry S_{hq} in the main body of the table gives the expected number of observed selections of category/alternative h which are predicted by the model to be selections of category/alternative q. For example, S_{12} is the number of observed selections of category 1 which are predicted by the model to be

Table 2.2
Prediction success table

		Predicted response category selection				Observed count	Observed share
		$q = 1$	2	... R			
Observed response category selection	$h = 1$	S_{11}	S_{12}	... S_{1R}		S_{1+}	S_{1+}/S_{++}
	$h = 2$	S_{21}	S_{22}	... S_{2R}		S_{2+}	S_{2+}/S_{++}
	.	.	.	.		.	.
	.	.	.	.		.	.
	$h = R$	S_{R1}	S_{R2}	... S_{RR}		S_{R+}	S_{R+}/S_{++}
Predicted count		S_{+1}	S_{+2}	... S_{+R}		S_{++}	1.0
Predicted share		S_{+1}/S_{++}	S_{+2}/S_{++}	... S_{+R}/S_{++}		1.0	
Proportion successfully predicted		S_{11}/S_{+1}	S_{22}/S_{+2}	... S_{RR}/S_{+R}		$(S_{11}+ \ldots +S_{RR})/S_{++}$	
Prediction success index σ_q (not normalized)		$\dfrac{S_{11}}{S_{+1}} - \dfrac{S_{+1}}{S_{++}}$	$\dfrac{S_{22}}{S_{+2}} - \dfrac{S_{+2}}{S_{++}}$	... $\dfrac{S_{RR}}{S_{+R}} - \dfrac{S_{+R}}{S_{++}}$		Overall success index $\sigma = \sum\limits_{q=1}^{R} \dfrac{S_{+q}}{S_{++}} \sigma_q$	

[†] In some applications, the reader should be aware that the ρ^2 statistic is defined as $\rho^2 = 1 - (\log_e \Lambda(0))$ where $\log_e \Lambda(0)$ is the maximized log likelihood when all the parameters in the model (*including the constant term*) are set to zero. The new maximized log likelihood, $\log_e \Lambda (0)$, is often referred to (see McFadden 1979, 307) as that produced by an *equal shares* model, whereas $\log_e \Lambda(C)$ is referred to (see Stopher and Meyburg 1979, 333) as that produced by a *market shares* model. The two versions of ρ will not produce the same numerical values (the equal shares version will always be larger), and it should be noted that it is the market shares version of equation [2.71] to which McFadden's statement about values between 0.2 and 0.4 representing a 'good fit' applies. The version in equation [2.71] is by far the most widely used in current practice and it is generally regarded as being preferable (except in the case to be discussed in Section 2.11 and Example 2.9 where there is no constant term in the model). However, it should be noted that certain computer programs compute only the market shares version and certain published applications report *only* the market shares version.

In addition, the reader should be aware that sometimes another version of ρ^2 will be reported. This is known as $\bar{\rho}^2$ and it represents the ρ^2 adjusted for degrees of freedom (for further details see Example 2.9).

selections of category 2, and S_{11} is the number of observed selections of category 1 correctly predicted by the model as selections of category 1. S_{hq} is defined as[†]

$$S_{hq} = \sum_{i=1}^{N} D_{hi}\hat{P}_{q|i} \qquad\qquad [2.72]$$

where $\hat{P}_{q|i}$ is the estimated or predicted probability that category q will be selected by individual i or at locality i, and D_{hi} is a dummy variable which equals 1 if category h is observed to be selected by individual i or at locality i, 0 otherwise. The prediction success table is shown in its general form appropriate for any number of response categories. In the dichotomous (2–category) case which we are concerned with in this section of Chapter 2, it clearly reduces to a 2×2 table.

Surrounding the main body of the table are the row and column sums (termed observed and predicted counts) which can be expressed as proportions of the grand total (S_{++}) and are termed observed and predicted shares. Below these figures are two rows of summary statistics. The first of these is termed the 'proportion successfully predicted'. This gives the proportion of the individuals predicted to select a certain category who actually select that category[‡]. It also gives the overall proportion, $S_{11} + \ldots + S_{RR}/S_{++}$, successfully predicted. The problem with this measure, however, is that the proportion successfully predicted for a given response category will depend upon, and vary in accordance with, the share of selections obtained by that category. A better measure, therefore, would take this into account and correct for the share of selections obtained by the response category. Such a measure is reported in the next row and is termed the *prediction success index*. It is defined as

$$\sigma_q = \frac{S_{hq}}{S_{+q}} - \frac{S_{+q}}{S_{++}} \qquad h = q \qquad\qquad [2.73]$$

As noted above, S_{hq}/S_{+q} $(h = q)$ is the proportion of the individuals predicted to select a certain category who actually select that category, whilst S_{+q}/S_{++} is the proportion which could be expected to be successfully predicted solely on the basis of the category's predicted share. If σ_q equals zero it follows, therefore, that the model does no better in prediction for category q than a simple reponse-category-share hypothesis. Alternatively, σ_q can usefully be viewed (McFadden 1979) as measuring the net contribution of the explanatory variables to the model's prediction success (i.e. the net contribution of all the terms in the model other than the constant term). The index, σ_q, will usually be non-negative with a maximum value of $1 - (S_{+q}/S_{++})$. If an index

---

[†] Alternative definitions are possible. McFadden (1979, 307) suggests the alternative of assuming that the category with the highest predicted probability is selected. This would amount to redefining [2.72] as

$$S_{hq} = \sum_{i=1}^{N} D_{hi}D^*_{qi}$$

where $D^*_{qi} = 1$ if $\hat{P}_{q|i} > \hat{P}_{g|i}$ $\quad g = 1, \ldots R \quad g \neq q$
$\qquad\qquad = 0$ otherwise

[‡] Alternatively: the proportion of localities at which a certain category is predicted to occur at which that category actually occurs.

normally lying between 0 and 1 is desired, the index [2.73] can be normalized by dividing through by $1 - (S_{+q}/S_{++})$.

An overall prediction success index can be formed by summing the σ_q's over the R alternatives and weighting each σ_q by S_{+q}/S_{++}. This gives

$$\sigma = \sum_{q=1}^{R} (S_{+q}/S_{++})\sigma_q \qquad\qquad [2.74]$$

Alternatively, we can rewrite [2.74] as

$$\sigma = \sum_{q=1}^{R} (S_{+q}/S_{++}) \left[\frac{S_{hq}}{S_{+q}} - \frac{S_{+q}}{S_{++}} \right] = \sum_{q=1}^{R} \left[\frac{S_{hq}}{S_{++}} - \left(\frac{S_{+q}}{S_{1+}} \right)^2 \right] \qquad [2.75]$$

where $h = q$

Again, this index will usually be non-negative with a maximum value of $1 - \sum_q (S_{+q}/S_{++})^2$ and it can be normalized to give an index usually lying within the range 0 to 1.

2.7.2 *Residuals*

In conventional regression modelling, considerable attention is given to the discrepancies between the observed values of the response variable and values estimated or predicted by the model. These discrepancies or residuals are fundamental to the least squares estimation of the regression model, to the testing of the validity of the assumptions (homoscedasticity, no autocorrelation, etc.) of the model, and to the process of model refinement. In the case of our dichotomous logistic/logit models estimated by maximum likelihood[†], analogous residuals can be defined as

$$\hat{\varepsilon}_i = Y_i - \hat{P}_{1|i} \qquad\qquad [2.76]$$

where Y_i, the observed value of the response variable, equals 1 or 0. Alternatively, they can be standardized (Cox 1970, 96) to have a mean of zero and a unit variance as follows

$$\hat{\varepsilon}_i = \frac{Y_i - \hat{P}_{1|i}}{\sqrt{[\hat{P}_{1|i}(1 - \hat{P}_{1|i})]}} \qquad\qquad [2.77]$$

Following conventional regression modelling practice (see for example Chatterjee and Price 1977), it is valuable to consider graphing these residuals against explanatory variables already in the model, against other variables which might potentially be included in the model as

---

[†] In the case of grouped data and weighted least squares estimation (which as we have noted is very rarely used for cell (d) problems), residuals can be defined as

$$\hat{\varepsilon}_g = Y_g - \hat{P}_{1|g}$$

Other definitions of residuals which are useful in the case of logistic/logit models are possible and include the 'components of deviance'. These definitions lie outside the scope of this chapter. They are considered in Chapter 8, and more fully in Wrigley and Dunn (1984).

explanatory variables, and against the estimated values of the response variable. Such graphs or plots allow us to examine the adequacy of the fitted model, and they usefully supplement the information derived from the overall goodness-of-fit measures and from the more formal joint, partial joint, and separate tests of significance. In addition, where appropriate locational information is available, geographers should supplement these plots with maps of the residuals. If systematic patterning is detected in either the residual plots or maps, it suggests that an inappropriately specified model has been fitted and/or that one of the implicit assumptions of the model has been violated. Recognition of such deficiency in the fitted model will then enable us to refine the model by means such as: including additional explanatory variables; altering the form in which the explanatory variables are included in the model (e.g. making the model non-linear in the explanatory variables but still linear in the parameters); giving the model a structure which incorporates spatial or temporal lags, etc. Likewise, the importance of anomalous ill-fitted data points (outliers) can also be determined from visual inspection of the residual plots, and model refinement which may involve the exclusion of these outliers can be undertaken.

Although visual inspection of residual plots and maps is an extremely useful aid in testing the adequacy of, and refining, a postulated model, it is dependent upon and limited by the capacity of the human eye to detect in a consistent manner patterning amongst the plotted or mapped residual values. For this reason, in the conventional regression model context, test statistics have been developed to detect various types of residual patterning, and formal tests of significance based upon these statistics are now widely used. Particularly important for geographers are those spatial autocorrelation test statistics (see Cliff and Ord 1972, 1973, 1981) which have been developed specifically for regression residuals.

Although analogous test statistics would be an extremely valuable addition to the residual plotting and mapping suggested for our logistic/logit models, certain difficulties exist which have resulted in there being little progress so far in developing such statistics. One difficulty (Cox 1970, 98) is that although the crude residuals [2.76] can be scaled (as in [2.77]) to have zero mean and unit variance, they typically have a very non-normal distribution. As the response variable can only take the values 0 or 1, residuals close to the value zero will not occur except for extreme values of $\hat{P}_{1|i}$. In such regions, where there is a high or low probability of the first response category being selected, the residuals have a very skew distribution. Thus, if the probability of selecting the first response category is high, the residuals are either small and positive or large and negative. If the probability of selecting the first response category is low, residuals are either small and negative or large and positive.

A second difficulty is that the majority of test statistics for residual patterning which exist have been developed for normal-theory linear models. As such, they are not immediately applicable to the residual plots and maps from our logistic/logit models, and there have been very few attempts, as yet, to generalize them.[†] In the case of spatial

---

[†] Since completion of this text in early 1983 some important developments have occurred. These are summarized in Wrigley (1984).

autocorrelation statistics, for example, the Cliff–Ord regression residual statistic which is familiar to geographers and to many environmental scientists was developed for normal-theory linear models. Since the development of this statistic, generalized procedures for evaluating spatial autocorrelation have been suggested (Hubert et al. 1981; Cliff and Ord 1981). These are based upon a matrix-comparison, randomization inference strategy,[†] and they can usefully be extended to nominal or ordinal data structures. Unfortunately, however, although these generalized procedures may appear to provide a means of assessing spatial patterning in the residual maps from our logistic/logit model they are, in fact, developed for, and limited at present to, spatially distributed variables other than residuals. The reason for this is that considerable care must be taken in developing test statistics for residuals because of the correlation between residuals introduced by the need to estimate and use a set of common β regression parameters. Because of this, it is well known for example, that residuals ($\hat{\varepsilon}_i$) can show evidence of correlation and unequal variance even though the population or 'true' error terms are uncorrelated and homoscedastic (see Theil 1971, 96; Cliff and Ord 1973, 92). This can have serious implications if it is not recognized, and it must be allowed for in developing appropriate test statistics and inference strategies.

It appears, therefore, that currently there are few appropriate and/or widely adopted formal test procedures for assessing patterning in the plotted or mapped residual values from our cell (d) logistic/logit models. At this stage in the text, reliance is therefore placed upon visual assessment of the residual plots and maps. In Chapter 8, however, we reconsider this issue and discuss other diagnostic tools for assessing and refining the fit of logistic/logit models.

EXAMPLE 2.5   _Oil and gas exploration in south-central Kansas (continued)_

In Example 2.4 we concluded by fitting a simplified model which contained only two explanatory variables: thickness of the Mississippian B zone and the shale content of the Mississippian B zone. We can now consider the fit of that model using the ρ^2 measure, the prediction success table and residual plots.

Substituting the log likelihood values given in [2.70] into the general ρ^2 definition [2.71] gives

$$\rho^2 = 1 - \frac{-58.0695}{-71.8419} = 0.192 \tag{2.78}$$

This is slightly below the range of ρ^2 values (0.2–0.4) which McFadden (1979) has suggested represent a 'good' fit. It indicates that the fit of the simplified model is reasonable but by no means outstanding. This should not surprise us as the two-explanatory variable model is a very simple specification. As we have noted previously, a more fully specified model would undoubtedly include other explanatory variables

---

[†] It is of interest to note that this strategy has also been extended to provide an alternative approach to assessing overall goodness-of-fit of the logistic/logit models for cell (d) problems. See Costanzo et al. 1982.

Table 2.3
Prediction success table for
two-explanatory-variable
model [2.70]

		Predicted response category selection		Observed count	Observed share
		Producing	Dry		
Observed response category selection	Producing	13.8	19.2	33.0	0.266
	Dry	19.2	71.8	91.0	0.734
Predicted count		33.0	91.0	124.0	1.0
Predicted share		0.266	0.734	1.0	
Proportion successfully predicted		0.418	0.789	0.690	
Prediction success index (not normalized)		0.152	0.055	0.081	
Prediction success index (normalized)		0.207	0.207	0.207	

(e.g. indicating lateral variations in the subsurface characteristics in the areas surrounding the wells), and on the evidence of this simplified model any such extended model would be likely to produce a good fit.

A second method of assessing the overall goodness-of-fit of the model is to construct the prediction success table shown in Table 2.3. This shows that, overall, the model correctly predicts 69 per cent of the observed response categories (i.e. 69 per cent of the observed 'producing' and dry wells are correctly predicted). The model appears to be more successful in predicting the occurrence of dry wells than 'producing' wells (78.9 per cent as compared to 41.8 per cent). However, when these figures are adjusted to take account of the differing predicted shares of the response categories (26.6 per cent for the 'producing' wells, 73.4 per cent for the dry wells) this apparent difference disappears, and the normalized prediction success index shows that the model outperforms a simple response-category-share hypothesis equally well for both 'producing' and dry wells.

Although they provide us with useful information such overall goodness-of-fit measures are ultimately somewhat crude methods of examining the adequacy of the fitted model. As in most statistical model building, there will often be no effective substitute for careful inspection of the residuals. As an illustration of this, consider Fig. 2.7, which shows the standardized residuals from the two-explanatory-variable model [2.70] plotted against the local geological structure variable; the explanatory variable which in Example 2.4 we deleted from our original three-explanatory-variable model.

Readers accustomed to using similar plots in conventional regression analysis must take considerable care when interpreting residual plots like that shown in Fig. 2.7, because of the dichotomous nature of the response variable and the non-normal distribution of residuals which this produces[†]. Despite such problems, however, in the case of Fig. 2.7

[†] Cox (1970, 98) suggests that it will often be valuable in such plots to combine adjacent residuals. This will help reduce systematic effects caused by the dichotomous nature of the response variable, and will clarify the pattern. For example, groups of four adjacent residuals can be taken, and the combined

Figure 2.7
Plot of standardized re-
siduals, $\hat{\varepsilon}_{i'}$ from model
[2.70] against local geolo-
gical structure variable X_i.

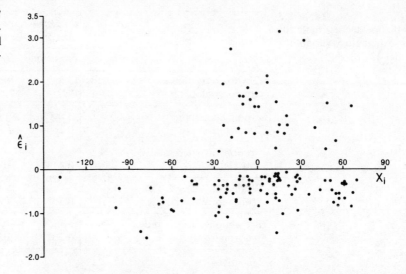

it does appear that a fairly pronounced residual patterning exists. In particular, it appears that positive residuals *only* occur in those localities in which the local geological structure variable (which, it will be recalled from Example 2.1, is measured by the residual from a first-order trend surface of the top of the Mississippian B) takes values close to zero. In localities where the local geological structure variable takes large negative or positive values (particularly large negative values), negative residuals predominate. In conventional terms, this suggests that the model is tending to 'underpredict'[†] the probability of a well being a 'producing' well in localities where the local geological structure variable takes values close to zero, and is tending to 'overpredict'[†] that probability in localities where the local geological structure variable takes large negative or positive values. In conventional regression analysis, the type of residual plot shown in Fig. 2.7 would normally (e.g. see Bibby 1977, 52) be taken to indicate the need to consider reintroducing into model [2.70] the omitted local geological structure variable in a new form. This form would be a non-linear, quadratic, X^2, form rather than its original linear form.

The conclusion we have drawn from the residual plot can be confirmed and elaborated by a consideration of Fig. 2.8; a map of the residuals. In this case, rather than simply inspect a residual map in which all positive and negative residual values are plotted and then contoured, it is more effective to exploit the knowledge we derived from Fig. 2.7 and to superimpose a map of positive residuals over a map of

---

residual is then defined as $\sqrt{4}$ times the average of the four adjacent residuals. In Fig. 2.7, for example, the first four residuals are -0.179, -0.879, -0.409 and -1.409, and the combined residual is $\sqrt{4} \times (-2.876/4) = -1.438$. The factor $\sqrt{4}$ is included to make the variance approximately one. As an exercise, the reader is encouraged to try this procedure on Fig. 2.7.

[†] The terms 'underpredict' and 'overpredict' must be interpreted cautiously in this logistic/logit model context. Any prediction for a locality in which $Y_i = 1$, i.e. in which a 'producing' well is observed will in almost all circumstances be an underprediction, and can never be an overprediction, as $\hat{P}_{1|i}$ has an upper bound of 1. Similarly, any prediction for a locality in which $Y_i = 0$, i.e. in which a dry well is observed, will in almost all circumstances be an overprediction, and can never be an underprediction, as $\hat{P}_{1|i}$ has a lower bound of 0. More important is the relative distribution of the positive and negative residuals, and the distribution within each of the two classes of residuals. In this case, these distributions suggest a clear deficiency in the fitted model.

Figure 2.8
Map of positive residuals (shown by dots) from model [2.70] superimposed upon contour map of local geological structure variable.

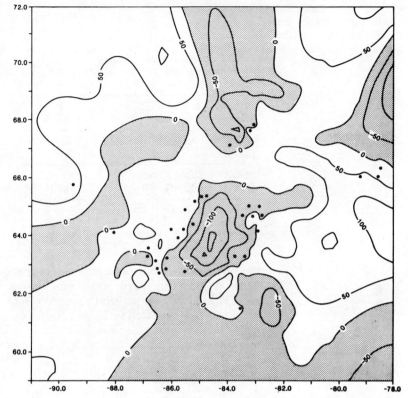

the omitted variable: local geological structure. From Fig. 2.8 we can then see that the positive residuals tend to follow the zero-value contour on the local geological structure map. In other words, the residuals show a distinct spatial patterning, and the model [2.70] tends to underestimate the probability of a well being a 'producing' well along the flanks of local positive structures.

Given the conclusions which we have drawn from an inspection of Figs. 2.7 and 2.8, we can now reintroduce into our model the local geological structure variable, X_{i3}, but now in quadratic, X^2, form. In this case, $x_i'\hat{\boldsymbol{\beta}}$ becomes:

$$x_i'\hat{\boldsymbol{\beta}} = -0.7469 \quad +0.0746X_{i2} \quad -0.0006X_{i3}^2 \quad -8.2262X_{i4} \qquad \textbf{[2.79]}$$
$$\phantom{x_i'\hat{\boldsymbol{\beta}} =} (0.9698) \quad\quad (0.0227) \quad\quad\quad (0.0002) \quad\quad\quad\quad (3.1159)$$
$$\phantom{x_i'\hat{\boldsymbol{\beta}} =} [-0.7702] \quad [3.2892] \quad\quad [-3.0215] \quad\quad\quad [-2.6401]$$

$$\log_e \Lambda(C) = -71.8419 \qquad\qquad \log_e \Lambda(\hat{\boldsymbol{\beta}}) = -50.5221$$

Comparing [2.79] with [2.70], it can be seen that the inclusion of the local geological structure variable in a X^2 form gives rise to a significant improvement in the fit of the model. The overall ρ^2 goodness-of-fit measure increases from 0.192 in the case of the previous specification [2.70] to 0.297 in the case of [2.79], and this is now well within the range (0.2–0.4) of values which McFadden (1979) has suggested represent a 'good' fit. In addition, it can be seen from the new prediction success table, Table 2.4, that the model now correctly predicts 73.4 per cent of

Table 2.4
Prediction success table for
the reformulated model
[2.79]

		Predicted response category selection		Observed count	Observed share
		Producing	Dry		
Observed response	Producing	16.50	16.50	33.0	0.266
category selection	Dry	16.47	74.53	91.0	0.734
Predicted count		33.0	91.0	124.0	1.0
Predicted share		0.266	0.734	1.0	
Proportion successfully predicted		0.500	0.819	0.734	
Prediction success index (not normalized)		0.234	0.085	0.125	
Prediction success index (normalized)		0.319	0.319	0.319	

the observed response categories, and it shows a significant improvement in the normalized prediction success index (and hence in the model's ability to outperform a simple response-category-share hypothesis) over that reported in Table 2.3. Inspection of the 't statistic' of the newly included variable shows it to be significant at the conventional 5 per cent level, and we can reasonably reject the null hypothesis that $\beta_3 = 0$, and suggest that local geological structure is significantly but non-linearly related to the odds of a well being a 'producing' rather than a 'dry' well. This inference can readily be confirmed using the maximized log likelihood values of [2.70] and [2.79] and a partial joint test like that employed in Example 2.4. This is left as an exercise for the reader.

By careful consideration of the fit of our provisional model, we have been able to improve it significantly. In so doing, an important principle has been demonstrated, namely that many of our statistical models are, as Cox (1970, 94) notes, 'provisional working bases for the analysis rather than rigid specifications to be accepted uncritically'. Having produced the improved model [2.79], our task may not be completed. Further residual analysis and model refinement may yet be possible, as shown in Chapter 8 and by Wrigley and Dunn (1984).

EXAMPLE 2.6   *Resource evaluation in Newfoundland*

In the previous example, estimated or predicted values of the response variable have been used to construct the prediction success table and to define the residuals which were used in model evaluation and refinement. Another use of such estimated or predicted probabilities is illustrated in this example; a use which some environmental scientists appear to have found particularly valuable.

In research conducted for the Canadian Geological Survey, Chung (1978) used a logistic/logit model of the occurrence of volcanogenic massive sulphide (VMS) deposits in Newfoundland, in which the probability of occurrence of VMS deposits was taken to be a function of several geological variables. For the purposes of the investigation, the

Figure 2.9
Known VMS deposits in
Newfoundland at time of
research (adapted from
Chung 1978).

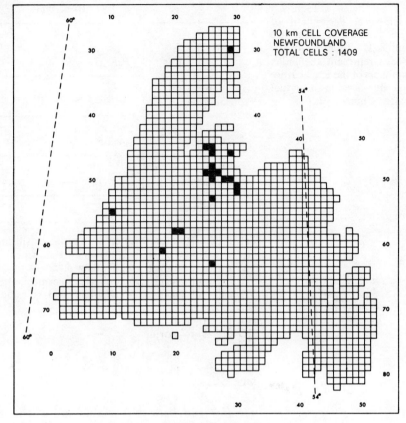

Figure 2.9
Known VMS deposits in
Newfoundland at time of
research (adapted from
Chung 1978).

contains known occurrences of VMS deposits ●

island of Newfoundland was subdivided into the 1,409 grid-squares of size 10 km by 10 km shown in Fig. 2.9, and all the VMS deposits discovered in Newfoundland by the time of the research were found to be located in just twenty one of the 1,409 grid-squares. Furthermore, it was found that volcanic rocks of four geologic ages (Lower Silurian, Middle-Upper Ordovician, Lower Ordovician, and Cambrian) hosted all the known VMS deposits. Hence, it was considered possible to model the occurrence of VMS deposits as a function of seventeen explanatory variables, defined as the seventeen litho-age units shown in Table 2.5. In each of the 1,409 grid-squares, a regular spatial sample grid of 400 points was then superimposed, and the number of sample points (maximum 400 per grid-square) at which each of the seventeen litho-age units shown in Table 2.5 was present in each grid-square was recorded. Only 307 of the 1,409 grid-squares were found to contain at least one of the seventeen litho-age units.

Chung then estimated a logistic/logit model using the maximum likelihood procedure, and the fitted model had the form

$$\hat{P}_{1|i} = \frac{e^{x'_i\hat{\boldsymbol{\beta}}}}{1 + e^{x'_i\hat{\boldsymbol{\beta}}}} \quad \text{or} \quad \log_e \frac{\hat{P}_{1|i}}{\hat{P}_{2|i}} = x'_i\hat{\boldsymbol{\beta}} \qquad [2.80]$$

Table 2.5
Geological description of the seventeen litho-age units. The numbers in the boxes represent the index numbers of the explanatory variables used in the model (after Chung 1978).

Rock types

	Pel	Gw	Va	V	Vb	Gr (Pre-Silurian)	Um (Pre-Silurian)	GD
Lower Silurian	2	—	3	5	—			
Middle-Upper Ordovician	4	6	—	7	—	8	9	10
Lower Ordovician	11	12	—	13	14			
Cambrian	15	16	18	—	17			

(Age is indicated along the left vertical axis.)

Pel : Pelite (marine with or without minor sillstone and carbonate)
Gw : Greywacke - pelite, turbidite: deep water
Va : Acid volcanics greater than 50%
V : Basic volcanics; minor acid volcanics
Vb : Basic volcanics; no reported acid volcanics
Gr : Granite
Um : Ultramafic
GD : Gabbro-dioritic

where

$$x'_i\hat{\boldsymbol{\beta}} = \hat{\beta}_1 + \hat{\beta}_2 X_{i2} + \ldots + \hat{\beta}_{18} X_{i18}$$
$$= -4.560 - 7.504 X_{i2} + \ldots + 0.004 X_{i18} \qquad [2.81]$$

$\hat{P}_{1|i}$ = predicted probability of a VMS deposit occurring in grid-square i.

X_{i2} = number of sample points (max. 400) at which litho-age unit 2 (see Table 2.5) is present in grid-square i.

$\quad \vdots \qquad\qquad\qquad\qquad \vdots \qquad\qquad \vdots$

X_{i18} = number of sample points (max. 400) at which litho-age unit 18 is present in grid-square i.

He then noted that the average $\hat{P}_{1|i}$ in the twenty-one grid-squares containing known VMS deposits was 0.26, and proceeded to map those cells where *no* VMS deposits had yet been discovered but which had $\hat{P}_{1|i}$ values greater than 0.26. The implication was that these grid-squares were the areas of Newfoundland most suited to further exploration for VMS deposits.

Unfortunately, closer inspection of Chung's data reveals that several of the explanatory variables produce very sparse columns in the X data matrix (e.g. X_{15} has only three non-zero values out of 307 entries in that column of the data matrix, X_{11} has only five non-zero values, X_2 has only fourteen non-zero values, etc.). Consequently (and remembering that the first column of the data matrix, associated with the constant term, is a column of ones), these variables are likely to cause multicollinearity problems. For this reason, we can remove these offending explanatory variables from Chung's model and re-estimate his model with five original variables (X_2, X_{11}, X_{15}, X_{17} and X_{18}) removed. The fitted model now has the form [2.80] where,

$$x'_i\hat{\beta} = \hat{\beta}_1 + \hat{\beta}_3 X_{i3} + .. + \hat{\beta}_{14} X_{i14} + \hat{\beta}_{16} X_{i16}$$
$$= -4.0248 + 0.0100 X_{i3} + \ldots + 0.0120 X_{i14} + 0.0039 X_{i16} \qquad [2.82]$$

For this new model, the average $\hat{P}_{1|i}$ in the twenty-one grid-squares containing known VMS deposits is 0.217 and the largest $\hat{P}_{1|i}$ is 0.47. Figure 2.10 shows a map of the grid-squares where *no* VMS deposits had been discovered at the time of Chung's research but which had $\hat{P}_{1|i}$ values greater than 0.217, or between 0.108 and 0.217. Once again, the implication is that these are the areas of Newfoundland most suited to further exploration for VMS deposits. It must be stressed, however, that there are two limitations to the value of this map (Fig. 2.10). First, the estimated values used to produce this map are merely *point* estimates of the probabilities. There is no attempt made in Chung's paper, or in

Figure 2.10
Grid-squares containing no known VMS deposits but with high or moderately high predictions of VMS occurrence.

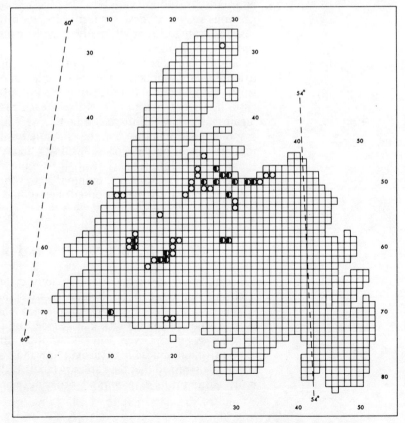

estimated probability >0.217 ◑   estimated probability >0.108 and <0.217 ○

other environmental science applications, to derive confidence intervals and to give an interval prediction of the probabilities. Second, the map is based upon the information which existed at the time of Chung's research. Any subsequent discovery of VMS deposits would change the data matrix. This would necessitate re-estimation of the model, and would perhaps also change the explanatory variables which enter the model. Nevertheless, in certain circumstances, such maps can be a useful aid to the environmental scientist, and it should be noted that a similar map could also be constructed for the Kansas oil and gas well data.

2B   MULTIPLE RESPONSE CATEGORIES: THE POLYTOMOUS CASE

2.8 Introduction

So far in this chapter we have considered the case in which we have a dichotomous (2-category, 2-alternative) response variable. Clearly, however, there are far more cases in which geographers and environmental scientists will wish to deal with multiple-category, multiple-alternative response variables. In this section we will generalize our previous results, focusing specifically upon the situation in which the response categories or alternatives can be considered to have no natural ordering or ranking.

To aid our discussion, it is useful to think in terms of a specific example. For simplicity, we will once again consider our example of a survey of opinions about the siting of a proposed mental hospital. In this case, however, instead of classifying each respondent as 'in favour' or 'against' the mental hospital scheme, we will assume initially that the respondents' opinions can be classified into three exhaustive and mutually exclusive categories: 'against', 'in favour' or 'undecided' about the scheme. Once again, we wish to build a simple statistical model which relates respondents' opinions about the scheme to a measure of their proximity to the proposed site of the mental hospital.

2.9 The extended linear logit and logistic models

We can begin to extend our previous dichotomous models by first arbitrarily denoting the probability that respondent i is 'in favour' of the scheme as $P_{1|i}$, the probability that respondent i is 'against' the scheme as $P_{2|i}$, and the probability that respondent i is 'undecided' about the scheme as $P_{3|i}$. Given our assumption that these categories are exhaustive and mutually exclusive, they must sum to one. Furthermore, it must be stressed that the category labels 1, 2 and 3 are arbitrary; there is no assumption of a natural ordering amongst the categories.

2.9.1   *The linear logit model*

In the case of our dichotomous logit model [2.15], the reader will recall that we basically had a model in which the odds of choosing one particular response category over the other response category ($P_{1|i}/P_{2|i}$) was some function of the explanatory variable X. In our extended mental hospital example we now have three response categories instead of the previous two and, therefore, we can now define three distinct sets of odds $P_{1|i}/P_{3|i}$, $P_{2|i}/P_{3|i}$ and $P_{1|i}/P_{2|i}$ and express each as some function of the explanatory variable X. Instead of the one linear logit model

$$\log_e \frac{P_{1|i}}{P_{2|i}} = \alpha + \beta X_i \qquad [2.83]$$

which we had in the dichotomous response variable case, it would appear that, in the three-response category case, we now potentially have three linear logit models

$$\log_e \frac{P_{1|i}}{P_{3|i}} = \alpha_{13} + \beta_{13} X_i$$

$$\log_e \frac{P_{2|i}}{P_{3|i}} = \alpha_{23} + \beta_{23} X_i \qquad [2.84]$$

$$\log_e \frac{P_{1|i}}{P_{2|i}} = \alpha_{12} + \beta_{12} X_i$$

where the subscripts on the parameters α and β have been introduced to denote which set of log-odds a particular equation refers to, and where the equations are linked together by the necessity for the sum of the individual probabilities to equal one. Fortunately, however, it can be demonstrated that one of these equations is redundant, and that the number of parameters to be estimated can be reduced from six to four.

To see why one equation and two parameters in [2.84] are redundant, we can take any pair of response categories (say, categories 1 and 3: 'in favour' and 'undecided') and show that[†]

$$\log_e \frac{P_{1|i}}{P_{3|i}} = \log_e \frac{P_{1|i}}{P_{2|i}} + \log_e \frac{P_{2|i}}{P_{3|i}} \qquad [2.85]$$

Writing these equations out fully implies:

$$\log_e \frac{P_{1|i}}{P_{3|i}} = \alpha_{13} + \beta_{13} X_i$$

$$= \log_e \frac{P_{1|i}}{P_{2|i}} + \log_e \frac{P_{2|i}}{P_{3|i}} = \alpha_{12} + \alpha_{23} + (\beta_{12} + \beta_{23}) X_i \qquad [2.86]$$

[†] $\log_e \dfrac{P_{1|i}}{P_{2|i}} + \log_e \dfrac{P_{2|i}}{P_{3|i}} = \log_e \left[\dfrac{P_{1|i}}{P_{2|i}} \times \dfrac{P_{2|i}}{P_{3|i}} \right] = \log_e \dfrac{P_{1|i}}{P_{3|i}}$

or in terms of the parameters alone

$$\alpha_{13} = \alpha_{12} + \alpha_{23} \quad \text{and} \quad \beta_{13} = \beta_{12} + \beta_{23} \qquad\qquad [2.87]$$

or alternatively

$$\alpha_{12} = \alpha_{13} - \alpha_{23} \quad \text{and} \quad \beta_{12} = \beta_{13} - \beta_{23} \qquad\qquad [2.88]$$

This implies that the parameters of [2.84] are linked by two constraints, and that the parameters of the third equation in [2.84] can be calculated once we know the parameters of the first two equations. In turn, this implies that the parameters of the third equation do not need to be estimated directly, and that the third equation is redundant.

Next, we can usefully redefine all the parameters in [2.84] as follows:

$$
\begin{aligned}
\alpha_{13} &= \alpha_1 - \alpha_3 & \beta_{13} &= \beta_1 - \beta_3 \\
\alpha_{23} &= \alpha_2 - \alpha_3 & \beta_{23} &= \beta_2 - \beta_3 \qquad\qquad [2.89]\\
\alpha_{12} &= \alpha_1 - \alpha_2 & \beta_{12} &= \beta_1 - \beta_2
\end{aligned}
$$

and write the model system [2.84] as:

$$\log_e \frac{P_{1|i}}{P_{3|i}} = (\alpha_1 - \alpha_3) + (\beta_1 - \beta_3)X_i$$

$$\log_e \frac{P_{2|i}}{P_{3|i}} = (\alpha_2 - \alpha_3) + (\beta_2 - \beta_3)X_i \qquad\qquad [2.90]$$

$$\log_e \frac{P_{1|i}}{P_{2|i}} = (\alpha_1 - \alpha_2) + (\beta_1 - \beta_2)X_i$$

Since the parameters of the third equation can be calculated once we know the parameters of the first two equations, the third equation need not be estimated and can be dropped from the model system. Furthermore, the right-hand sides of the first two equations in [2.90] show that it is only the difference of the α's and β's from those of an arbitrary base or anchor category (in our case, category 3 – 'undecided') which matter. Therefore, we can arbitrarily impose the constraints

$$\alpha_3 = \beta_3 = 0 \qquad\qquad [2.91]$$

without loss.

Instead of the single linear logit equation [2.83] for each individual which we had in the dichotomous response variable case, in the extended three-category response variable case, we now, therefore, have a system of two-linear logit equations of the form

$$L_{13|i} = \log_e \frac{P_{1|i}}{P_{3|i}} = \alpha_1 + \beta_1 X_i$$

$$\qquad\qquad [2.92]$$

$$L_{23|i} = \log_e \frac{P_{2|i}}{P_{3|i}} = \alpha_2 + \beta_2 X_i$$

This can be summarized, if required, as

$$L_{r3|i} = \log_e \frac{P_{r|i}}{P_{3|i}} = \alpha_r + \beta_r X_i \qquad r = 1, 2 \qquad \text{[2.93]}$$

2.9.2 The logistic model

In the dichotomous case, the reader will recall that the logistic model (which it must be remembered is simply a re-expression of the linear logit model, and vice versa), took the form (see equations [2.14] and [2.17]),

$$P_{1|i} = \frac{e^{\alpha + \beta X_i}}{1 + e^{\alpha + \beta X_i}} \quad \text{and} \quad P_{2|i} = \frac{1}{1 + e^{\alpha + \beta X_i}} \qquad \text{[2.94]}$$

In the case of our three-response category example, these equations generalize in a very straightforward manner to the form

$$P_{1|i} = \frac{e^{\alpha_1 + \beta_1 X_i}}{1 + e^{\alpha_1 + \beta_1 X_i} + e^{\alpha_2 + \beta_2 X_i}}$$

$$P_{2|i} = \frac{e^{\alpha_2 + \beta_2 X_i}}{1 + e^{\alpha_1 + \beta_1 X_i} + e^{\alpha_2 + \beta_2 X_i}}$$

$$\text{and} \quad P_{3|i} = \frac{1}{1 + e^{\alpha_1 + \beta_1 X_i} + e^{\alpha_2 + \beta_2 X_i}} \qquad \text{[2.95]}$$

In developing the generalization, we use the parameter notation utilized in [2.92] and exploit the results established in the previous section. In particular, it can be seen that only four parameters require estimation, and that we exploit the arbitrary constraints [2.91].

Under the assumption that the arbitrary constraints [2.91] are to be imposed, the three equations in [2.95] can be summarized in one general expression as

$$P_{r|i} = \frac{e^{\alpha_r + \beta_r X_i}}{\sum\limits_{s=1}^{3} e^{\alpha_s + \beta_s X_i}} \qquad r = 1, 2, 3 \qquad \text{[2.96]}$$

where $P_{r|i}$ represents the probability that the rth response category will be selected by the ith individual or at the ith locality. To move from equations [2.96] to those in [2.95] the reader should recall that $\alpha_3 = \beta_3 = 0$ by the imposition of the arbitrary constraints [2.91], and that $e^0 = 1$ by definition. Hence, when $r = 3$, the numerator of equation [2.96] becomes $e^{0 + 0X_i} = 1$. Similarly, in the denominator of equation [2.96], when $s = 3$, $e^{0 + 0X_i} = 1$ and the denominator becomes $e^{\alpha_1 + \beta_1 X_i} + e^{\alpha_2 + \beta_2 X_i} + 1$.

2.9.3   *General forms*

Having seen the extension of our dichotomous logit and logistic models to the three-response category case, it simply remains for us to note that the same principles of extension hold when generalizing our models from three categories to any number of response categories. For example, in the case of R response categories, where R stands for any integer number (in most empirical examples R will be small), our three-category logit model [2.92] with only one explanatory variable generalizes to a system of $R-1$ linear logit equations of the form:

$$\log_e \frac{P_{1|i}}{P_{R|i}} = \alpha_1 + \beta_1 X_i$$

$$\vdots \qquad \vdots$$

$$\log_e \frac{P_{R-1|i}}{P_{R|i}} = \alpha_{R-1} + \beta_{R-1} X_i$$

[2.97]

This can be summarized as

$$\log_e \frac{P_{r|i}}{P_{R|i}} = \alpha_r + \beta_r X_r \qquad r = 1, 2, \ldots R-1$$

[2.98]

In the same case, our three-category logistic model generalizes to a system of R non-linear equations of the form

$$P_{1|i} = \frac{e^{\alpha_1 + \beta_1 X_i}}{1 + \sum_{s=1}^{R-1} e^{\alpha_s + \beta_s X_i}}$$

$$\vdots \qquad \vdots$$

$$P_{r|i} = \frac{1}{1 + \sum_{s=1}^{R-1} e^{\alpha_s + \beta_s X_i}}$$

[2.99]

which, assuming the imposition of the usual arbitrary constraints relating to the base category (in this case $\alpha_R = \beta_R = 0$), can be summarized (see Mantel 1966; Cox 1970, 105; Mantel and Brown 1973; Wrigley 1975; Schmidt and Strauss 1975a) as

$$P_{r|i} = \frac{e^{\alpha_r + \beta_r X_i}}{\sum_{s=1}^{R} e^{\alpha_s + \beta_s X_i}} \qquad r = 1, 2, \ldots R$$

[2.100]

When there are several explanatory variables (often of both ϕ and ψ types – see Examples 2.2, 2.3, Section 2.11 and Appendix 2.1) rather than the single explanatory variable of our previous discussion, we can simply employ vector notation expressions $x'_{ri}\boldsymbol{\beta}_r$, $r = 1, 2, \ldots R$, similar to that which we have used in previous sections, and substitute these general expressions into the equations above (see the final part of Appendix 2.1 for a further discussion of these expressions). In the case of [2.98] and [2.99], and assuming the imposition of the usual arbitrary constraints relating to the base category (in this case $\boldsymbol{\beta}_R = \boldsymbol{0}$), this gives the multiple explanatory variable, multiple response category models

$$\log_e \frac{P_{r|i}}{P_{R|i}} = x'_{ri}\boldsymbol{\beta}_r \qquad r = 1, 2, \ldots R-1 \qquad\qquad \textbf{[2.101]}$$

and

$$P_{r|i} = \frac{e^{x'_{ri}\boldsymbol{\beta}_r}}{1 + \sum\limits_{s=1}^{R-1} e^{x'_{si}\boldsymbol{\beta}_s}} \qquad r = 1, 2, \ldots R \qquad\qquad \textbf{[2.102]}$$

2.10 Estimating the parameters of the extended models

Earlier in this chapter we considered the relative merits of weighted least squares and direct maximum likelihood procedures as possible strategies for estimating the parameters of dichotomous logistic/logit models. We noted that for problems in cell (d) of Table 1.1 (categorical response but continuous explanatory variables) it is highly unlikely that sufficient repetitions of explanatory variable values will be available to facilitate the use of the weighted least squares procedure. For this reason, in addition to its own intrinsic merits, we saw that the direct maximum likelihood procedure has become the almost universally employed estimation procedure for the dichotomous logistic/logit models used for the cell (d) problems of Table 1.1.

When we move from the two-category to the multiple-response category case, exactly the same arguments apply to our choice of estimation procedure and, once again, it is the direct maximum likelihood procedure which is almost universally employed. For this reason, discussion of how the weighted least squares procedure can be generalized to deal with our extended models will be deferred until the more appropriate context of Chapter 4, and we will concentrate in this section upon generalizing the maximum likelihood procedure.

In our earlier discussion of the dichotomous case we saw that the maximum likelihood procedure begins with the derivation of an expression for the likelihood of observing the pattern of response category choices in our data set. Assuming that the response category choices are viewed as independent drawings from a binomial distribution, and assuming that we have arbitrarily ordered the observations so that the N_1 choices of the first response category come prior to the $N - N_1$ choices of the second response category, we saw that the likelihood in the dichotomous case is

$$\Lambda = \prod_{i=1}^{N_1} P_{1|i} \prod_{i=N_1+1}^{N} P_{2|i} \qquad\qquad \textbf{[2.103]}$$

In the three-response category case, and with related assumptions of ordering and independence, this likelihood expression generalizes very simply to the form

$$\Lambda = \prod_{i=1}^{N_1} P_{1|i} \prod_{i=N_1+1}^{N_2} P_{2|i} \prod_{i=N_2+1}^{N} P_{3|i} \qquad [2.104]$$

where N is the total number of observations in the data set, N_1 is the number of choices of the first response category, $N_2 - N_1$ is the number of choices of the second response category, and $N - N_2$ is the number of choices of the third response category. Assuming, as in our mental hospital scheme example, that we have only one explanatory variable, we can then substitute the expressions for $P_{1|i}$, $P_{2|i}$ and $P_{3|i}$ from [2.95] into [2.104] and this gives:

$$\Lambda = \prod_{i=1}^{N_1} \frac{e^{\alpha_1 + \beta_1 X_i}}{1 + \sum_{s=1}^{2} e^{\alpha_s + \beta_s X_i}} \prod_{i=N_1+1}^{N_2} \frac{e^{\alpha_2 + \beta_2 X_i}}{1 + \sum_{s=1}^{2} e^{\alpha_s + \beta_s X_i}}$$

$$\prod_{i=N_2+1}^{N} \frac{1}{1 + \sum_{s=1}^{2} e^{\alpha_s + \beta_s X_i}} \qquad [2.105]$$

Once again, it can be seen that the value of the likelihood depends upon the values of the unknown parameters α_1, β_1, α_2 and β_2, and we estimate these parameters by taking as estimates those values which maximize the value of the likelihood. In practice, however, rather than maximize the likelihood [2.105] it proves to be more convenient to maximize the logarithm of the likelihood,

$$\log_e \Lambda = \sum_{i=1}^{N_1} \alpha_1 + \beta_1 X_i + \sum_{i=N_1+1}^{N_2} \alpha_2 + \beta_2 X_i - \sum_{i=1}^{N} \log_e \left(1 + \sum_{s=1}^{2} e^{\alpha_s + \beta_s X_i}\right) \qquad [2.106]$$

instead. The maximization can be achieved by taking the first partial derivatives of [2.106] with respect to α_1, β_1, α_2 and β_2 and setting these partial derivatives equal to zero (the so-called first-order conditions for a maximum):

$$\frac{\partial \log_e \Lambda}{\partial \alpha_1} = N_1 - \sum_{i=1}^{N} \frac{e^{\alpha_1 + \beta_1 X_i}}{1 + \sum_{s=1}^{2} e^{\alpha_s + \beta_s X_i}} = 0$$

$$\vdots \qquad \qquad \vdots \qquad \qquad [2.107]$$

$$\frac{\partial \log_e \Lambda}{\partial \beta_2} = \sum_{i=N_1+1}^{N_2} X_i - \sum_{i=1}^{N} X_i \frac{e^{\alpha_2 + \beta_2 X_i}}{1 + \sum_{s=1}^{2} e^{\alpha_s + \beta_s X_i}} = 0$$

Evaluating these derivatives at $\hat{\alpha}_1$, $\hat{\beta}_1$, $\hat{\alpha}_2$ and $\hat{\beta}_2$, the solutions of this set of simultaneous equations are the required maximum likelihood parameter estimators $\hat{\alpha}_1$, $\hat{\beta}_1$, $\hat{\alpha}_2$ and $\hat{\beta}_2$. In practice, this is done by resorting to a computer which can be programmed to find a numerical solution by means of some form of iterative procedure. Once again, this turns out to be much less complex than it sounds and many of the package programs (see Ch.7) which permit maximum likelihood estimation of the parameters of dichotomous logistic/logit models also permit similar estimation of the extended multiple-category models. Consistent estimates of the asymptotic variances of $\hat{\alpha}_1, \hat{\beta}_1, \hat{\alpha}_2$ and $\hat{\beta}_2$ will normally be computed automatically as part of the iterative search procedure employed by such programs, and this will be accomplished by taking the diagonal elements of the inverse of the so-called 'information matrix' (see Section 2.4, and Schmidt and Strauss 1975a, 485).

Having seen the extension of the maximum likelihood procedure from the two-response category to the three-response category case, it is sufficient for us to note that the same principles of extension hold when generalizing the procedure from three to any number, R, categories. For example, in the case of R categories with appropriate assumptions of ordering of observations and independence of response category choices, the likelihood becomes

$$\Lambda = \prod_{i=1}^{N_1} P_{1|i} \prod_{i=N_1+1}^{N_2} P_{2|i} \cdots \prod_{i=N_{(R-1)}+1}^{N} P_{R|i} \qquad [2.108]$$

When there are several explanatory variables rather than the single explanatory variable of our previous discussion, we can simply employ the vector notation expressions $x'_{ri}\beta_r$, $r = 1, 2, \ldots R$ and substitute these, where appropriate, into the equations above. For example, equation [2.106] becomes

$$\log_e \Lambda = \sum_{i=1}^{N_1} x'_{1i}\beta_1 + \sum_{i=N_1+1}^{N_2} x'_{2i}\beta_2 - \sum_{i=1}^{N} \log_e \left(1 + \sum_{s=1}^{2} e^{x'_{si}\beta_s}\right) \qquad [2.109]$$

EXAMPLE 2.7   _Oil and gas exploration in south-central Kansas (continued)_

As a first example of the application of multiple response category logistic/logit models, it is appropriate to return to our Kansas oil and gas exploration data. In our previous uses of this data set we concentrated upon fitting and refining a dichotomous model which attempted to account for the probability that a well penetrating the B division of the Mississippian Osage Series would be a 'producing' rather than a dry well. In fact, the data set is somewhat richer than this use would indicate. In contains information not only on whether a well is 'producing' or dry but also on the type of production, i.e. oil or gas. We can, therefore, fit to these data a three-category logistic/logit model in which we consider the determinants of whether a well is an 'oil-producing' well, a 'gas-producing' well, or a dry well.

The specification of our three-category model draws upon the results of Example 2.5. We include, as explanatory variables, measures of the thickness of the Mississippian B zone and its shale content (X_2 and X_4), and also a measure of local geological structure (X_3) entered into the model in a non-linear, quadratic form. Using the maximum likelihood procedure to estimate the parameters of this model and assuming the usual base category parameter constraints (i.e. $\beta_3 = 0$), the fitted model takes the form:

$$\hat{P}_{r|i} = \frac{e^{x'_{ri}\hat{\beta}_r}}{1 + \sum_{s=1}^{2} e^{x'_{si}\hat{\beta}_s}} \qquad \text{or} \qquad \log_e \frac{\hat{P}_{r|i}}{\hat{P}_{3|i}} = x'_{ri}\hat{\beta}_r \qquad \qquad [2.110]$$

where $r = 1 = $ oil; $\quad r = 2 = $ gas; $\quad r = 3 = $ dry.[†]

$$x'_{ri}\hat{\beta}_1 = -0.5825 \quad + 0.0762X_{i2} \quad - 0.0015X_{i3}^2 \quad - 9.9619X_{i4}$$
$$\quad\quad (0.7457) \quad\quad (0.0282) \quad\quad\quad (0.0006) \quad\quad\quad (3.8413)$$
$$\quad\quad [-0.7811] \quad\; [2.7008] \quad\;\; [-2.6842] \quad\;\; [-2.5934]$$

$$\qquad\qquad\qquad\qquad\qquad\qquad\qquad\qquad\qquad\qquad [2.111]$$

$$x'_{ri}\hat{\beta}_2 = -2.6389 \quad + 0.0764X_{i2} \quad - 0.0003X_{i3}^2 \quad - 5.5941X_{i4}$$
$$\quad\quad (0.8514) \quad\quad (0.0287) \quad\quad\quad (0.0002) \quad\quad\quad (2.1952)$$
$$\quad\quad [-3.0994] \quad\; [2.6600] \quad\;\; [-1.2718] \quad\;\; [-2.5483]$$

$$\log_e \Lambda(C) = -92.8460 \qquad \log_e \Lambda(\hat{\beta}) = -67.5632$$

For clarity of presentation it is often valuable to use acronyms instead of single letters for the variable names and response category types, and to write out the model in a full linear logit format. In this case, [2.110] becomes

$$\log_e \frac{\hat{P}_{\text{OIL}|i}}{\hat{P}_{\text{DRY}|i}} = -0.5825 + 0.0762\text{THK}_i - 0.0015\,\text{LGS}_i^2 - 9.9619\,\text{SHALE}_i$$
$$\qquad\qquad\quad (0.7457) \quad (0.0282) \quad\quad\quad (0.0006) \quad\quad\quad (3.8413)$$
$$\qquad\qquad\quad\; [0.7811] \quad [2.7008] \quad\quad [-2.6842] \quad\quad [-2.5934]$$

$$\qquad\qquad\qquad\qquad\qquad\qquad\qquad\qquad\qquad\qquad [2.112]$$

$$\log_e \frac{\hat{P}_{\text{GAS}|i}}{\hat{P}_{\text{DRY}|i}} = -2.6389 + 0.0764\,\text{THK}_i - 0.0003\,\text{LGS}_i^2 - 5.5941\,\text{SHALE}_i$$
$$\qquad\qquad\quad (0.8514) \quad (0.0287) \quad\quad\quad (0.0002) \quad\quad\quad (2.1952)$$
$$\qquad\qquad\quad [-3.0994] \quad [2.6600] \quad\quad [-1.2718] \quad\quad [-2.5483]$$

This simple model indicates that in this area of south-central Kansas the determinants of whether a well which penetrates the Mississippian B zone is either oil-producing or gas-producing are similar in nature.[‡] The parameter estimates have the same signs in both equations; hence the odds that a well will be oil-producing rather than dry, or gas-producing rather than dry, are affected in similar qualitative ways (positively or negatively) by changes in the values of the explanatory variables. However, it appears that in the case of the shale content and local geological structure variables such effects are likely to be quantitatively different. Shale content of the Mississippian B zone appears to exert a negative influence which is almost twice as strong on the odds of a well being oil-producing rather than dry, than it is on the odds of a well being gas-producing rather than dry. Similarly, variations in local geological structure appear to have a much stronger influence (though

---

[†] In this particular case where there are no type ϕ variables, the subscript r in x'_{ri} is not strictly necessary, i.e., $x'_{ri} = x'_i$ for all r. See Appendix 2.1.

[‡] The reader interested in the substantive content of this example should treat these results cautiously. The data set includes only 124 wells, and of these only a small number can be classified as gas-producing (11) and oil-producing (22). This example is for illustration purposes only. Most empirical applications of multiple-category logistic/logit models will be to data sets with much greater numbers in any particular response category.

one which is of the same qualitative type) on the odds of a well being oil-producing than on the odds of a well being gas-producing. In contrast, variations in the thickness of the Mississipian B zone appear to have an identical influence on the odds of a well being oil-producing or gas-producing.

It will be recalled that the compact form of systems of linear logit equations such as [2.97] is achieved by exploiting the fact that certain equations and certain parameters in the underlying equation system are redundant. Of course, should we wish to write out any of these redundant equations, it is a simple matter to do so (see Schmidt and Strauss 1975a, for further illustration). For example, in this case, we might consider it valuable to write out the equation for the odds of oil-producing wells over gas–producing wells, i.e. the equation $\log_e$ $(P_{1|i}/P_{2|i})$. To do this we simply utilize the parameter constraints $(\boldsymbol{\beta}_3 = \boldsymbol{0})$ which hold. This gives (see [2.90])

$$\log_e \frac{P_{1|i}}{P_{2|i}} = x'_{ri}(\boldsymbol{\beta}_1 - \boldsymbol{\beta}_2) \qquad [2.113]$$

or in terms of the equations of the fitted model [2.112]

$$\log_e \frac{P_{\text{OIL}|i}}{P_{\text{GAS}|i}} = 2.0564 - 0.0002 \text{ THK}_i - 0.0012 \text{ LGS}_i^2 - 4.3678 \text{ SHALE}_i \qquad [2.114]$$

This equation confirms our previous statements about the relative influence of the shale content of the Mississippian B zone and the local geological structure on the odds of a well being oil-producing and the odds of a well being gas-producing. However, it does it in a more direct fashion by considering explicitly the odds of a well being oil-producing rather than gas-producing.

Substituting the log-likelihood values given in [2.111] into the ρ^2 definition gives the overall goodness-of-fit measure for the model

$$\rho^2 = 1 - \frac{-67.5632}{-92.8460} = 0.2723 \qquad [2.115]$$

This is within the range of values which McFadden has suggested represent a 'good' fit, and the prediction success table (Table 2.6) shows that the model correctly predicts 69.3 per cent of the observed response categories. However, inspection of the normalized prediction success index shows that the fit of the model is significantly poorer for gas-producing wells than it is for oil-producing or dry wells (0.096 compared to 0.296 and 0.325). This result should not surprise us as we have noted already that there is a significantly smaller number of gas-producing wells in the data set than oil-producing or dry wells. This small size magnifies the significance of minor variations within the gas-producing sample of wells and makes the modelling of the determinants of gas-production relatively more difficult than for the other categories. Also it implies that when (as in the case of Example 2.5) we modelled the determinants of the combined 'producing' well sample, oil-producing wells dominated the sample and the particular explanatory variable specification adopted tended to reflect their dominance. Hence, when the sample of 'producing' wells is divided into two groups but the same explanatory variable specification as in the

Table 2.6
Prediction success table for
model [2.110–2.111]

		Predicted response category selection			Observed count	Observed share
		Oil	Gas	Dry		
Observed response	Oil	9.248	2.807	9.944	22.0	0.1774
category selection	Gas	2.672	1.931	6.397	11.0	0.0887
	Dry	10.068	6.239	74.694	91.0	0.7339
Predicted count		21.988	10.977	91.035	124.0	1.0
Predicted share		0.1773	0.0885	0.7342	1.0	
Proportion successfully predicted		0.421	0.176	0.820	0.693	
Prediction success index (not normalized)		0.243	0.087	0.086	0.114	
Prediction success index (normalized)		0.296	0.096	0.325	0.271	

combined sample is used, the fit, not surprisingly, is better for the larger oil-producing well sample than for the smaller gas-producing well sample.

The poorer fit of the model for gas-producing wells than for oil-producing or dry wells is reflected in the parameter estimates and standard errors of the models [2.112]. On the basis of the separate ('t') tests, the parameter associated with the local geological structure variable in the second equation (i.e. $\log_e \hat{P}_{GAS}/\hat{P}_{DRY}$) is not significantly different from zero, suggesting that local geological structure (particularly the importance of location along the flanks of a local positive structure around the zero-value contour line on the local geological structure map) is not as important a determinant in the case of gas-producing wells, as it is in the case of oil-producing wells. However, this result could well be an artifact of the small sample size of the gas-producing wells. It would, therefore, be unwise to remove this variable from the second equation, particularly as it has the same sign as in the first equation indicating a consistent relationship in qualitative terms.[†]

2.11 Types of explanatory variables and response categories

2.11.1 Types of explanatory variables

Earlier in this chapter (in Examples 2.2 and 2.3) we discussed differences in the types of explanatory variables which are found in

---

[†] If we remove the local geological structure variable from both equations in [2.111], and refit the model, the maximized log likelihood value of the fitted model, $\log_e \Lambda (\hat{\boldsymbol{\beta}})$, becomes −78.7773. As an exercise, the reader should conduct a partial joint test, to check whether this deletion is valid. The appropriate null hypothesis would be $H_0 : \beta_{31} = \beta_{32} = 0$.

typical applications of dichotomous logistic/logit models. In general, we noted that the explanatory variables which enter our logistic/logit models can be partitioned into two broad groups; a first group of variables (which we termed X^ϕ variables) which represent attributes of the choice alternatives/response categories, and a second group (which we termed X^ψ variables) which represent attributes of the choice maker/respondent i or the locality i at which the choice is made or the response category is measured. (In terms of this grouping, the explanatory variables in Example 2.7 will readily be seen to be X^ψ variables). Unfortunately, the reader will not always find the recognition of types of explanatory variables as simple as it is in the case of Example 2.7; but will find the whole issue to be surrounded by (some people would suggest that it is confused by) a specialized terminology which has developed in transportation science: one of the major fields of application of logistic/logit models with which geographers and environmental scientists are likely to come into contact. In this section we will attempt to introduce and make sense of this terminology, and to develop a general expression for $x'_{ri}\boldsymbol{\beta}_r$ which distinguishes the various types of explanatory variables (and associated parameters) which the reader will encounter in his wider reading.

To aid our understanding it is useful to think in terms of a simple hypothetical example. Let us imagine that we wish to investigate mode of travel on journey to work trips for a small sample of 100 commuters in one particular city, and that our commuters divide into three groups: those who usually travel by car to work; those who usually travel by bus to work; those who usually travel by metro to work. For each commuter in the sample, we have information on a number of variables: his usual mode of travel to work; his income (I); his age (AG); the time (T) it takes him to travel to work by his usual mode and the alternative modes; the cost (C) of travelling to work by his usual mode and the alternative modes; the number of cars per licensed driver in his household (CAV); the number of bus transfers required if travelling to work by bus (BTR).

The appropriate form of the three-category logit model for this example (where the codes, 1, 2 and 3 indicate CAR, BUS, METRO respectively) is the system of linear logit models

$$\log_e \frac{P_{1|i}}{P_{3|i}} = \beta_{11} + \beta_2(T_i^1 - T_i^3) + \beta_3(C_i^1 - C_i^3) + \beta_{41}I_i + \beta_{51}AG_i + \beta_{61}CAV_i$$

[2.116]

$$\log_e \frac{P_{2|i}}{P_{3|i}} = \beta_{12} + \beta_2(T_i^2 - T_i^3) + \beta_3(C_i^2 - C_i^3) + \beta_{42}I_i + \beta_{52}AG_i + \beta_{71}BTR_i$$

The reader will notice immediately that the variables are treated in somewhat different ways. Some of the variables (I and AG) are included in both equations in the same manner as the variables in our previous example (2.7). They also have similar parameters to those in Example 2.7; that is to say, the parameters β_4 and β_5 vary with respect to the response category (transport mode) and are written as β_{41}, β_{42} and β_{51}, β_{52} to indicate this. In contrast, some of the variables (T and C) are entered (like those in Examples 2.2 and 2.3) as differences between the attributes of response categories (travel modes) and have associated parameters β_2 and β_3 which remain constant across response categories. Finally, some of the variables (CAV and BTR) are entered in a similar manner to the socio-economic variables I and AG but are included in only one of the equations.

In the literature of transportation science, the members of the second of these groups of variables (i.e. T and C) are termed 'generic variables' (GVs) and this terminology has become widely adopted in the field of study known as discrete choice modelling (which we will consider in detail in Part 4). The characteristics of a generic variable are that it varies in value across all response categories and has an associated generic parameter which remains *constant* in value across all response categories. In our example, travel time and travel cost are generic variables. In the first response category (CAR), travel time equals the in-car travel time, plus the time taken to walk to and from the parked car, and travel cost equals the fuel cost plus the parking charges, etc. In the second response category (BUS), travel time equals the in-bus travel time, plus the time spent walking to and from the bus stop, waiting at the bus stop, and making any necessary transfers. Travel cost equals the bus fare. Similarly, in the third response category (METRO), travel time equals the in-metro travel time, plus time spent walking to and from the metro station, waiting at the metro station, and making any necessary transfers. Both generic variables have associated parameters β_2 or β_3 which remain constant and, as a result, these parameters do not have a second subscript (unlike β_{41}, β_{42}, etc.) which indicates the response category.

In practical modal choice research, generic variables of this type are often decomposed into their constituent parts. For example, travel time is often decomposed into two or three generic variables such as: in-vehicle travel time and out-of-vehicle travel time; or walk time, waiting time and in-vehicle time. When such decomposed variables are used, a generic variable may take the value zero for one alternative (e.g. waiting time for the car alternative). It should be stressed, however, that in this circumstance the zero value is a natural measured zero value, rather than the type of 'assigned' zero value we will discuss below.

Members of the first and third groups of variables we distinguished above (i.e. I, AG, and CAV, BTR) are referred to in the transportation science and discrete choice modelling literature as 'alternative specific variables' (ASVs). Alternative specific variables are variables which do not vary in value across all response categories (response alternatives) and which, therefore, take an 'assigned' value of zero for certain response categories (alternatives) in the choice set. ASVs can be said to have an identifiable correspondence with particular response categories (alternatives), and they have associated alternative specific parameters which are specific to particular response categories (alternatives), e.g. β_{41} is specific to response category 1, β_{42} is specific to response category 2, etc. In our example it is clear that CAV and BTR are specific to response categories CAR and BUS respectively, and take the value zero elsewhere. They thus have associated parameters β_{61} and β_{72} respectively. The general socio-economic variables I and AG, however, are introduced in such a way that they are specific to $R-1$ response categories where R is the total number of response categories in the choice set (in our case $3-1=2$). This is done by introducing each of these variables into the model as a series of $R-1$ alternative-specific variables. Each of these ASVs takes the appropriate income or age value for a specific response category and takes the value zero for all other categories. This produces $R-1$ alternative specific parameters (β_{41} and β_{42}) for the income variable, and $R-1$ alternative specific parameters (β_{51} and β_{52}) for the age variable, and it is directly equivalent to the system assumed in our original discussion on multiple-category models, and to the situation in Example 2.7.[†]

In the transportation science and discrete choice modelling literature, the constant terms β_{11} and β_{12} are referred to as 'alternative specific constants' (ASCs). Normally, a series of $R-1$ ASCs will be introduced when constants are used in the model. However, it is not necessary to introduce all $R-1$ ASCs and any number from 1 to $R-1$ can, in fact, be introduced. For example, in our case we could introduce just β_{12} and remove β_{11}.

For some readers[‡] who are familiar with matrix formulations of the general linear model, it may help their understanding of the previous discussion if the system of linear logit equations [2.116] is written out as a vector and matrix expression. This is shown below, where for simplicity $L_{13|i}$ denotes the logit $\log_e(P_{1|i}/P_{3|i})$, $L_{23|i}$ denotes the logit $\log_e(P_{2|i}/P_{3|i})$; $T_i^{13} = (T_i^1 - T_i^3)$, and $C_i^{23} = (C_i^2 - C_i^3)$. In all cases, $i = 1, \ldots 100$.

$$
\begin{bmatrix}
L_{13|1} \\
L_{13|2} \\
L_{13|3} \\
\cdot \\
\cdot \\
\cdot \\
L_{13|99} \\
L_{13|100} \\
L_{23|1} \\
L_{23|2} \\
L_{23|3} \\
\cdot \\
\cdot \\
\cdot \\
L_{23|99} \\
L_{23|100}
\end{bmatrix}
=
\begin{bmatrix}
1 & 0 & T_1^{13} & C_1^{13} & I_1 & 0 & AG_1 & 0 & CAV_1 & 0 \\
1 & 0 & T_2^{13} & C_2^{13} & I_2 & 0 & AG_2 & 0 & CAV_2 & 0 \\
1 & 0 & T_3^{13} & C_3^{13} & I_3 & 0 & AG_3 & 0 & CAV_3 & 0 \\
\cdot & & \cdot & \cdot & \cdot & & \cdot & & \cdot & \cdot \\
\cdot & & \cdot & \cdot & \cdot & & \cdot & & \cdot & \\
\cdot & & \cdot & \cdot & \cdot & & \cdot & & \cdot & \\
1 & 0 & T_{99}^{13} & C_{99}^{13} & I_{99} & 0 & AG_{99} & 0 & CAV_{99} & 0 \\
1 & 0 & T_{100}^{13} & C_{100}^{13} & I_{100} & 0 & AG_{100} & 0 & CAV_{100} & 0 \\
0 & 1 & T_1^{23} & C_1^{23} & 0 & I_1 & 0 & AG_1 & 0 & BTR_1 \\
0 & 1 & T_2^{23} & C_2^{23} & 0 & I_2 & 0 & AG_2 & 0 & BTR_2 \\
0 & 1 & T_3^{23} & C_3^{23} & 0 & I_3 & 0 & AG_3 & 0 & BTR_3 \\
\cdot & & \cdot & \cdot & & \cdot & & \cdot & & \cdot \\
\cdot & & \cdot & \cdot & & \cdot & & \cdot & & \cdot \\
\cdot & & \cdot & \cdot & & \cdot & & \cdot & & \cdot \\
0 & 1 & T_{99}^{23} & C_{99}^{23} & 0 & I_{99} & 0 & AG_{99} & 0 & BTR_{99} \\
0 & 1 & T_{100}^{23} & C_{100}^{23} & 0 & I_{100} & 0 & Ag_{100} & 0 & BTR_{100}
\end{bmatrix}
\begin{bmatrix}
\beta_{11} \\
\beta_{12} \\
\beta_2 \\
\beta_3 \\
\beta_{41} \\
\beta_{42} \\
\beta_{51} \\
\beta_{52} \\
\beta_{61} \\
\beta_{72}
\end{bmatrix}
\quad [2.117]
$$

Careful consideration of this matrix and vector formulation will demonstrate to the reader: (a) how ASVs take an 'assigned' value of zero for certain response categories; (b) how the socio-economic

---

[†] In some transportation science examples, the reader will encounter the socio-economic variable, income, entered as an interactive influence on one of the generic variables, e.g. Travel Cost/Income. This creates a new generic variable, and the interpretation is that individuals with different incomes value travel cost savings differently.

[‡] If it does not help you, just ignore this paragraph and move on immediately to the next.

variables such as income and age are introduced as a series of $R-1$ (in our case 3 -1) ASVs; (c) how the generic variables do not take an assigned value of zero for any of the response categories; (d) how the ASCs are normally introduced as a series of $R-1$ binary variables.

In addition to classifying our explanatory variables into generic and alternative specific, we can also classify them into the two broad groups, X^ϕ and X^ψ variables which we have distinguished in previous sections. X^ϕ variables represent attributes of the choice alternatives/response categories, and X^ψ variables represent attributes of the choice maker/ respondent i or the locality i at which the choice is made or the response category is measured. In our examples, the variables T, C and BTR are, therefore, X^ϕ variables (attributes of the choice alternatives), whereas the variables I, AG and CAV are X^ψ variables (attributes of the choice maker or his household).

Combining the generic/alternative-specific and X^ϕ/X^ψ classifications, we are now in a position to state a general expression for the $x'_{ri}\boldsymbol{\beta}_r$ function which we adopted in the general forms of the multiple-category logistic/logit models, equations [2.101] and [2.102]. Where there are K parameters in the function, we can write this general expression as:

$$x'_{ri}\boldsymbol{\beta}_r = \beta_{1r} + \sum_{k=1}^{K_I} \beta_k (X^\phi_{rik} - X^\phi_{Rik}) + \sum_{k=K_1+1}^{K_2} \beta_{kr} X^\phi_{rik} + \sum_{k=K_2+1}^{K} \beta_{kr} X^\psi_{ik} \quad \textbf{[2.118]}$$

$$\text{(ASC)} \quad \text{(Generic, type } \phi) \quad \text{(ASV, type } \phi) \quad \text{(ASV, type } \psi)$$

where K_1, K_2 represent arbitrary sub-totals (see also equation [A2.28] in Appendix 2.1). In any particular model, not all of these types of explanatory variables will necessarily be included, and the reader must become accustomed to recognizing different combinations of types. To facilitate this, at the end of this section, examples will be presented which illustrate a range of typical combinations.

2.11.2 *Types of response categories*

Linked to the distinction between generic and alternative specific variables is a specialized terminology which has also developed in transportation science and discrete choice modelling to describe two important types of response categories which are encountered in practical research. The first of these is what is known as the 'ranked' set of response categories/choice alternatives. A 'ranked' set is one in which the response categories/choice alternatives are the same for all individuals or at all localities in the sample. For example, in the case of the mode choice illustration above, $r = 1$ is the car alternative for all respondents, $r = 2$ is the bus alternative and $r = 3$ is the metro alternative for all respondents. In fact, all our previous examples have involved 'ranked' sets of response categories/alternatives. For example, $r = 1 = $ oil-producing, $r = 2 = $ gas-producing, $r = 3 = $ dry, at all loca- lities in our Kansas oil and gas exploration example. We have seen above that a 'ranked' set of categories/alternatives can be modelled using any combination of generic or alternative specific variables.

The second type of response categories/choice alternatives are known as 'unranked'. In this case the response categories or alternatives available to each individual in the sample (or at all localities) are not

necessarily the same, i.e. the alternatives $r = 1$, $r = 2$, $r = 3$ may have
a totally different identity from individual to individual. For example,
the alternatives may be the shops or shopping centres available to an
individual. Clearly, if the sample includes individuals from different
areas of a city or region the available shopping alternatives will differ
from individual to individual (e.g. for individual A shopping alternatives
1 and 2 may be shopping areas x and y respectively, i.e.
$r = 1$ = shopping area x, $r = 2$ = shopping area y; whereas for
individual B the shopping alternatives 1 and 2 may be shopping areas v
and z, i.e. $r = 1$ = shopping area v, $r = 2$ = shopping area z). In this
case of 'unranked' categories/alternatives, only generic variables can
meaningfully enter our models. For example, shopping centres can be
described in terms of attributes such as: range of goods and services
offered; general attractiveness; ease of parking, etc; and these would
represent generic variables. However, as the alternatives (shopping
centres) available differ from individual to individual, it is not
meaningful to introduce alternative specific variables into our model.

In certain cases a mixture of 'ranked' and 'unranked' response
categories/alternatives will occur. For example, in a particular city the
shopping centres available may differ from individual to individual
depending upon the individual's location within the city, but *all*
individuals may have one, two or more common alternatives (e.g. the
central business district (CBD) and a major inner city shopping centre)
within their choice-sets. In this case a mixture of generic and alternative
specific variables can be used to model the common alternatives (the
embedded 'ranked' subset with the wider 'unranked' set) whilst generic
variables alone are used for the unranked alternatives.

Further discussion of generic/alternative-specific variables and
ranked/unranked alternatives is to be found in Domencich and
McFadden (1975, 117–19), Richards and Ben-Akiva (1975, 30–3), and
Hensher and Johnson, (1981a, 118–27). We will now turn to a series of
examples which will illustrate these features.

EXAMPLE 2.8   *Shopping trip destination choice in Pittsburgh*

Domencich and McFadden (1975, 170–3) investigated the destination
choices on shopping trips of a small, relatively homogeneous sample of
respondents drawn from the southern suburban corridor of the
Pittsburgh area. The shopping destinations available differed from
respondent to respondent. The choice alternatives were, therefore,
'unranked' and it was necessary to express their attributes in generic
terms rather than in terms specific to each alternative. The model they
fitted had the form:

$$\log_e \frac{\hat{P}_{r|i}}{\hat{P}_{R|i}} = -1.06(\hat{IP}_i^r - \hat{IP}_i^R) + 0.844(E_i^r - E_i^R) \qquad [2.119]$$
$$\qquad\qquad (0.28) \qquad\qquad (0.23)$$

where,

$\hat{P}_{r|i}/\hat{P}_{R|i}$ is the predicted odds that respondent i will choose to
shop at destination r rather than destination R.

$\hat{I}P_i^r$ is the estimated 'inclusive price' of travel to destination r for respondent i. This estimate is derived using parameter estimates from a previous shopping mode of travel model (see Domencich and McFadden 1975, 172 for details).

E_i^r is a measure of the shopping opportunities available at destination r for respondent i (defined as the retail employment at r as a percentage of the total retail employment in the Pittsburgh region).

Standard errors are shown in parentheses below the parameter estimates.

It can be seen from model [2.119] that both generic variables are highly significant determinants of the choice of shopping destination in this area. As the inclusive price of travel to a given destination r increases relative to the household's alternative destination, so the odds of travelling to destination r fall. On the other hand, the greater the relative retail trade opportunities at r the greater the odds of travelling to that destination to shop.

Models such as [2.119] which are composed entirely of generic variables are often referred to in the transportation science literature as 'abstract-alternative' (or in the case of mode choice 'abstract-mode') models. Since there are no ASVs present in the model that relate to any specific alternative, but only generic variables which measure attributes common to all alternatives, an abstract-alternative (generic-variables-only) model can be applied for forecasting purposes to situations significantly different from those used for model estimation. Such a model is, therefore, particularly suited to the evaluation of systems not currently in use. In Part 4 we will consider an example of such a model used for forecasting purposes.

EXAMPLE 2.9   *Shopping trip destination choice in West Yorkshire*

For purposes of comparison, and to give a British dimension to these examples, we will consider a similar (but not identical) generic-variables-only model of the choice of destination on shopping trips in West Yorkshire. This model is taken from a much wider study of mode and destination choice on work, social/recreational, and shopping trips conducted by Southworth (1981). The study is notable for its regional rather than intra-urban scale, for its large sample size, and for certain aspects of its design (including its approach to spatial sampling and the household diary survey employed).

Within the West Yorkshire region, Southworth selected the fourteen sample areas shown in Fig. 2.11. These sample areas include a selection of the major inter-urban and suburb-to-urban-centre flows within the region. Each of the fourteen sample areas contains between three and six of the census wards used in the 1970 M62 Transpennine Motorway Economic Impact Study, and within each of these wards household travel diaries were completed for 7 days by a maximum of thirty-three households. The shopping trips generated by these households within the 7-day period were then used to estimate the shopping trip destination choice model.

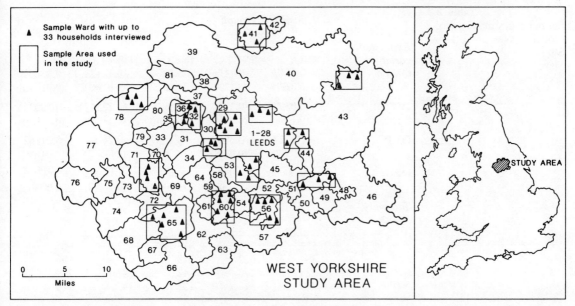

Figure 2.11
The sample design underlying Southworth's study (adapted from Southworth, 1981)

Southworth estimated his shopping trip destination choice model for each of three household income groupings (INC1 = less than £1,000 p.a., INC2 = £1,000–1,999 p.a., INC3 = £2,000 or more p.a.) and two household structure groupings (HS2 = household with 1 employed resident, HS1 and HS3 = households with 0 or more than 1 employed residents respectively). In order to display Southworth's results, we can use an explicit linear logit equation form of the type employed in model [2.119]; one equation of the type [2.119] for each of the household groupings. Alternatively, we can choose to present the results in the much more compact form shown in Table 2.7.

From the parameter estimates given in Table 2.7 we can draw similar general conclusions to those of the previous example (2.8). That is to say, as the cost of travel to a given shopping destination r (in terms of both monetary expenditure and time) increases relative to the household's alternative destination so the odds of travelling to shopping destination r will fall. On the other hand, the greater the relative attractiveness of shopping destination r, the greater the odds (with one exception) of travelling to that destination. In this example, however, the generic variable which in Example 2.8 measures the 'inclusive price' of travel to a destination has been decomposed into three separate variables. This decomposition, together with the segmentation of the sample into different household income and household structure groupings, provides valuable additional information. For example, the parameter estimates associated with monetary expenditure on travel (OPTC) have a consistent trend across the household income groups and are least strong for the highest income group (INC 3). In contrast, the in-vehicle travel time (IVT) parameter estimates have the opposite trend across income groups, and are at their maximum for the highest income group households. This suggests that the most affluent households will substitute monetary loss to compensate for travel time savings. Another feature which this decomposition reveals is that in the

Table 2.7
Shopping destination choice
model. West Yorkshire
Study

Explanatory variables	*Household groupings*				
	INC1	INC2	INC3	HS 1+3	HS 2
OVT	−0.1317	−0.1135	−0.1804	−0.12123	−0.08090
	(0.0227)	(0.1720)	(0.0248)	(0.02981)	(0.01674)
IVT	−0.0899	−0.1720	−0.1940	−0.23547	−0.29035
	(0.1066)	(0.0974)	(0.1116)	(0.14773)	(0.09176)
OPTC	−2.2127	−0.3907	−0.1389	−0.26004	−0.21182
	(0.0379)	(0.0848)	(0.2094)	(0.04702)	(0.04306)
DEST	−0.00006	+0.00003	+0.00009	+0.00002	+0.00006
	(0.00001)	(0.00001)	(0.00001)	(0.00002)	(0.00001)
Summary Statistics					
$\bar{\rho}^2$	0.63	0.66	0.65	0.60	0.68
t	517	792	415	422	967
t_1	389	622	373	340	698
t_2	128	170	42	82	269
N	200	314	122	145	400

where

OVT = out-of-vehicle travel time to a given shopping destination by a given mode of travel

IVT = in-vehicle travel time to a given shopping destination by a given mode of travel

OPTC = monetary expenditure (or out-of-pocket travel cost) from a respondent's residence to a given shopping destination by a given mode of travel

DEST = aggregate 'pull' or attraction of a given destination spatial zone for shopping purposes.

destination-choice context, where varying trip distance is of concern, in-vehicle travel time (IVT) is relatively more important than out-of-vehicle travel time (OVT) in influencing choice. This contrasts with Southworth's mode-of-travel-choice results (which are not reported here but can be seen in Southworth's original report). These results show that, in the mode-choice context, out-of-vehicle travel time is regarded as significantly more onerous than in-vehicle travel time, and OVT is therefore a more important determinant of choice in this context than IVT.

Below the parameter estimates in Table 2.7 a series of summary statistics are presented. N indicates the number of households in the overall sample which belong to that particular household grouping, and t_1 and t_2 indicate the number of private transport and public transport trips made over the 7-day trip-diary monitoring period by the households in that group. The first of the summary statistics is $\bar{\rho}^2$, the ρ^2 goodness-of-fit statistic adjusted for degrees of freedom. In general, the definition of $\bar{\rho}^2$ is

$$\bar{\rho}^2 = 1 - \frac{\log_e \Lambda(\hat{\boldsymbol{\beta}})/(\sum_{i=1}^{N} R_i - 1) - K}{\log_e \Lambda(C)/(\sum_{i=1}^{N} R_i - 1)}$$

[2.120]

where,

R_i is the number of response categories (alternatives) available to the individual (household) i or at the ith locality

K is the number of parameters in the model.

However, in this case where there are only generic variables and no alternative specific constants, Southworth appropriately uses the alternative version of equation [2.120] where the *equal shares* maximized log likelihood value $\log_e \Lambda\,(0)$ is substituted for the *market-shares* log likelihood value $\log_e \Lambda\,(C)$ (see footnote, page 50 for further details on this issue).

EXAMPLE 2.10   *Shopping destination and mode of travel choice in San Francisco*

In the two previous examples we have considered cases in which the choice alternatives were 'unranked' and in which the fitted models contained only generic variables. We will now consider an example in which the choice alternatives are a 'ranked' set and in which both generic and alternative specific variables appear in the fitted model.

The example is taken from a much wider study of shopping choice behaviour in the San Francisco Bay Area by McCarthy (1980), and it concerns an attempt to model shopping destination choice and choice of mode of travel on shopping trips for a small sample of 135 respondents in the Mission District area of central San Francisco. The model employed was a simultaneous (or joint) destination-mode-choice model,[†] and the choice alternative set (a 'ranked' set) consisted of five alternative mode-destination combinations: (1) car travel to the downtown San Francisco shopping area; (2) bus travel to the downtown San Francisco shopping area; (3) car travel to the Mission Street shopping area; (4) bus travel to the Mission Street shopping area; (5) walk or bicycle travel to the Mission Street shopping area.

Writing the model fitted by McCarthy in the same general form as in Example 2.9 (i.e. not in explicit linear logit equation form) gives the results shown in Table 2.8. The results indicate that individuals in this central city environment are sensitive in their destination-mode choices to both travel cost and out-of-vehicle walk time. As both the cost and walking time associated with a given destination-mode combination increase relative to the alternative combination so the odds of selecting that particular destination-mode combination will decrease. On the other hand, the greater the relative attraction of a particular shopping area and the greater the relative safety of a shopping trip to that shopping area, the greater will be the odds of selecting a destination-mode combination which includes that shopping area. The car availability parameter has the correct sign (indicating an increase in the odds of selecting a destination-mode combination which involves the use of a car as car availability in the household increases) but is not significant at conventional levels if a separate 't statistic' test is employed.

[†] See Ch.10 for a discussion of the role of simultaneous models in the modelling of decision structures.

Table 2.8
Simultaneous destination-mode-choice model, Mission District, San Francisco

Explanatory variables	Parameter estimates
BUSCON – alternative specific constant taking value 1 for bus travel alternatives, 0 otherwise	−1.163 (0.55)
SFCON – alternative specific constant taking value 1 for downtown San Francisco shopping area alternatives, 0 otherwise	2.110 (0.44)
WOVT – out-of-vehicle walk time: generic variable	−0.078 (0.04)
OPTC – round trip out-of-pocket travel costs from a respondent's residence to a given shopping destination: generic variable	−0.870 (0.46)
SATT – generalized measure of shopping area attraction: generic variable	4.478 (1.09)
TSAFE – generalized measure of the safety of a shopping trip to a given shopping destination: generic variable	5.935 (2.86)
CAV – alternative specific car availability variable taking value of number of cars per driver in household (max 1) for car travel alternatives, 0 otherwise	0.439 (0.65)

$\bar{\rho}^2 = 0.349$

One particularly interesting aspect of this study is that McCarthy derives the generalized measures SATT and TSAFE from a factor analysis of attitudinal data. This allows him to calculate the impact of changes in the component elements of the generalized measures (e.g. SATT is derived from five component elements: variety of goods; reliability of retailers; store-to-store accessibility within the shopping area; price levels within the shopping area; opening hours) on an individual's choice of destination. These impact measures (elasticities) are not reported here but they indicate that, within this central city environment, personal as well as vehicular safety is an important feature in an individual's choice of destination. Also, they indicate the importance of variety of goods, store-to-store accessibility, and price of goods in determining the choice of shopping destination.

EXAMPLE 2.11   *Shopping destination and mode of travel choice in Eindhoven*

In the previous examples we have considered cases in which the choice alternatives are either 'ranked' or 'unranked'. In this final example we will consider a case in which there is a mixture of 'ranked' and 'unranked' alternatives.

The example is taken from a much wider study of travel demand in the Eindhoven region of the Netherlands conducted by Richards and Ben-Akiva (1975). Within this region (see Fig. 2.12), two municipalities, Best, and Son and Breugel, were selected for further study. Although both are a similar distance from Eindhoven, they differ

Figure 2.12
The Eindhoven study area (adapted from Richards and Ben-Akiva 1975, 52)

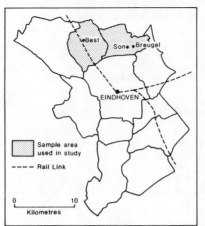

considerably in average income, proportion of local employment, and in the nature of public transport connections to Eindhoven.

The part of the study we will concentrate on concerns an attempt to develop a simultaneous (or joint) shopping-destination-mode-choice model of the type we discussed in Example 2.10. Unlike that San Francisco example, however, the respondents in the Eindhoven study are drawn from a much wider geographical area and their available destination-mode choice alternatives are not necessarily the same in number or kind. As a result, the destination-mode alternatives must, in principle, be treated as 'unranked' and that implies the use of a generic-variables-only model. However, it is possible in this case to regard certain elements in the set of choice alternatives as being 'ranked'. For example, the Eindhoven central shopping area is a shopping alternative common to all respondents and, similarly, the travel modes – car, public transport, moped, bicycle and walk – can be regarded as an embedded 'ranked' subset. In fact, it is only the local area shopping alternatives which differ from respondent to respondent and which are strictly 'unranked' alternatives. As a result, a model which contains a mixture of generic and alternative specific variables can be developed.

The form of the model fitted by Richards and Ben-Akiva (one of many specifications they tried) is shown in Table 2.9. In the model shown in Table 2.9, in-vehicle travel time (IVT) is a generic variable which equals the time spent in or on the vehicle for all mechanical modes (car, public transport, moped, bicycle) and which takes a natural zero value for the walk mode. The other variables are all alternative specific (taking 'assigned' zero values for all alternatives other than those indicated in brackets), though it is possible to regard the shopping destination attraction variable EMPLOK as being a generic variable across all the 'unranked' shopping destination alternatives.

The parameter estimates have the expected signs. The negative signs of IVT, WOVT, POVT and WSOVT indicate that as the relative in-vehicle or out-of-vehicle travel times of a particular destination-mode combination increase relative to the alternative combinations, so the odds of selecting that particular destination-mode combination will decrease. On the other hand, the positive signs of EMPLOK and

Table 2.9
Simultaneous destination-mode choice model, Eindhoven study area. Richards and Ben-Akiva (1975. 122–23 Model 3)

Explanatory variables (alternative specific to)	Parameter estimates
MOPCON – alternative specific constant (moped)	−1.5163 (0.2920)
IVT – in-vehicle travel time	−0.1502 (0.0132)
WOVT – total walking time of trip (walk)	−0.2122 (0.0177)
POVT – out-of-vehicle travel time (car, moped, bicycle)	−0.4325 (0.0511)
WSOVT – walking time to and from public transport stop or station (public transport)	−0.2605 (0.0246)
CAV×$\log_e$ IVT – car availability; number of cars per licensed driver, times $\log_e$ IVT (car)	0.5734 (0.2197)
BCAV – bicycle availability; number of bicycles per person over 5 years of age (bicycle)	0.7254 (0.2325)
EMPLOK – shopping destination attraction measured by retail employment (all shopping destinations except Eindhoven central area)	0.0344 (0.0017)
EMPEIND – shopping destination attraction measured by retail employment (Eindhoven central area)	0.0027 (0.0002)

$\bar{\rho}^2 = 0.46$

EMPEIND indicate that the greater the relative attraction of a particular shopping destination, the greater will be the odds of selecting a destination-mode combination that includes that shopping destination. The bicycle and car availability variables have the expected positive signs indicating an increase in the odds of selecting a destination-mode combination which involves the use of these modes as bicycle/car availability in the household increases. The unusual form of the car availability variable is an attempt to specify the coefficient of car availability as a function of trip length (where trip length is proxied by IVT).

Appendix 2.1   A comparison of alternative derivations of logistic/logit models

The dichotomous case: a general appreciation

In the derivation of our dichotomous logistic/logit models we have implicitly assumed (up until Examples 2.2 and 2.3) that all our explanatory variables are type ψ variables. That is to say, we have assumed that all our explanatory variables represent attributes of the choice maker/respondent i or the locality i at which the choice is made or the response category is measured. The implication of this assumption is that, for a given individual i, the selection of each response category/choice alternative is related to the *same* explanatory variables and values of those explanatory variables (i.e. the explanatory variable values are common to both category/alternative selections). Put rather more formally,

$$P_{1|i} = F(z'_i\boldsymbol{\beta}) \qquad\qquad [A2.1]$$

$$P_{2|i} = 1 - P_{1|i} = 1 - F(z'_i\boldsymbol{\beta}) \qquad\qquad [A2.2]$$

where, for purposes of this appendix, z_i represents the vector of explanatory variable values for individual i. Expressions [A2.1] and [A2.2] imply that the probability that the ith repondent will select the first response category is a function of the same explanatory variable values and parameters as the probability that the ith respondent will select the second response category. If we specify the arbitrary function in [A2.1] to be the cumulative logistic probability function this produces, as we saw in equations [2.14] and [2.17], the expressions:

$$P_{1|i} = \frac{e^{z'_i\boldsymbol{\beta}}}{1 + e^{z'_i\boldsymbol{\beta}}} \qquad\qquad P_{2|i} = \frac{1}{1 + e^{z'_i\boldsymbol{\beta}}} \qquad\qquad [A2.3]$$

Let us now make the alternative assumption that all our explanatory variables are type ϕ variables (attributes of the choice alternatives/response categories). In this case, the selection of each response category/choice alternative is related to the particular attributes of each category/alternative. More formally, we can say:

$$P_{1|i} = F(z'_{1i}\boldsymbol{\beta}, z'_{2i}\boldsymbol{\beta}) \qquad P_{2|i} = 1 - F(z'_{1i}\boldsymbol{\beta}, z'_{2i}\boldsymbol{\beta}) \qquad [A2.4]$$

where z_1 is the set of explanatory variables measuring the attributes of category/alternative 1 and z_2 is the set of explanatory variables measuring the attributes of category/alternative 2. If we now employ the cumulative logistic probability function we get[†]

$$P_{1|i} = \frac{e^{z'_{1i}\boldsymbol{\beta}}}{e^{z'_{1i}\boldsymbol{\beta}} + e^{z'_{2i}\boldsymbol{\beta}}} \qquad\qquad P_{2|i} = \frac{e^{z'_{2i}\boldsymbol{\beta}}}{e^{z'_{1i}\boldsymbol{\beta}} + e^{z'_{2i}\boldsymbol{\beta}}} \qquad\qquad [A2.5]$$

or by dividing through the numerator and denominator of both of these equations by $e^{z'_{2i}\boldsymbol{\beta}}$

$$P_{1|i} = \frac{e^{(z_{1i}-z_{2i})'\boldsymbol{\beta}}}{1 + e^{(z_{1i}-z_{2i})'\boldsymbol{\beta}}} \qquad\qquad P_{2|i} = \frac{1}{1 + e^{(z_{1i}-z_{2i})'\boldsymbol{\beta}}} \qquad\qquad [A2.6]$$

In other words, the probabilities of selecting either of the two categories/alternatives can be seen to be related to the differences between the attributes of the two choice alternatives or response categories.

[†] These logistic expressions appear different from those adopted in [A2.3] in that they have a 'symmetric' rather than 'asymmetric' form (see Stopher and Meyburg 1979, 320). However, the reader should not be concerned about this superficial difference, for on dividing through the numerator and denominator of the 'symmetric' expressions [A2.5] we obtain the 'asymmetric' expressions [A2.6]. This demonstrates the equivalence of the 'symmetric' and 'asymmetric' logistic forms, and the reader must expect to encounter both forms in his wider reading.

What we have just observed in terms of logistic models can also be seen if we re-express [A2.3] and [A2.6] in equivalent linear logit form. In this case the type ψ explanatory variables equations [A2.3] become

$$\log_e \frac{P_{1|i}}{P_{2|i}} = z'_i \boldsymbol{\beta} \qquad\qquad [A2.7]$$

and the type ϕ explanatory variables equations [A2.6] become

$$\log_e \frac{P_{1|i}}{P_{2|i}} = (z_{1i} - z_{2i})' \boldsymbol{\beta} \qquad\qquad [A2.8]$$

This, in essence, is the contrast between the models in Examples 2.1 and 2.2. When all the explanatory variables are type ϕ variables, it follows that they are included in the dichotomous logistic/logit models as *differences* between the attributes of the two choice alternatives/ response categories. In practice, this use of type ϕ variables and the associated derivation of logistic/logit models is typical of research in transportation science and the field of study which has become known as discrete choice modelling (which we will consider in detail in Part 4). The use of type ψ variables only, is more typical of work in statistics, econometrics and biometrics, and it is the approach we have chosen to introduce logistic/logit models in Chapter 2.

The dichotomous case: further refinement

Although the derivation above provides a useful summary of the essence of the contrast between models which include only type ψ variables and those which include only type ϕ variables, it is not sufficient as it stands. Before we can proceed further we need to reconsider the issue and add some extra refinement. This will involve us in the use of some of the concepts discussed in Sections 2.9 to 2.11 and the reader should have considered those sections of Chapter 2 before proceeding with this appendix.

Clearly, the two cases we discussed above should be equivalent when $z'_{1i} = z'_{2i} = z'_i$. In other words, they should be equivalent when the attributes of choice alternative/response category 1 and the particular values of those attributes are identical to those of alternative/category 2, and when the probability of selecting each alternative/category is thus a function of a common set of explanatory variable values. Unfortunately, in this situation we run into problems, for if we substitute identical z'_{1i} and z'_{2i} into the expressions [A2.6] they simply cancel out in the difference form $(z'_{1i} - z'_{2i})$. The only way to overcome this problem is to make the further assumption that when the probability of selecting each alternative/category is a function of a common set of explanatory variable values, the parameters in expression [A2.4] (and by implication those in [A2.1] and [A2.2] also) are *alternative specific*. In other words, in this situation (given that $z'_{1i} = z'_{2i} = z'_i$) we must assume that [A2.1], [A2.2] and [A2.4] can, more precisely, be written as

$$P_{1|i} = F(z'_i \boldsymbol{\beta}_1, \ z'_i \boldsymbol{\beta}_2) \qquad P_{2|i} = 1 - F(z'_i \boldsymbol{\beta}_1, \ z'_i \boldsymbol{\beta}_2) \qquad [A2.9]$$

where $\boldsymbol{\beta}_1$ are parameters specific to alternative/category 1 and $\boldsymbol{\beta}_2$ are parameters specific to alternative/category 2. If we now employ the cumulative logistic probability function we get, in place of expression [A2.5],

$$P_{1|i} = \frac{e^{z'_i \boldsymbol{\beta}_1}}{e^{z'_i \boldsymbol{\beta}_1} + e^{z'_i \boldsymbol{\beta}_2}} \qquad P_{2|i} = \frac{e^{z'_i \boldsymbol{\beta}_2}}{e^{z'_i \boldsymbol{\beta}_1} + e^{z'_i \boldsymbol{\beta}_2}} \qquad \text{[A2.10]}$$

or by dividing through the numerator and denominator of both these equations by $e^{z'_i \boldsymbol{\beta}_2}$

$$P_{1|i} = \frac{e^{z'_i (\boldsymbol{\beta}_1 - \boldsymbol{\beta}_2)}}{1 + e^{z'_i (\boldsymbol{\beta}_1 - \boldsymbol{\beta}_2)}} \qquad P_{2|i} = \frac{1}{1 + e^{z'_i (\boldsymbol{\beta}_1 - \boldsymbol{\beta}_2)}} \qquad \text{[A2.11]}$$

The equivalent linear logit model expression is then

$$\log_e \frac{P_{1|i}}{P_{2|i}} = z'_i (\boldsymbol{\beta}_1 - \boldsymbol{\beta}_2) \qquad \text{[A2.12]}$$

In both the logistic and linear logit expressions it can be seen that it is only the difference of the $\boldsymbol{\beta}$ parameters from those of the arbitrary base, 'anchor', or denominator category (in our case category 2) which matters. We can, therefore, arbitrarily impose the constraints

$$\boldsymbol{\beta}_2 = 0 \qquad \text{[A2.13]}$$

without loss. The logistic and linear logit expressions then become

$$P_{1|i} = \frac{e^{z'_i \boldsymbol{\beta}_1}}{1 + e^{z'_i \boldsymbol{\beta}_1}} \qquad P_{2|i} = \frac{1}{1 + e^{z'_i \boldsymbol{\beta}_1}} \qquad \text{[A2.14]}$$

and

$$\log_e \frac{P_{1|i}}{P_{2|i}} = z'_i \boldsymbol{\beta}_1 \qquad \text{[A2.15]}$$

The subscript 1 on the parameters is clearly redundant as no other parameters enter the models and there is only one category/alternative other than the base or denominator category. For convenience, the subscript can be dropped, producing the expressions we have seen previously in [A2.3] and [A2.7].

It appears, therefore, that if we wish to reconcile the two cases discussed above, and the expressions in [A2.3] and [A2.6] we must be prepared to make further assumptions. We must assume that when the probability of selecting each choice alternative/response category is a function of identical explanatory variable values, the associated parameters are alternative specific and the parameters of the arbitrary base or denominator category are constrained to zero. As we have demonstrated above, these are implicit assumptions which underlie the dichotomous logistic and logit model expressions [A2.3] and [A2.7] which we employed in the early parts of Chapter 2.

Once these implict assumptions which underlie the type ψ case models [A2.3] and [A2.7] are understood, it is a relatively simple matter to extend our previous discussion to consider the case in which both types of explanatory variables (ψ and ϕ) are included in the same model (see Example 2.3). In this case the probability of selecting each choice alternative/response category is a function of the particular values of the type ϕ attributes of the alternatives/categories and also of the common set of type ψ explanatory variable values. More formally, we can combine [A2.4] and [A2.9] and write this as

$$P_{1|i} = F(z'_{1i}\boldsymbol{\beta}+z'_i\boldsymbol{\beta}_1, \; z'_{2i}\boldsymbol{\beta}+z'_i\boldsymbol{\beta}_2)$$

$$P_{2|i} = 1-F(z'_{1i}\boldsymbol{\beta}+z'_i\boldsymbol{\beta}_1, \; z'_{2i}\boldsymbol{\beta}+z'_i\boldsymbol{\beta}_2)$$

[A2.16]

If we now employ the cumulative logistic probability function we get, in place of expressions [A2.5] or [A2.10],

$$P_{1|i} = \frac{e^{z'_{1i}\boldsymbol{\beta}+z'_i\boldsymbol{\beta}_1}}{e^{z'_{1i}\boldsymbol{\beta}+z'_i\boldsymbol{\beta}_1} + e^{z'_{2i}\boldsymbol{\beta}+z'_i\boldsymbol{\beta}_2}} \qquad P_{2|i} = \frac{e^{z'_{2i}\boldsymbol{\beta}+z'_i\boldsymbol{\beta}_2}}{e^{z'_{1i}\boldsymbol{\beta}+z'_i\boldsymbol{\beta}_1} + e^{z'_{2i}\boldsymbol{\beta}+z'_i\boldsymbol{\beta}_2}} \qquad \text{[A2.17]}$$

or dividing through the numerator and denominator of these equations by $e^{z'_{2i}\boldsymbol{\beta}+z'_i\boldsymbol{\beta}_2}$ we get

$$P_{1|i} = \frac{e^{(z_{1i}-z_{2i})'\boldsymbol{\beta}+z'_i(\boldsymbol{\beta}_1-\boldsymbol{\beta}_2)}}{1 + e^{(z_{1i}-z_{2i})'\boldsymbol{\beta}+z'_i(\boldsymbol{\beta}_1-\boldsymbol{\beta}_2)}} \qquad P_{2|i} = \frac{1}{1 + e^{(z_{1i}-z_{2i})'\boldsymbol{\beta}+z'_i(\boldsymbol{\beta}_1-\boldsymbol{\beta}_2)}} \qquad \text{[A2.18]}$$

We then impose the usual arbitrary constraints [A2.13] on the parameters of the base or denominator category, and drop the unnecessary subscript on $\boldsymbol{\beta}_1$. This gives

$$P_{1|i} = \frac{e^{[(z_{1i}-z_{2i})'+z'_i]\boldsymbol{\beta}}}{1 + e^{[(z_{1i}-z_{2i})'+z'_i]\boldsymbol{\beta}}} \qquad P_{2|i} = \frac{1}{1 + e^{[(z_{1i}-z_{2i})'+z'_i]\boldsymbol{\beta}}} \qquad \text{[A2.19]}$$

The equivalent linear logit expression is then

$$\log_e \frac{P_{1|i}}{P_{2|i}} = [(z_{1i}-z_{2i})'+z'_i]\boldsymbol{\beta} \qquad \text{[A2.20]}$$

It should be noted that [A2.19] and [A2.20] include the models of the two original cases ([A2.3], [A2.6], [A2.7] and [A2.8]) as special cases. If, for convenience, we now rewrite the right-hand side of [A2.20] in the shorthand form

$$[(z_{1i}-z_{2i})'+z'_i]\boldsymbol{\beta} = x'_i\boldsymbol{\beta} \qquad \text{[A2.21]}$$

this gives us the general expression used in the multiple explanatory variable dichotomous logistic/logit models in Chapter 2 (see Section 2.5). Quite clearly, therefore, the general expression $x'_i\boldsymbol{\beta}$ in these sections of Chapter 2 should be taken to encompass the case in which both types of explanatory variables (ψ and ϕ) are included in the same model (as in Example 2.3), and also both of the special cases: (a) in which all the explanatory variables are type ψ (as in Example 2.1); (b) in which all the explanatory variables are type ϕ and are included in difference form (as in Example 2.2).

The multiple-category case

Before considering the multiple-category models, it is useful to see that we can rewrite the dichotomous models [A2.17] as

$$P_{1|i} = \frac{e^{z_{1i}\boldsymbol{\beta}+z'_i\boldsymbol{\beta}_1}}{\sum\limits_{s=1}^{2} e^{z'_{si}\boldsymbol{\beta}+z'_i\boldsymbol{\beta}_s}}$$

[A2.22]

$$P_{2|i} = \frac{e^{z'_{2i}\boldsymbol{\beta}+z'_i\boldsymbol{\beta}_2}}{\sum\limits_{s=1}^{2} e^{z'_{si}\boldsymbol{\beta}+z'_i\boldsymbol{\beta}_s}}$$

As we have seen above, if we then divide through the numerator and denominator of both of these equations by $e^{z'_{2i}\boldsymbol{\beta}+z'_i\boldsymbol{\beta}_2}$, impose the constraints [A2.13] on the parameters of the base category and drop the unnecessary subscript on $\boldsymbol{\beta}_1$, we produce expressions [A2.19].

The multiple-category model equivalent to [A2.22] can then be written

$$P_{r|i} = \frac{e^{z'_{ri}\boldsymbol{\beta}+z'_i\boldsymbol{\beta}_r}}{\sum\limits_{s=1}^{R} e^{z'_{si}\boldsymbol{\beta}+z'_i\boldsymbol{\beta}_s}} \qquad r = 1, 2, \ldots R$$

[A2.23]

Arbitrarily selecting the Rth category as the base category, we can then divide through the numerator and denominator of [A2.23] by $e^{z'_{Ri}\boldsymbol{\beta}+z'_i\boldsymbol{\beta}_R}$ and impose the constraints $\boldsymbol{\beta}_R = 0$. This produces the multiple-category logistic models (see equation [2.99]

$$P_{r|i} = \frac{e^{(z_{ri}-z_{Ri})'\boldsymbol{\beta}+z'_i\boldsymbol{\beta}_r}}{1+\sum\limits_{s=1}^{R-1} e^{(z_{si}-z_{Ri})'\boldsymbol{\beta}+z'_i\boldsymbol{\beta}_s}} \qquad r = 1, 2, \ldots R-1$$

[A2.24]

$$P_{R|i} = \frac{1}{1+\sum\limits_{s=1}^{R-1} e^{(z_{si}-z_{Ri})'\boldsymbol{\beta}+z'_i\boldsymbol{\beta}_s}}$$

[A2.25]

The equivalent multiple-category linear logit model is then system of $R-1$ linear equations which can be summarized as

$$\log_e \frac{P_{r|i}}{P_{R|i}} = (z_{ri}-z_{Ri})'\boldsymbol{\beta}+z'_i\boldsymbol{\beta}_r \qquad r = 1, 2, \ldots R-1$$

[A2.26]

As in the dichotomous case, it should be noted that models [A2.24], [A2.25] and [A2.26] include the models in which all the explanatory variables are type ψ or type ϕ as special cases. Also we can now, for convenience, rewrite the right-hand side of [A2.26] in the shorthand form

$$(z_{ri}-z_{Ri})'\boldsymbol{\beta}+z_i'\boldsymbol{\beta}_r = x_{ri}'\boldsymbol{\beta}_r \qquad \text{[A2.27]}$$

This gives us the general expression used in the multiple-category logistic/logit models (e.g in equations [2.101] and [2.102] and subsequently in Chapters 2 and 3). Clearly, in these sections of Chapters 2 and 3 the general expression $x_{ri}'\boldsymbol{\beta}_r$ should be understood, therefore, to encompass the case in which both types of explanatory variables (ψ and ϕ) are included in the same model, and the special cases in which only type ψ variables or type ϕ variables are included.

Finally, it should be noted that in equations [A2.23] to [A2.27] we have adopted the simple concept that the expression $(z_{ri}-z_{Ri})'\boldsymbol{\beta}$ represents the type ϕ variables entered in difference form, and the expression $z_i'\boldsymbol{\beta}_r$ represents the type ψ variables with their associated alternative specific parameters. However, as discussed in Section 2.11, some type ϕ variables are entered into our models as alternative specific rather than generic variables. Thus, the expression $z_i'\boldsymbol{\beta}_r$ in [A2.23] to [A2.27] for type ψ variables must be supplemented by the addition of $z_{ri}'\boldsymbol{\beta}_r$ to represent the type ϕ variables which are entered in alternative specific form. The expression $(z_{ri}-z_{Ri})'\boldsymbol{\beta}$, on the other hand, can be taken to represent any generic variables with the associated generic parameters, and generic variables will always be type ϕ variables. For this reason, in Chapters 2 and 3 we normally find it useful to use the shorthand expression $x_{ri}'\boldsymbol{\beta}_r$ derived in [A2.27], with the addition of $z_{ri}'\boldsymbol{\beta}_r$, but sometimes to write out the shorthand expression as (see equation [2.118]),

$$x_{ri}'\boldsymbol{\beta}_r = \beta_{1r} + \sum_{k=2}^{K_1}\beta_k(X_{rik}^{\phi} - X_{Rik}^{\phi}) + \sum_{k=K_1+1}^{K_2}\beta_{kr}X_{rik}^{\phi} + \sum_{k=K_2+1}^{K}\beta_{kr}X_{ik}^{\psi} \qquad \text{[A2.28]}$$

(ASC)   (Generic, type ϕ)   (ASV, type ϕ)   (ASV, type ψ)

The shorthand expression $x_{ri}'\boldsymbol{\beta}_r$ used in Chapters 2 and 3 (and again| in Part 4 of the book) must be understood to imply any combination of the types of explanatory variables and parameters shown on the right-hand side of [A2.28].

CHAPTER 3
Categorical response variable, mixed explanatory variables

The previous chapter was by necessity rather long. In it we considered the limitations of conventional regression models when we try to extend such models to the problems of cell (d) of Table 1.1 (see Section 1.2), and we considered the characteristics of alternative models; logistic, linear logit, probit, which overcome these difficulties. In this chapter we move from cell (d) of Table 1.1 to cell (e). Conceptually this is a much smaller step than our previous move from cells (a)–(c) to cell (d). The principles of parameter estimation, hypothesis testing, goodness-of-fit measurement, etc, which we established in Chapter 2 continue to hold, as does the typology of explanatory variables and response categories, etc. The essence of the move from cell (d) to (e) is that we now have some mixture of continuous and categorical explanatory variables, and we incorporate these additional categorical explanatory variables by adopting the well-known principles used in the extension of the conventional regression models of cell (a) to the 'dummy' variable regression models of cell (b).

3.1 Dummy variables in conventional regression models

Although it is assumed in the book that the reader is familiar with the classical regression model and some of its most important extensions, it may be valuable at this point to present the reader with an *aide-mémoire* in the form of a brief summary of the principles involved in the incorporation of categorical ('dummy') explanatory variables into conventional regression models. (Readers well acquainted with this material should move straight on to the next section.) To do this we will assume that we have a simple urban housing market regression model relating a dependent variable Y (a measure of household overcrowding–

average number of persons per room) to an explanatory variable X (household income)

$$Y_i = \beta_1 + \beta_2 X_i + \varepsilon_i \qquad \text{[3.1]}$$

In any particular urban housing market the relationship between Y and X might be summarized by the regression line shown in Fig. 3.1(a). That is to say, average number of persons per room might be seen to decline as household income rises.

Let us next assume that we wish to introduce into this simple model a categorical ('dummy') explanatory variable D indicating the racial characteristics of the household, and that the variable is a dichotomous variable which takes the value 1 if the head of the household is 'non-white' and 0 if the head of the household is 'white'. There are three ways of introducing this categorical variable into our basic model.

First, it can be introduced in a simple additive manner

$$Y_i = \beta_1 + \beta_2 X_i + \beta_3 D_i + \varepsilon_i \qquad \text{[3.2]}$$

As can be seen in Fig. 3.1(b), this model allows the intercept (constant) terms for the two household groups to differ, but the slopes of the regression lines remain the same. The dummy variable parameter β_3 is interpreted as the *difference* between the intercepts of the two categories of household (i.e. between 'white' households and 'non-white' households), and by testing whether this β_3 parameter is statistically significantly different from zero, we can test whether or not any apparent difference between the intercepts is merely a random fluctuation of no substantive importance.

Fig. 3.1
An illustration of three different ways of introducing a dummy explanatory variable into a simple regression model

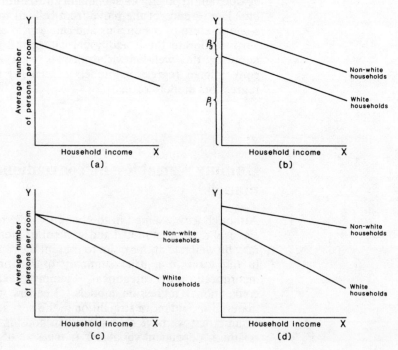

Second, it can be introduced in an 'interactive' manner in which the original explanatory variable is multiplied by the dummy variable and the resulting variable is added into the model in the usual way

$$Y_i = \beta_1 + \beta_2 X_i + \beta_3 D_i X_i + \varepsilon_i \qquad \text{[3.3]}$$

As can be seen in Fig.3.1(c), this model allows the slopes of the regression lines for the two household groups to differ but the intercepts remain the same. The dummy variable parameter β_3 is now interpreted as the *difference* between the regression line slopes of the two household groups, and any apparent difference can be examined by testing whether the β_3 parameter is statistically significantly different from zero.

Third, the dummy variable can be introduced in both the previous manners.

$$Y_i = \beta_i + \beta_2 X_i + \beta_3 D_i + \beta_4 D_i X_i + \varepsilon_i \qquad \text{[3.4]}$$

As can be seen in Fig. 3.1(d) this model allows the regression relationships for the two household groups to differ both in terms of the intercept and the slope. It is virtually equivalent[†] to fitting separate regression relationships to each household group.

In the case of multiple-category explanatory variables the same principles continue to hold. Instead of a single dummy variable ($D_i = 1$ or 0) for the case of a two-category explanatory variable, we introduce $J - 1$ dummy variables ($D_{i1} = 1$ or 0, $D_{i2} = 1$ or 0, . . . $D_{iJ-1} = 1$ or 0) when we have a J–category explanatory variable, and we interpret the intercept and slope terms associated with these dummy variables as differences from the base level or 'anchor' category J. To illustrate this, let us extend the previous example and assume that we now recognize three household groups, 'West Indian', 'Indian' and 'British'. In this case two dummy variables

$$
\begin{aligned}
D_{i1} &= 1 \quad \text{if head of household is 'West Indian'} \\
&= 0 \quad \text{otherwise} \\[4pt]
D_{i2} &= 1 \quad \text{if head of household is 'Indian'} \\
&= 0 \quad \text{otherwise}
\end{aligned}
\qquad \text{[3.5]}
$$

are introduced into the basic model [3.1]. The three ways of doing this produce the models:

$$Y_i = \beta_1 + \beta_2 X_i + \beta_3 D_{i1} + \beta_4 D_{i2} + \varepsilon_i \qquad \text{[3.6]}$$

$$Y_i = \beta_1 + \beta_2 X_i + \beta_3 D_{i1} X_i + \beta_4 D_{i2} X_i + \varepsilon_i \qquad \text{[3.7]}$$

$$Y_i = \beta_1 + \beta_2 X_i + \beta_3 D_{i1} + \beta_4 D_{i2} + \beta_5 D_{i1} X_i + \beta_6 D_{i2} X_i + \varepsilon_i \qquad \text{[3.8]}$$

and Figs. 3.2(a) and 3.2(b) illustrate the 'different intercepts – same

---

[†] The slight difference is that in model [3.4] the variance of the error term is assumed to be the same in both household groups, whereas if two regression models were used the error variance would be allowed to differ in the two household groups.

Fig. 3.2
An illustration of two different ways of introducing a three-category explanatory variable into a simple regression model

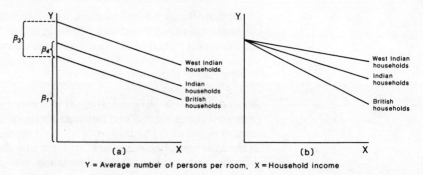

Y = Average number of persons per room,   X = Household income

slopes' and 'same intercept – different slopes' implication of models [3.6] and [3.7] respectively.

It is important to draw the reader's attention to two aspects of this multiple-category example.

First, as in this example, we must always introduce into our basic model one less dummy variable than the number of categories, otherwise we would create perfect linear dependence between our explanatory variables, and consequently be unable to obtain least squares estimates of the regression parameters. For example, if a third dummy variable ($D_{i3} = 1$ if head of household is 'British' and 0 otherwise) was introduced in our previous example, then for each household i, $D_{i1} + D_{i2} + D_{i3} = 1$. As a result it can readily be seen that:

$$D_{i3} = 1 - D_{i1} - D_{i2} \qquad [3.9]$$

which clearly shows that the third dummy variable is redundant and, if introduced, would create perfect linear dependence between the explanatory variables and, hence, perfect multicollinearity.

Second, it would not be correct to use a single variable D_i taking the values: 0 for 'British', 1 for 'Indian' and 2 for 'West Indian' in place of the two dummy variables D_{i1} and D_{i2} defined above. The reason is that such a variable would imply that the difference in regression slope and/or intercept between British and Indian households is identical to that between Indian and West Indian households. Unless we know *a priori* that this is the case, we are not justified in making this assumption.

Further details of the use of dummy variables in conventional regression models are to be found in Kmenta (1971, 409–23), Chatterjee and Price (1977, 74–100), Pindyck and Rubinfeld (1976, 77–84), Wonnacott and Wonnacott (1979, 100–9).

3.2 Categorical explanatory variables in logistic/ logit models

The dummy variable principles discussed in the previous section can be transferred directly to the logistic/logit models of Chapter 2 (and by implication to probit models also). To see this, we will once again take our example (see Sections 2.1 and 2.8) of a model which relates respondent's opinions about a proposed mental hospital scheme to a measure of their proximity to the proposed site of the mental hospital. In this case, however, we assume that we wish to add an extra explanatory variable into our model. This extra variable is a two-category (dummy) variable which indicates the tenure group to which the respondent belongs. ($D_i = 1$ = member of home owner household, $D_i = 0$ = member of a private renting or local authority renting household).

We can first consider the two-category or dichotomous response variable case, where the two response categories are 'in favour' or 'against' the scheme. In this case the logistic or linear logit models with the additional dummy variable generalize to the form:

$$P_{1|i} = \frac{e^{x'_i\beta}}{1+e^{x'_i\beta}} \quad \text{or} \quad \log_e \frac{P_{1|i}}{P_{2|i}} = x'_i\beta \tag{3.10}$$

where

$$x'_i\beta = \beta_1 + \beta_2 X_i + \beta_3 D_i \tag{3.11}$$

$$\text{or} \quad x'_i\beta = \beta_1 + \beta_2 X_i + \beta_3 D_i X_i \tag{3.12}$$

$$\text{or} \quad x'_i\beta = \beta_1 + \beta_2 X_i + \beta_3 D_i + \beta_4 D_i X_i \tag{3.13}$$

depending upon which of the three ways discussed in the previous section we choose to introduce the dummy variable.

In the three-category response variable case where the three response categories are 'against', 'in favour' or 'undecided' about the scheme (and assuming the imposition of the usual constraints relating to the base category i.e. $\beta_3 = 0$), the logistic and linear logit models with the additional dummy variable generalize to the forms:

$$P_{r|i} = \frac{e^{x'_{ri}\beta}}{1 + \sum_{s=1}^{2} e^{x'_{si}\beta_s}} \quad r = 1, 2, 3 \tag{3.14}$$

and

$$\log_e \frac{P_{r|i}}{P_{3|i}} = x'_{ri}\beta_r \quad r = 1, 2 \tag{3.15}$$

where[†]

$$x'_{ri}\beta_r = \beta_{1r} + \beta_{2r} X_i + \beta_{3r} D_i \tag{3.16}$$

$$\text{or} \quad x'_{ri}\beta_r = \beta_{1r} + \beta_{2r} X_i + \beta_{3r} D_i X_i \tag{3.17}$$

$$\text{or} \quad x'_{ri}\beta_r = \beta_{1r} + \beta_{2r} X_i + \beta_{3r} D_i + \beta_{4r} D_i X_i \tag{3.18}$$

depending on which of the three possible methods we select as the

---

[†] In this particular case where there are no type ϕ variables, the subscript r in x'_{ri} is not strictly necessary, i.e. $x'_{ri} = x'_i$ for all r. See Appendix 2.1.

means of introducing the dummy variable. The multiple response category generalization, where R is any arbitrary integer number, then follows immediately. We simply replace the number 3 by R in equations [3.14] and [3.15] and assume that $r = 1, 2 \ldots R - 1$ in equation [3.15].

When our extra explanatory variable is a multiple-category rather than two-category variable (for example, where we recognize three or more tenure groups rather than the two of our previous examples) we incorporate two or more dummy variables into our basic model. To illustrate this let us assume that we now recognize three tenure groups: home owner households, private renting households and local authority renting households. In this case, two dummy variables:

$D_{i1} = 1$ if respondent is member of a home owner household
     $= 0$ otherwise

$D_{i2} = 1$ if respondent is member of a private renting household
     $= 0$ otherwise

would be introduced into the basic two-category or multiple-category logistic or linear logit models in any of the three ways discussed in the previous section (see equations [3.6]–[3.8]).

If the proportion of categorical variables in the 'mixed' set of explanatory variables is sufficiently high, it is possible, in certain circumstances, to obtain a sufficient number of repetitions of each combination of values of the explanatory variables to facilitate the use of the weighted least squares parameter estimation procedure discussed in Section 2.4.1. However, in practice, the necessary combination of a large sample size plus a sufficient proportion of categorical variables for WLS estimation is rarely encountered. Consequently, the parameters of the type of mixed explanatory variable logistic/logit models which we have considered in this section are in almost all cases estimated using the direct maximum likelihood procedure discussed in Section 2.4.2.

3.3 Dummy variables within the general typology of explanatory variables

In Chapter 2 (Section 2.11) we discussed a typology of explanatory variables based upon the combination of the generic/alternative-specific and X^ϕ/X^ψ classifications. This enabled us to write a general expression for the $x'_{ri}\beta_r$ function which we had adopted in the general forms of the multiple-category logistic/logit models, equations [2.101] and [2.102]. This expression took the form (see also equation [A2.29] in Appendix 2.2)

$$x'_{ri}\beta_r = \beta_{1r} + \sum_{k=2}^{K_1}\beta_k(X^\phi_{rik} - X^\phi_{Rik}) + \sum_{k=K_1+1}^{K_2}\beta_{kr}X^\phi_{rik} + \sum_{k=K_2+1}^{K}\beta_{kr}X^\psi_{ik} \qquad [3.19]$$

(ASC)   (Generic, type ϕ)   (ASV, type ϕ) (ASV, type ψ)

In terms of this typology, the categorical (dummy) explanatory variables which we have added to our logistic/logit models in this chapter can take the form of either generic variables or alternative-specific (ASV) variables, and as ASVs they can be of either type ϕ (i.e.

attributes of the choice alternatives/response categories) or type ψ (i.e. attributes of the choice maker i or the locality i at which the choice is made). In almost all cases, however, dummy explanatory variables will occur as ASVs. As a result, dummy variables are sometimes referred to in the transportation science literature as alternative-specific dummy variables (ASDVs). As shown in equations [3.15] to [3.18] many of the dummy variables (particularly those of type ψ) will be introduced in such a way that they are specific to $R-1$ response categories (alternatives) where R is the total number of response categories (alternatives) in the choice set and category R acts as the base or 'anchor' response category. Others, however, (particularly those of type ϕ) will be introduced in such a way that they are specific to just one or two response categories (alternatives).

3.4 A range of illustrative examples

As the 'mixed' explanatory variable logistic/logit models of cell (e) of Table 1.1 are so similar to those of cell (d) and involve indentical principles of parameter estimation, hypothesis testing, goodness-of-fit measurement, etc. it will be sufficient to refer the reader to the discussion of these issues in Chapter 2, and to devote the remainder of this short chapter to a range of illustrative examples of the type of cell (e) models which the reader will encounter in the literature. It will be seen in these examples that although there are three potential methods of introducing categorical (dummy) explanatory variables into our models (in simple additive form, in interactive form, and in both simple additive and interactive form) it is the first of these methods (equivalent to equations [3.11] and [3.16]) which dominates current practice.

EXAMPLE 3.1   *Determinants of housing tenure in Sydney*

Hensher (1980) reports an illustrative investigation of the determinants of housing tenure (renting versus owner-occupancy) for a small group of 274 households in the outer south-western suburbs of Sydney, Australia. The models he fitted were of a simple dichotomous logistic/logit form and all the explanatory variables were type ψ (i.e. attributes of the choice maker/household i or the locality i at which the choice is made) alternative-specific variables. The fitted model reported by Hensher takes the form:

$$\hat{P}_{1|i} = \frac{e^{x'_i\hat{\beta}}}{1 + e^{x'_i\hat{\beta}}} \quad \text{or} \quad \log_e \frac{\hat{P}_{1|i}}{\hat{P}_{2|i}} = x'_i\hat{\beta} \qquad [3.20]$$

where:

$x'_i\hat{\beta} = $ 1.640 − 0.148 DAYSAV$_i$ − 0.130 NUMCAR$_i$ − 0.975 SEX$_i$
(0.859) (0.097)　　　　(0.096)　　　　(0.439)

−0.223 NOEMPL$_i$ + 0.699 NOPENS$_i$ + 0.484 ADULT2$_i$ + 0.595 YOUNG$_i$
(0.151)　　　　(0.411)　　　　(0.178)　　　　(0.209)

−0.359 GPRWKN$_i$ + 1.128 SHA$_i$ − 0.010 LRES$_i$ 　　　　　　　　　[3.21]
(0.094)　　　　(0.333)　　　(0.003)

$\hat{P}_{1|i}$ = predicted probability that respondent i will belong to a 'renting' household.

DAYSAV = number of days per week the car is available for commuting in household.

NUMCAR = number of cars in household.

SEX = (1 = male, 0 = female)

NOEMPL = number of employed persons in household (full-time and part-time).

NOPENS = number of persons receiving a pension in household.

ADULT2 = number of adults aged 18–30 in household.

YOUNG = number of young people aged 14–17 in household.

GPRWKN = gross weekly personal income.

SHA = renting authority (1 = state housing, 0 = other).

LRES = length of residence in current dwelling.

The ρ^2 goodness-of-fit measure for this model is 0.25.

Care must be exercised when interpreting the results of this model as the data set used in this study had been collected for purposes other than the study of the determinants of housing tenure and thus the explanatory variables available for inclusion in the model were not particularly appropriate. However, for our purposes, it does show two categorical (dummy) explanatory variables (SEX and SHA) introduced alongside the continuous explanatory variables in the simple additive form described in equation [3.11] above. In addition, it does demonstrate, as might be expected *a priori*, that households with higher incomes and with greater length of residence in their current dwelling are more likely to be owner occupiers than renters.

Another reason for displaying this model is that Hensher went on to stratify his data set on the basis of the type of organization (state housing authority versus private landlords) from which 'renting' households rented their dwelling and he then fitted separate models to the two renting groups. In other words, he stratified the data set on the basis of variable SHA and fitted separate models for the two situations: renting from a state housing authority versus owner-occupation; renting from private landlords versus owner-occupation. This produces results which are very similar but not quite identical (see footnote, p. 93) to those which would have been derived had Hensher introduced the SHA dummy variable into [3.21] in *both* a simple additive and an interactive fashion (i.e. in the form described in equation [3.13] above). We will see again that this practice of fitting separate models is much more common for cell (e) logistic/logit models than the alternative of introducing dummy variables in both simple additive and interactive fashion and estimating a single 'full' model.

The fitted models for the stratified data set take the form of model [3.20] where:

(a) *Renting from state housing authority versus owner occupation*

$$x_i'\beta = \begin{array}{l} 2 \cdot 845 \quad - 0 \cdot 143 \text{ DAYSAV}_i \quad - 0 \cdot 027 \text{ NUMCAR}_i \quad + 1 \cdot 382 \text{ SEX}_i - \\ (1 \cdot 210) \quad (0 \cdot 124) (0 \cdot 125) (0 \cdot 617) \end{array}$$

$$\begin{array}{l} 0 \cdot 381 \text{ NOEMPL}_i \quad + 2 \cdot 238 \text{ NOPENS}_i \quad + 0 \cdot 697 \text{ ADULT2}_i \quad + 0 \cdot 818 \text{ YOUNG}_i \\ (0 \cdot 210) (0 \cdot 888) (0 \cdot 289) (0 \cdot 272) \end{array}$$

$$\begin{array}{l} - 0 \cdot 419 \text{ GPRWKN}_i \quad + 0 \cdot 013 \text{ LRES}_i \\ (0 \cdot 136) (0 \cdot 005) \end{array}$$

[3.22]

(b) *Renting from private landlords versus owner-occupation*

$$x_i'\beta = 1.903 - 0.189 \text{ DAYSAV}_i - 0.401 \text{ NUMCAR}_i + 0.886 \text{ SEX}_i +$$
$$ (1.464) \quad (0.189) \qquad\quad (0.279) \qquad\qquad (0.763)$$

$$0.453 \text{ NOEMPL}_i - 0.762 \text{ NOPENS}_i + 0.181 \text{ ADULT2}_i -$$
$$(2.715) \qquad\qquad (0.952) \qquad\qquad (0.270)$$

$$0.371 \text{ YOUNG}_i - 0.281 \text{ GPRWKN}_i - 0.012 \text{ LRES}_i$$
$$(0.598) \qquad\qquad (0.146) \qquad\qquad (0.005)$$ **[3.23]**

Once again, considerable care must be exercised when interpreting these results. In both situations households with higher incomes and with greater length of residence in their current dwelling are shown to be more likely to be owner occupiers than renters. However, three variables (NOEMPL, NOPENS, YOUNG) have different signs in the two models. In the state housing authority case, the odds of renting rather than owning diminish as the number of employed persons increases, as the number of pensioners decreases and as the number of young persons falls. In contrast, in the private landlords case, the odds of renting rather than owning diminish as the number of employed persons decreases, and as the number of pensioners and young persons increases. In this latter case, however, the three variables are not statistically significantly different from zero, and thus it is not possible to draw policy-relevant conclusions. Nevertheless, the results do indicate that the model [3.21] in which the dummy variable is introduced in just a simple additive fashion may not be a satisfactory specification. The results also indicate that it may be necessary in any future analysis of this data set to consider estimating a three-alternative, polytomous, logistic/logit model in which the two renting class are separated (we will see in Ch. 11 that there are also other possible models which may prove more suitable than the three-alternative logistic/logit model).

EXAMPLE 3.2   *Occupational attainment in the United States*

In the previous example we considered a two-category, dichotomous model. We now turn to a multiple-category model in which the categories are a 'ranked' set (see Section 2.11.2) and all the explanatory variables (continuous and categorical) are alternative-specific variables of type ψ.

The example is taken from the work of Schmidt and Strauss (1975a) who estimated a set of multiple-category logit models of occupational attainment in the United States, using race, sex, educational attainment and labour market experience as explanatory variables. In this study the models were estimated using information on the characteristics of a random sample of 1,000, 934 and 1,000 full-time workers in the years 1960, 1967 and 1970 respectively, and the models were used to predict the odds that individuals would be in one of five occupational groups: 'professional' (PROF); 'white collar' (WC); 'craft' (CFT); 'blue collar' (BC); 'menial' (MEN). The multiple-category logit model fitted to the 1967 data takes the form:

$$\log_e (\hat{P}_{\text{BC}|i}/\hat{P}_{\text{MEN}|i}) = 1.203 - 0.081 ED_i - 0.026 EX_i + 0.670 R_i +$$
$$\phantom{\log_e (\hat{P}_{\text{BC}|i}} (0.704) \quad (0.048) \qquad (0.010) \qquad (0.331)$$
$$1.179 S_i$$
$$(0.255)$$

$$\log_e (\hat{P}_{\mathrm{CFT}|i}/\hat{P}_{\mathrm{MEN}|i}) = -4.319 + 0.115ED_i - 0.007EX_i + 0.818R_i +$$
$$\phantom{\log_e (\hat{P}_{\mathrm{CFT}|i}/\hat{P}_{\mathrm{MEN}|i}) =} (0.971)\quad (0.054)\qquad (0.012)\qquad (0.431)$$
$$\phantom{\log_e (\hat{P}_{\mathrm{CFT}|i}/\hat{P}_{\mathrm{MEN}|i}) = -4.319 +} 3.793S_i$$
$$\phantom{\log_e (\hat{P}_{\mathrm{CFT}|i}/\hat{P}_{\mathrm{MEN}|i}) = -4.319 +} (0.567)$$

$$\log_e (\hat{P}_{\mathrm{WC}|i}/\hat{P}_{\mathrm{MEN}|i}) = -2.821 + 0.274ED_i - 0.033EX_i + 1.778R_i -$$
$$\phantom{\log_e (\hat{P}_{\mathrm{WC}|i}/\hat{P}_{\mathrm{MEN}|i}) =} (0.837)\quad (0.054)\qquad (0.0108)\qquad (0.433)$$
$$\phantom{\log_e (\hat{P}_{\mathrm{WC}|i}/\hat{P}_{\mathrm{MEN}|i}) = -2.821 +} 0.597S_i$$
$$\phantom{\log_e (\hat{P}_{\mathrm{WC}|i}/\hat{P}_{\mathrm{MEN}|i}) = -2.821 +} (0.268)$$

[3.24]

$$\log_e (\hat{P}_{\mathrm{PROF}|i}/\hat{P}_{\mathrm{MEN}|i}) = -6.098 + 0.477ED_i - 0.012EX_i + 1.066R_i +$$
$$\phantom{\log_e (\hat{P}_{\mathrm{PROF}|i}/\hat{P}_{\mathrm{MEN}|i}) =} (0.866)\quad (0.056)\qquad (0.011)\qquad (0.412)$$
$$\phantom{\log_e (\hat{P}_{\mathrm{PROF}|i}/\hat{P}_{\mathrm{MEN}|i}) = -6.098 +} 1.077S_i$$
$$\phantom{\log_e (\hat{P}_{\mathrm{PROF}|i}/\hat{P}_{\mathrm{MEN}|i}) = -6.098 +} (0.280)$$

where ED_i = years of schooling of individual i
EX_i = years of labour market experience of individual i (calculated as age minus ED minus 5)
R_i = race of individual i (1 if white, 0 if black)
S_i = sex of individual i (1 if male, 0 if female)

It can be seen that the model includes two continuous explanatory variables and two categorical (dummy) explanatory variables.

In the interpretation of the results of the model particular interest centres upon the race and sex variables[†]. Significant non-zero parameter estimates can be taken to indicate that race and sex affect an individual's occupational attainment, even when differences in education and experience are taken into account. That is to say, non-zero parameter estimates indicate differential access to certain occupations depending on the individual's race and sex (i.e. race and sex discrimination). The parameter estimates in [3.24] and those not reported for 1960 and 1970 are consistent with this interpretation.

In the case of the race variable, the parameter estimates indicate that if occupation groups are ordered as follows:
'menial'
'blue collar'
'craft'
'professional'
'white collar'
then, other things being held constant, being black (white) implies that the individual is more likely to be in the groups higher (lower) on this list. In other words, being black makes it more likely that the individual will be in the 'menial' or 'blue collar' occupational groups.

In the case of the sex variable, the parameter estimates indicate that if the occupational groups are ordered as follows:
'white collar'
'menial'
'professional'
'blue collar'
'craft'

[†] The parameter estimates of the education variable indicate, as expected *a priori*, that more education increases the likelihood that the individual will be in the more prestigious occupational groups. The labour market experience parameter estimates, on the other hand, are much less significant and/or consistent in pattern, though over the three years they do indicate that 'professionals' have more labour market experience than 'blue collar', workers, holding race, sex and educational attainment constant.

then, other things being held constant, being female (male) implies that the indiviual is more likely to be in occupational groups higher (lower) on this list, i.e. females are more likely to be in the 'white collar' or 'menial' occupational groups.

In model [3.24] it can be seen that the dummy variables, race and sex, have been introduced in the simple additive form described in equation [3.16] above. As we have noted above, this allows only the constant terms to be shifted by race and sex. In other words, the parameter estimates of the education and experience variables are implicitly assumed to be the same for all individuals regardless of race or sex. Also, the parameter estimate of the race dummy variable is implicitly assumed to be the same for both sexes and vice-versa. To test the validity of these implicit assumptions Schmidt and Strauss extended the analysis and refitted models similar to [3.24] to new samples stratified by race and sex.

In the case of the race variable, they generated further random samples of 1,000 whites and 1,000 blacks from the 1967 data files and fitted model [3.24] without the race dummy variable. The results they obtained are given below (where for convenience of expression and to facilitate comparison, the subscript i and the standard errors have been omitted).

(a) *1967: sample of whites only*

$$\log_e \hat{P}_{BC}/\hat{P}_{MEN} = \quad 2.013 - 0.059ED - 0.033EX + 0.750S$$
$$\log_e \hat{P}_{CFT}/\hat{P}_{MEN} = -2.965 + 0.018ED - 0.016EX + 4.459S$$
$$\log_e \hat{P}_{WC}/\hat{P}_{MEN} = -1.537 + 0.310ED - 0.016EX - 0.866S \qquad [3.25]$$
$$\log_e \hat{P}_{PROF}/\hat{P}_{MEN} = -5.231 + 0.532ED - 0.007EX + 0.423S$$

(b) *1967: sample of blacks only*

$$\log_e \hat{P}_{BC}/\hat{P}_{MEN} = -0.103 + 0.010ED - 0.033EX + 2.249S$$
$$\log_e \hat{P}_{CFT}/\hat{P}_{MEN} = -4.876 + 0.122ED - 0.011EX + 4.087S$$
$$\log_e \hat{P}_{WC}/\hat{P}_{MEN} = -3.885 + 0.319ED - 0.031EX + 1.151S \qquad [3.26]$$
$$\log_e \hat{P}_{PROF}/\hat{P}_{MEN} = -6.804 + 0.497ED - 0.020EX + 0.876S$$

The differences in the parameter estimates for the education, experience and sex variables are not large and with a couple of minor exceptions the parameters have the same signs and give rise to similar substantive interpretations. As a result, Schmidt and Strauss conclude that the differences are not sufficiently large to invalidate the usefulness of the single-sample, both-racial-group model [3.24] with the simple additive dummy variable to represent racial differences.

The stratification procedure which Schmidt and Strauss adopt to question the validity of the dummy variable specification is similar to that we saw used by Hensher in the previous example. Consequently, it produces results which are somewhat similar to those we would expect if Schmidt and Strauss had introduced the race dummy in both a simple additive and an interactive fashion (i.e. in the form described in equation [3.18] above). In this case, however, the difference is that

Schmidt and Strauss have a data base rich enough to allow the drawing of new samples stratified in terms of the dummy variables. This is likely to be a rare situation in most empirical research.

Finally, geographers will be interested to note that the richness of the data base also enabled Schmidt and Strauss to introduce a regional dimension into their analysis. As in the case of the race and sex dummy variables, they did this by drawing from the 1967 data files additional random samples of 1,000 individuals in each of four Census regions (Northeast, North Central, South, West) and estimating models of the form [3.24]. Using the parameter estimates from the four regional models and from the combined United States sample model [3.24], they were then able to evaluate the probabilities of being in each of the five occupational groups in each of the four Census regions. These probabilities were evaluated at the regional sample means for education and experience and for all four permutations of race and sex, and were derived by plugging the parameter estimates and sample means into an equation of the form [3.14]. The results are shown in Table 3.1. The regional variation observed is due to both regional differences in parameter estimates and to regional differences in average education and labour market experience.

Table 3.1
Predicted probabilities of being in each occupational group given average education and experience and based upon parameter estimates from regional and combined U.S. sample models (after Schmidt and Strauss 1975a, 480)

Race-sex combination	Region	'Menial'	'Blue collar'	'Craft'	'White collar'	'Professional'
Black female	US	0.366	0.258	0.015	0.221	0.140
	NE	0.211	0.245	0.014	0.379	0.150
	NC	0.257	0.277	0.007	0.296	0.164
	S	0.515	0.273	0.006	0.098	0.108
	W	0.349	0.217	0.009	0.190	0.234
Black male	US	0.151	0.346	0.283	0.050	0.170
	NE	0.156	0.340	0.201	0.139	0.164
	NC	0.111	0.405	0.122	0.137	0.226
	S	0.215	0.452	0.147	0.048	0.139
	W	0.180	0.435	0.126	0.045	0.216
White female	US	0.140	0.192	0.013	0.499	0.156
	NE	0.112	0.206	0.017	0.422	0.223
	NC	0.136	0.235	0.013	0.433	0.183
	S	0.156	0.241	0.019	0.384	0.200
	W	0.114	0.086	0.015	0.535	0.250
White male	US	0.067	0.299	0.284	0.131	0.219
	NE	0.081	0.278	0.247	0.158	0.236
	NC	0.054	0.315	0.216	0.184	0.231
	S	0.047	0.288	0.343	0.136	0.185
	W	0.073	0.217	0.263	0.158	0.289

Combined U.S. sample: average ED = 11.509, average EX = 24.203
Northeast sample: average ED = 11.552, average EX = 24.730
North Central sample: average ED = 11.458, average EX = 24.245
South sample: average ED = 10.807, average EX = 24.571
West sample: average ED = 12.219, average EX = 23.265

EXAMPLE 3.3   *Housing choice in Pittsburgh*

In Example 3.2 we considered a cell (e) 'mixed' explanatory variable model in which all the explanatory variables (both continuous and dummy variables) took the form of alternative-specific variables of type ψ. We will now consider an example of a cell (e) model in which all the explanatory variables take the form of generic variables of type ϕ. (i.e. variables which represent attributes of the choice alternatives/response categories).

The example comes from the work of Quigley (1976) who estimated a set of multiple-category logistic/logit models of household choice amongst various types of residential housing in the Pittsburgh Metropolitan Area. The models were estimated using information gathered in 1967 on 25,000 dwelling units in the Pittsburgh Metropolitan Area and on the housing choices made by approximately 3,000 'renting' households who made residential location decisions within the 7-year period, 1960–67. Eighteen types of rental housing were defined in the study using proxies for residential density, quality and interior size, and the models employed by Quigley took the form:

$$\log_e \frac{P_{r|i}}{P_{R|i}} = \beta_1(CW_i^r - CW_i^R) + \beta_2(APT_i^r - APT_i^R) + \beta_3(BR_i^r - BR_i^R)$$

[3.27]

$$+ \beta_4(AG_i^r - AG_i^R) + \beta_5(MPR_i^r - MPR_i^R) + \beta_6(ST_i^r - ST_i^R)$$

where

CW^r = (1 if housing type r is a common-wall unit, 0 otherwise)
APT^r = (1 if housing type r is an apartment unit, 0 otherwise)
BR^r = number of bedrooms in housing type r
AG^r = (1 if housing type r was built before 1930)
MPR^r = effective monthly cost (price) of consuming housing type r
ST^r = number of units (i.e. stock) of housing of type r in the sample.

Quigley estimated the parameters of model [3.27] separately for each of thirty combinations of household income and size (six household income classes multiplied by five household size classes). For convenience, only a sample of these parameter estimates are reported in Table 3.2. However, these are sufficient to demonstrate the structuring of Quigley's tables of results, and to indicate some of the major findings (note that household income is in terms of 1967 prices).

On theoretical grounds, Quigley argued that *a priori* we should expect the following:
(a) $\hat{\beta}_5$ should be negative to ensure negative own-price elasticity.
(b) $\hat{\beta}_4$ should be negative since, other things being equal, households prefer higher quality dwelling units (age in this U.S. context is being used as a proxy for housing quality and obsolescence).
(c) $\hat{\beta}_3$ should be positive since, other things being equal, households should prefer more interior space.
(d) $\hat{\beta}_6$ should be positive, since households can obtain more information, at the same search cost, for housing types in greater supply.
(e) $\hat{\beta}_3$ should be greater in magnitude for larger households than for smaller households of the same income since, holding income constant,

Household size	Number of observations	$\hat{\beta}_1$ Common wall (CW)	$\hat{\beta}_2$ Apartment (APT)	$\hat{\beta}_3$ Number of bedrooms (BR)	$\hat{\beta}_4$ Structure age (AG)	$\hat{\beta}_5$ Relative monthly cost/price (MPR)	$\hat{\beta}_6$ Number of units (stock) (ST)
Income $3,000–$4,900							
1	104	1.023 (0.339)	2.438 (0.365)	−0.757 (0.214)	−0.650 (0.396)	−6.866 (1.248)	0.004 (0.001)
2	140	0.075 (0.268)	1.792 (0.272)	−0.261 (0.206)	−1.930 (0.400)	−2.170 (1.043)	0.007 (0.001)
3	95	−1.523 (0.332)	−0.276 (0.300)	1.687 (0.285)	−2.548 (0.447)	6.475 (1.408)	0.010 (0.001)
4	88	−1.054 (0.351)	−0.810 (0.338)	1.770 (0.340)	0.455 (0.379)	−6.364 (1.644)	0.006 (0.001)
5+	87	−1.278 (0.344)	−1.930 (0.445)	3.282 (0.368)	−0.247 (0.369)	−3.998 (1.379)	0.010 (0.001)
Income $5,000–$6,999							
1	91	0.150 (0.484)	2.564 (0.466)	−1.222 (0.381)	−3.049 (0.716)	−1.888 (1.573)	−0.009 (0.001)
2	223	0.291 (0.176)	0.874 (0.204)	−0.223 (0.043)	−0.223 (0.178)	−2.906 (0.730)	0.003 (0.0004)
3	224	−1.500 (0.212)	−0.849 (0.191)	2.020 (0.185)	−2.383 (0.264)	−4.465 (0.761)	0.010 (0.001)
4	194	−2.693 (0.266)	−0.903 (0.213)	3.823 (0.282)	−2.738 (0.309)	−4.140 (0.843)	0.014 (0.001)
5+	223	−1.591 (0.211)	−2.013 (0.275)	3.170 (0.236)	−0.893 (0.230)	−6.116 (0.925)	0.008 (0.001)

Table 3.2
Selected parameter estimates from Quigley's Pittsburgh housing choice model organized by household size and income class

we would expect that larger families demand larger dwelling units and more interior space (similarly $\hat{\beta}_1$ and $\hat{\beta}_2$ should be smaller in magnitude or more negative for larger households than for smaller households). (f) $\hat{\beta}_3$ should be greater in magnitude for higher income households than for lower income households of the same size since, holding family size constant, we would expect that higher income households are associated with greater consumption of higher quality, larger units with more interior space (similarly $\hat{\beta}_1$, $\hat{\beta}_2$ and $\hat{\beta}_4$ should be more negative for higher income households than for lower income households).

These expectations can be compared with the results shown in Table 3.2 (and the results for the other household size and income classes in Quigley's original paper). The conclusions which can be drawn are as follows:
(a) The price parameter estimates, $\hat{\beta}_5$, have the anticipated sign and most have 't' ratios (parameter estimates divided by standard errors) at the conventional levels of significance. This suggests that choice of

housing types for the overwhelming proportion of rental households is strongly influenced by relative prices. There is also a clear pattern in the magnitude of the price parameter estimates for households of different sizes. Within each income class, the magnitude of the $\hat{\beta}_5$ value increases with household size. It appears that larger households with greater demands for necessities are more responsive to relative prices in their choices among housing types.

(b) The structure age parameter estimates, $\hat{\beta}_4$, have the anticipated sign and significant 't' ratios in almost all equations.

(c) The number of bedrooms parameter estimates, $\hat{\beta}_3$, have the anticipated pattern across income classes and household sizes. For each income class the magnitude of $\hat{\beta}_3$ increases with household size, and there is a tendency for β_3 to increase with income level for a given household size.

(d) The common wall-unit and apartment unit parameter estimates, $\hat{\beta}_1$ and $\hat{\beta}_2$, have broadly the anticipated pattern across income classes and household sizes. Other things being equal, single detached housing units are preferred to common-wall units or apartments by larger households and households with higher incomes.

Having estimated the parameters of model [3.27] for each of the thirty combinations of household income and size, and derived the results which we have just discussed, Quigley went on to use the parameter estimates contained in Table 3.2 (and those for the other household income and size groups) to calculate a set of predicted probabilities. These were the probabilities of choice among housing types for otherwise identical households who differ only in the sense that their members are employed at different work places in the metropolitan area. The probabiities were calculated for households employed at four specific work sites: one located in the heart of the Pittsburgh CBD; one located in the inner city east of the CBD; one

Table 3.3 Quigley's predicted probabilities of housing type choice for otherwise identical households oriented to four different work sites in the Pittsburgh metropolitan area.

Type of dwelling	Work Places			
	CBD	Inner city	Central city	Suburbs
Four-person Families—Income $5,000–$6,999				
Common-wall units	0.51	0.54	0.50	0.42
Apartments	0.40	0.29	0.19	0.11
Single detached	0.09	0.17	0.30	0.47
One bedroom	0.16	0.13	0.13	0.14
Two bedrooms	0.63	0.63	0.61	0.57
Three bedrooms	0.21	0.23	0.26	0.28
Five-person Families—Income $5,000–$6,999				
Common-wall units	0.58	0.51	0.36	0.19
Apartments	0.28	0.15	0.07	0.02
Single detached	0.15	0.33	0.58	0.78
One bedroom	0.05	0.05	0.06	0.07
Two bedrooms	0.46	0.43	0.37	0.33
Three bedrooms	0.49	0.53	0.57	0.60

located in the outskirts of the central city; one located in the suburbs east of Pittsburgh. The predictions were derived, for each household size and income class and for a range of housing types, by plugging the parameter estimates and mean values of the explanatory variables into an equation of the form [3.14] in a similar fashion to the procedure employed in Example 3.2. In this context, it should be noted that different households face different effective prices (MPR) because of the interaction of contract housing prices and the accessibility costs to specific work places. It is this variation in MPR associated with employment at different work places in the metropolitan area which plays a crucial role in the derivation of the predicted probabilities. Table 3.3 shows an illustration of these predictions for just two of the thirty household income and size combinations.

EXAMPLE 3.4   *Work trip mode choice in Washington D.C. and the spatial transferability of models*

In the two previous examples we have considered cases in which the 'mixed' explanatory variables enter our cell (e) multiple-category logistic/logit models as either *all* alternative-specific variables or *all* generic variables. Much more common is the case where both generic and alternative specific variables are included in the same model. As we saw in Chapter 2, this case can only occur where the response categories (choice alternatives) form a 'ranked' set. We will see in this example that in this most commonly encountered case, the categorical (dummy) explanatory variables will normally take the form of alternative specific variables, whereas the continuous explanatory variables in the mixture will take the form of either generic or alternative specific variables.

The example we will consider is taken from the work of Atherton and Ben-Akiva (1976). Using a data set consisting of a sample of 1,114 respondents collected in Washington D.C. in 1968, they attempted to model the choice of mode of travel on work trips using a three-category logistic/logit model in which the choice alternatives (1 = 'car – driving alone', 2 = 'car – sharing the ride', 3 = public transport) form a 'ranked' set. Rather than write out their fitted model in explicit linear logit form they chose to present their results as shown in Table 3.4, i.e. in the type of form which we used in Examples 2.9, 2.10 and 2.11.[†]

[†] As an exercise, it may be useful for the reader to revise the discussion of types of explanatory variables in Section 2.11 (see also Appendix 2.1) and to notice that the results shown in Table 3.4 can be written in explicit linear logit form as:

$$\log_e \frac{\hat{P}_{1|i}}{\hat{P}_{3|i}} = -3.24 \quad -28.80(OPTC/INC_i^1 - OPTC/INC_i^3) \quad -0.0154(IVT_i^1 - IVT_i^3)$$
$$\phantom{\log_e \frac{\hat{P}_{1|i}}{\hat{P}_{3|i}} =} (0.47) \quad (12.74) \quad\quad\quad\quad\quad\quad\quad\quad (0.0058)$$
$$-0.160(OVT/DIST_i^1 - OVT/DIST_i^3) \; + 3.99AALD1_i \; +0.890BW_i$$
$$(0.039) \quad\quad\quad\quad\quad\quad\quad\quad (0.40) \quad\quad\quad (0.186)$$
$$-0.854CBD1_i \; + 0.000071DINC_i$$
$$(0.311) \quad\quad (0.000020)$$

$$\log_e \frac{\hat{P}_{2|i}}{\hat{P}_{3|i}} = -2.24 \quad -28.80(OPTC/INC_i^1 - OPTC/INC_i^3) \quad -0.0154(IVT_i^1 - IVT_i^3)$$
$$\phantom{\log_e \frac{\hat{P}_{2|i}}{\hat{P}_{3|i}} =} (0.40) \quad (12.74) \quad\quad\quad\quad\quad\quad\quad\quad (0.0058)$$
$$-0.160(OVT/DIST_i^1 - OVT/DIST_i^3) \; + 1.62AALD2_i \; +0.287GW_i$$
$$(0.039) \quad\quad\quad\quad\quad\quad\quad\quad (0.31) \quad\quad\quad (0.161)$$
$$-0.404CBD2_i \; + 0.000071DINC_i \; + 0.0983NWRS_i \; + 0.00065DTECA_i$$
$$(0.297) \quad\quad (0.000020) \quad\quad (0.0954) \quad\quad (0.00049)$$

Table 3.4
Work trip mode choice
model – Washington D.C.

Explanatory variables		Generic or alternatives specific to	Parameter estimates
ALT1	– Alternative specific constant=1 for 'car drive', 0 otherwise	1	−3.24 (0.47)
ALT2	– Alternative specific constant=1 for 'car share', 0 otherwise	2	−2.24 (0.40)
OPTC/INC	– Round trip out-of-pocket travel cost divided by household annual income	Generic	−28.80 (12.74)
IVT	– Round trip in-vehicle travel time	Generic	−0.0154 (0.0058)
OVT/DIST	– Round trip out-of-vehicle travel time divided by one-way distance	Generic	−0.160 (0.039)
AALD1	– Cars available per licensed driver	1	3.99 (0.40)
AALD2	– Cars available per licensed driver	2	1.62 (0.31)
BW	– Dummy variable indicating whether worker is head of household (1 = head of household. 0 otherwise)	1	0.890 (0.186)
GW	– Dummy variable indicating whether worker is employee of federal government (1=yes, 0=no)	2	0.287 (0.161)
CBD1	– Trip destination dummy variable (1=work place in CBD, 0 = otherwise)	1	−0.854 (0.311)
CBD2	– Trip destination dummy variable (1=work place in CBD, 0=otherwise)	2	−0.404 (0.297)
DINC	Household disposable income	1, 2	0.000071 (0.000020)
NWRS	Number of workers in household	2	0.0983 (0.0954)
DTECA	Employment density at the work zone (employees per commercial acre) times one way distance	2	0.00065 (0.00049)

The parameter estimates all have the expected signs and these correspond to the results achieved in many similar transportation studies. For example, the negative signs of the travel time and cost generic variables IVT, OVT/DIST, OPTC/INC are in accordance with what we have previously observed in Examples 2.9, 2.10 and 2.11, and the positive signs of the car availability parameter estimates are in accordance with the results in Examples 2.10 and 2.11.

In more detail we can note the following:

(a) The negative signs of the IVT, OVT/DIST and OPTC/INC parameter estimates indicate that as the relative in-vehicle travel time, out-of-vehicle travel time and out-of-pocket travel cost of a particular mode increase relative to an alternative mode so the odds of selecting that particular mode will decrease.

(b) The positive signs of the AALD and DINC parameter estimates indicate an increase in the odds of selecting the car modes 1 and 2 as car availability and household disposal income increases.

(c) The negative signs of the CBD parameter estimates indicate a decrease in the odds of selecting the car modes 1 and 2 if work trips terminate in the central city.

(d) The positive sign of the BW variable indicates an increase in the odds of selecting the 'car driving alone' mode if the worker is the head of the household.

(e) The positive signs of the GW and NWRS parameter estimates indicate an increase in the odds of selecting the 'car – sharing the ride' (i.e. car pooling) mode if the worker is a federal employee (large government organizations offer incentives to car pooling) and as the number of workers in the household increases (and thus, other things being equal, competition for household vehicles increases).

Following the estimation of this model using the Washington D.C. data, Atherton and Ben-Akiva went on to consider the extent to which the Washington model could be transferred to other cities. To this end they estimated similar models using work trip data sets collected in Los Angeles and New Bedford, Massachusetts. In the process of transferring the model they retained all the variables of the Washington model with the exception of the CBD and federal-government-worker variables. In the case of the CBD variables they argued that the congestion and inconvenience associated with driving into the CBD of a large, dense city such as Washington are real factors in choosing between car and public transit modes of travel, whereas, in a small city such as New Bedford, or a large, very diffuse city such as Los Angeles, the distinction between CBD and non-CBD trips would probably have

Table 3.5
Transferability of work trip mode choice model to different cities

Explanatory variable	Los Angeles		New Bedford	
	Parameter estimate	Standard error	Parameter estimate	Standard error
ALT1	−2.746	0.566	−2.198	0.830
ALT2	−1.830	0.463	−1.535	1.000
OPTC/INC	−24.37	11.77	−87.33	55.41
IVT	−0.0147	0.0065	−0.0199	0.0410
OVT/DIST	−0.186	0.046	−0.101	0.035
AALD1	3.741	0.520	2.541	0.692
AALD2	0.609	0.386	0.450	0.531
BW	0.810	0.247	1.026	0.272
DINC	0.000083	0.000036	0.000072	0.000056
NWRS	0.0810	0.1761	0.1874	0.1500
DTECA	0.00027	0.00012	0.00060	0.00078
Sample size	879		453	

much less effect on such a choice and thus the CBD variables should be omitted from the model. Similarly, they argued that the effects of large government organizations offering incentives to car pooling do not exist in either New Bedford or Los Angeles and thus the government worker variable should also be omitted. Using the same definitions as in Table 3.4, the results Atherton and Ben-Akiva achieved in Los Angeles and New Bedford are shown in Table 3.5. It can be seen that the parameter estimates are remarkably similar to those in Table 3.4 and Atherton and Ben-Akiva took this as empirical support for the spatial transferability of well-specified mode-choice models. (It should be noted, however, that subsequently Talvitie and Kirshner (1978) have cast some doubt on the validity of this inference and the reader should consult Galbraith and Hensher (1982) and McCarthy (1982) for up-to-date assessments of the current concensus of opinion on the spatial and temporal transferability of mode-choice models.)

3.5 Summary

We have seen in this chapter that categorical explanatory variables can be incorporated into our logistic/logit models by adopting the same principles as used in the extension of the conventional regression models of cell (a) of Table 1.1 to the 'dummy' variable regression models of cell (b). Principles of parameter estimation, hypothesis testing, goodness-of-fit measurement, etc., indentical to those which we discussed for the cell (d) models in Chapter 2, hold for our 'mixed' explanatory variable logistic/logit models.

Using a range of illustrative examples, we have seen that, in practice, only one of the three possible methods of introducing dummy variables into our models is widely used; the simple additive method. The interactive form of introducing dummy variables into our models is very rarely used as the sole means of including the dummy variables. Also, it appears that it is more common, in current practice, for the researcher to stratify his sample (or to draw new samples) on the basis of a dummy variable and to fit separate models to each stratum, rather than include the dummy variable in both a simple additive and interactive fashion and estimate a 'full' model on the whole sample. Finally, we have noted that perhaps the most typical example of a cell (e) model in the current literature is a model in which the response categories/choice alternatives form a 'ranked' set, and in which the categorical (dummy) variables take the form of alternative-specific variables (of type ψ or ϕ) and the continuous variables take the form of either generic or alternative-specific variables. This reflects the widespread use of such models in transportation science. In other contexts, models in which all the explanatory variables (both continuous and categorical) take the form of alternative-specific variables, or in which all take the form of generic variables, are to be found.

CHAPTER 4

Categorical response variable, categorical explanatory variables: the linear logit model approach

In this chapter we move from cell (e) of Table 1.1 to cell (f). Conceptually, this is a very small step but a number of changes in practice accompany it, and these can easily obscure the similarities with the approaches adopted in Chapters 2 and 3.

First, all variables in cell (f) problems (response and explanatory) are now categorical. This implies that instead of having, as in cells (d) and (e), sample data sets like that displayed in Table 1.7 (repeated below as Table 4.1 for convenience) we have sample data sets which can be represented in the form of Table 1.6 (repeated as Table 4.2 for convenience). In other words, instead of having a listing of individual cases, one observation for each distinct combination of values of the explanatory variables, we have grouped sets of observations or sub-populations where the groups or sub-populations are defined on the basis of the cross-classification of the categorical explanatory variables. Recalling the discussion of estimation procedures in Chapter 2, it should be clear to the reader that with grouped data in the form of Table 4.2 the

Table 4.1
Data table when explanatory variables continuous or mixed

Individual cases	Response categories				Total
	$r = 1$	2	. . .	R	
$i = 1$	$n_{1\mid1} = 1$ or 0	$n_{2\mid1} = 1$ or 0	. . .	$n_{R\mid1} = 1$ or 0	1
$i = 2$	$n_{1\mid2} = 1$ or 0	$n_{2\mid2} = 1$ or 0	. . .	$n_{R\mid2} = 1$ or 0	1
.	.	.		.	.
$i = N$	$n_{1\mid N} = 1$ or 0	$n_{2\mid N} = 1$ or 0	. . .	$n_{R\mid N} = 1$ or 0	1
					N

Note: each individual case i denotes a distinct combination of the values of the continuous or mixed explanatory variables.

Table 4.2
General form of contingency table when variables are divided into response and explanatory

Sub-population	Response categories				Total
	$r = 1$	2	...	R	
$g = 1$	$n_{1\|1}\ (f_{1\|1})$	$n_{2\|1}\ (f_{2\|1})$	...	$n_{R\|1}\ (f_{R\|1})$	$n_{+\|1}\ (1.0)$
$g = 2$	$n_{1\|2}\ (f_{1\|2})$	$n_{2\|2}\ (f_{2\|2})$	...	$n_{R\|2}\ (f_{R\|2})$	$n_{+\|2}\ (1.0)$
.	.	.		.	.
.	.	.		.	.
$g = G$	$n_{1\|G}\ (f_{1\|G})$	$n_{2\|G}\ (f_{2\|G})$	...	$n_{R\|G}\ (f_{R\|G})$	$n_{+\|G}\ (1.0)$
					$n_{+\|+} = N$

Note: $f_{r\|g} = n_{r\|g}/n_{+\|g}$ = observed proportion, $\sum_{r=1}^{R} f_{r\|g} = 1.0$

Each sub-population g denotes a distinct combination of values of the categorical explanatory variables.

non-iterative weighted least squares method now becomes a feasible parameter estimation procedure, so long as the size of each sub-population is sufficiently large.

Second, the problems of cell (f) lie in a zone of overlap or zone of transition between the domains of two major elements of the unified approach to categorical data analysis. That is to say, the problems of cell (f) lie between the domains of, on the one hand, the logistic, linear logit, probit models which we have considered in Chapters 2 and 3 and, on the other hand, the log-linear models which we will consider in Chapters 5 and 6. In terms of the cells of the bottom row of Table 1.1, we can imagine the influence of logistic, linear logit and probit models pushing towards the right from cells (d) and (e), and the influence of log-linear models pushing towards the left from cell (g). As a result, two types of models are now widely used to handle the problems of cell (f). These are either linear logit models whose parameters are estimated by a non-iterative weighted least squares procedure rather than the direct iterative maximum likelihood procedure we used in Chapters 2 and 3; or log-linear models whose parameters are estimated by an iterative proportional fitting procedure, an iterative weighted least squares procedure, or a Newton-Raphson procedure, and which will sometimes be written in linear logit model form.

For convenience, we will leave the second of these alternatives, log-linear models, until Chapters 5 and 6, and we will concentrate in this chapter upon the form and WLS estimation of the linear logit models used for the problems of cell (f).

4.1 Linear logit models for cell (f): basic forms

In the simple case of a two-category or dichotomous response variable, where we have several explanatory variables all of which are categorical, the standard linear logit model of Chapters 2 and 3 (see [2.49]) now takes the form:

$$\log_e \frac{P_{1|g}}{P_{2|g}} = x'_g \boldsymbol{\beta} \tag{4.1}$$

The letter g denotes the group or sub-population, and sub-populations are defined on the basis of the cross-classification of the categorical explanatory variables. For convenience, the left-hand side of [4.1] can be denoted $L_{12|g}$.

As we noted in Chapter 2, the probabilities on the left-hand side of [4.1] are unknown; that is to say, they are not directly observable quantities. Instead, we must approximate them by using the observed proportions, $f_{1|g}$ and $f_{2|g}$, as estimates. However, this approximation introduces an error and, consequently, the linear logit model [4.1] must be written in the form:

$$\bar{L}_{12|g} = \log_e \frac{f_{1|g}}{f_{2|g}} = x'_g \boldsymbol{\beta} + (\bar{L}_{12|g} - L_{12|g}) \tag{4.2}$$

where the term $(\bar{L}_{12|g} - L_{12|g})$ is introduced to take account of the fact that the left-hand side of [4.2] now contains only an estimate $(\bar{L}_{12|g})$ of the true logit $(L_{12|g})$. Under the assumption that the observed proportions, $f_{1|g}$ and $f_{2|g}$, are the results of independent random drawings from a binomial population with probabilities $P_{1|g}$ and $P_{2|g}$, it follows (Theil 1970, 137–8) that $\bar{L}_{12|g}$ is asymptotically normally distributed with mean:

$$E_A(\bar{L}_{12|g}) = L_{12|g} \tag{4.3}$$

and variance:

$$\text{Var}_A(\bar{L}_{12|g}) = 1/\left(n_{+|g} P_{1|g}(1 - P_{1|g})\right) \tag{4.4}$$

(where E_A means asymptotic expectation and Var_A means asymptotic variance). Furthermore, the error term $(\bar{L}_{12|g} - L_{12|g})$ has asymptotic properties:

$$E_A(\bar{L}_{12|g} - L_{12|g}) = 0$$

$$\text{Var}_A(\bar{L}_{12|g} - L_{12|g}) = 1/\left(n_{+|g} P_{1|g}(1 - P_{1|g})\right) \tag{4.5}$$

and it can be shown that the asymptotic properties of $\text{Var}_A(\bar{L}_{12|g} - L_{12|g})$ are not affected when we replace $P_{1|g}$ by its approximation $f_{1|g}$.

If we now take the expected value of both sides of the linear logit model [4.2] we get, because of [4.3] and [4.5], the result:

$$E_A(\bar{L}_{12|g}) = E_A\left[\log_e \frac{f_{1|g}}{f_{2|g}} \right] = x'_g \boldsymbol{\beta} \tag{4.6}$$

where

$$\log_e \frac{P_{1|g}}{P_{2|g}} = L_{12|g} = E_A(\bar{L}_{12|g}) \tag{4.7}$$

(i.e. the logit in the probabilities is the 'asymptotic expectation' of the logit in the observed proportions). As in the case of classical regression models where (as we saw in [2.1] and [2.2]) we can write the population regression model as either

$$Y_i = \beta_1 + \beta_2 X_{i2} + \beta_3 X_{i3} + \ldots + \beta_k X_{ik} + \varepsilon_i \qquad \text{[4.8]}$$

or $E(Y_i) = \beta_1 + \beta_2 X_{i2} + \beta_3 X_{i3} + \ldots + \beta_k X_{ik}$ [4.9]

we can, therefore, choose to write our linear logit model in the form of either [4.2] or [4.6]. In this chapter, for convenience, we will usually adopt the form [4.6].

When we extend our linear logit models from the dichotomous case to the polytomous, multiple-category case, the dichotomous linear logit model [4.6] generalizes[†] in the manner outlined in Section 2.9 to the following system of $R-1$ linear logit equations (where R denotes the number of response categories)

$$E_A(\bar{L}_{1R|g}) \quad = E_A \left[\log_e \frac{f_{1|g}}{f_{R|g}} \right] = x'_g \boldsymbol{\beta}_1$$

$$\qquad \vdots \qquad\qquad \vdots \qquad \vdots \qquad\qquad \text{[4.10]}$$

$$E_A(\bar{L}_{R-1\,R|g}) \; = E_A \left[\log_e \frac{f_{R-1|g}}{f_{R|g}} \right] = x'_g \boldsymbol{\beta}_{R-1}$$

This system of equations can be summarized as

$$E_A(\bar{L}_{rR|g}) = E_A \left[\log_e \frac{f_{r|g}}{f_{R|g}} \right] = x'_g \boldsymbol{\beta}_r \qquad r = 1, 2, \ldots R-1 \qquad \text{[4.11]}$$

The parameters of cell (f) linear logit models, both dichotomous [4.6] and polytomous [4.11] in form, are in almost all cases estimated using the least squares (WLS) procedure which we discussed in Section 2.4.1. In this chapter it will be necessary to generalize the WLS equations presented in Chapter 2. However, before we do that, it will be useful if the reader first gets a flavour of cell (f) linear logit models from a consideration of two simple illustrative examples.

EXAMPLE 4.1   *Preference for army camp location among American soldiers*

Table 4.3 presents data taken from a famous study of *The American Soldier* by Stouffer *et al.* (1949); data which have subsequently been analysed by many writers including Coleman (1964), Theil (1970) and Goodman (1972b). It can be seen that Table 4.3 is organized in the form of Table 4.2. The sub-populations are defined on the basis of the

---

[†] The assumptions discussed for the dichotomous case are also generalized. See footnote, p. 122–23.

Table 4.3
Preference for army camp location amongst American soldiers

| Sub-populations | | | Response categories | | |
| | | | r = 1 | r = 2 | |
Region of region	Race	Present camp	Preference for southern camp	Preference for northern camp	Total
g = 1: North	Negro	North	36 (0.085)	387 (0.915)	423 (1.0)
g = 2: North	Negro	South	250 (0.222)	876 (0.778)	1,126 (1.0)
g = 3: North	White	North	162 (0.145)	955 (0.855)	1,117 (1.0)
g = 4: North	White	South	510 (0.369)	874 (0.631)	1,384 (1.0)
g = 5: South	Negro	North	270 (0.414)	383 (0.586)	653 (1.0)
g = 6: South	Negro	South	1,714 (0.819)	379 (0.181)	2,093 (1.0)
g = 7: South	White	North	176 (0.628)	104 (0.372)	280 (1.0)
g = 8: South	White	South	869 (0.905)	91 (0.095)	961 (1.0)

cross-classification of three potential explanatory variables: (a) the soldier's region of origin (whether he comes from the north or the south); (b) the soldier's race (negro or white); (c) the location of the soldier's present camp (north or south). The response variable is the soldier's preference for army camp location, and it has two categories: (a) preference for camp in the north; (b) preference for camp in the south.

To the data of Table 4.3 we can fit a dichotomous linear logit model of the form [4.6] which relates the odds of expressing a preference for a southern army camp to the three categorical explanatory variables. The model has the form

$$E_A(\bar{L}_{12|g}) = E_A\left[\log_e \frac{f_{1|g}}{f_{2|g}}\right] = \beta_1 + \beta_2 X_{g2} + \beta_3 X_{g3} + \beta_4 X_{g4} \qquad \textbf{[4.12]}$$

where

$X_{g2} = \text{ORG}_g$ = 1 whenever sub-population g includes soldiers from the south, 0 from the north

$X_{g3} = \text{RACE}_g$ = 1 whenever sub-population g includes whites, 0 negroes

$X_{g4} = \text{PLOC}_g$ = 1 whenever sub-population g includes soldiers whose present camp is in the south, 0 if present camp in the north.

Theil (1970, 109) reports weighted least squares estimates of the parameters of [4.12]. They are (estimated standard errors in parentheses):

$$\hat{\beta}_1 = -2.74 \qquad \hat{\beta}_2 = 2.60 \qquad \hat{\beta}_3 = 0.76 \qquad \hat{\beta}_4 = 1.54 \qquad \textbf{[4.13]}$$
$$\phantom{\hat{\beta}_1 = }(0.08) \qquad \phantom{\hat{\beta}_2 = }(0.06) \qquad \phantom{\hat{\beta}_3 = }(0.06) \qquad \phantom{\hat{\beta}_4 = }(0.06)$$

The parameter estimates in [4.13] have signs which conform with our expectations about locational loyalty, inertia and racial attitudes in the United States. They indicate that if soldiers come from the south rather than the north, are white rather than negro, and are presently located in a southern rather than a northern camp, the odds of their expressing a preference for a southern camp are increased. The value of the constant term $(\hat{\beta}_1)$ can be taken to represent an estimate of the log-odds of expressing a preference for a southern camp amongst negroes of northern origin who are presently located in a northern camp. In this context it can be seen that in model [4.12] $E_A(\bar{L}_{12|g}) = \beta_1$ when $\text{ORG}_g = \text{RACE}_g = \text{PLOC}_g = 0$.

The model [4.12] is, of course, a very simple specification, and is included merely to give a brief illustration of dichotomous cell (f) linear logit models. It is a straightforward matter, however, to extend the model to take account of any interactions which might exist among the explanatory variables, and Theil (1970, 110–12) demonstrates this by including in the model a variable which takes account of interaction effects between the region of origin (ORG) and location of present camp (PLOC) variables.

In addition, the model [4.12] can serve a second useful function. It can be used to provide the reader with a brief introduction to the type of matrix expressions which will be encountered in subsequent sections of this chapter. To see this, let us display model [4.12] in explicit vector and matrix form as

$$
\begin{bmatrix}
E_A(\bar{L}_{12|1}) \\
E_A(\bar{L}_{12|2}) \\
E_A(\bar{L}_{12|3}) \\
E_A(\bar{L}_{12|4}) \\
E_A(\bar{L}_{12|5}) \\
E_A(\bar{L}_{12|6}) \\
E_A(\bar{L}_{12|7}) \\
E_A(\bar{L}_{12|8})
\end{bmatrix}
=
\begin{bmatrix}
X_{11} & X_{12} & X_{13} & X_{14} \\
X_{21} & X_{22} & X_{23} & X_{24} \\
X_{31} & X_{32} & X_{33} & X_{34} \\
X_{41} & X_{42} & X_{43} & X_{44} \\
X_{51} & X_{52} & X_{53} & X_{54} \\
X_{61} & X_{62} & X_{63} & X_{64} \\
X_{71} & X_{72} & X_{73} & X_{74} \\
X_{81} & X_{82} & X_{83} & X_{84}
\end{bmatrix}
\begin{bmatrix}
\beta_1 \\
\beta_2 \\
\beta_3 \\
\beta_4
\end{bmatrix}
\qquad \textbf{[4.14]}
$$

or $\qquad E_A(\bar{L}) = \mathbf{X}\boldsymbol{\beta}$ $\qquad\qquad\qquad\qquad\qquad\qquad$ **[4.15]**

The left-hand side of [4.14] is a column vector of observed logit values; one value for each sub-population g. The right-hand side is composed of a matrix $\mathbf{X}$ of explanatory variable values and a column vector $\boldsymbol{\beta}$ of parameters. Using the principles of matrix multiplication, any particular element in the logit vector is given by multiplying the elements of that particular row of the $\mathbf{X}$ matrix by elements in the $\boldsymbol{\beta}$ parameter vector. For example, the fifth element in the logit vector is given by

$$
E_A(\bar{L}_{12|5}) = [X_{51}\ X_{52}\ X_{53}\ X_{54}]
\begin{bmatrix}
\beta_1 \\
\beta_2 \\
\beta_3 \\
\beta_4
\end{bmatrix}
\qquad \textbf{[4.16]}
$$

or more compactly,

$$
E_A(\bar{L}_{12|5}) = X_{51}\beta_1 + X_{52}\beta_2 + X_{53}\beta_3 + X_{54}\beta_4 = \mathbf{x}'_5\boldsymbol{\beta} \qquad \textbf{[4.17]}
$$

Model [4.12] is thus simply [4.17] where g represents any particular sub-population ($g = 1, 2, \ldots 8$), i.e. model [4.12] can be seen to be

$$E_A(\bar{L}_{12|g}) = x'_g \beta \qquad g = 1, 2, \ldots 8 \qquad\qquad\qquad [4.18]$$

The matrix of explanatory variables $\mathbf{X}$ is known as the *design matrix*. In this case, because the explanatory variables (ORG, RACE, PLOC) are all dummy variables, it is composed entirely of zeros and ones (the X_1, constant term, is represented by a column of ones). Matrix $\mathbf{X}$ in [4.14] can therefore be written explicitly as

$$\mathbf{X} = \begin{bmatrix} 1 & 0 & 0 & 0 \\ 1 & 0 & 0 & 1 \\ 1 & 0 & 1 & 0 \\ 1 & 0 & 1 & 1 \\ 1 & 1 & 0 & 0 \\ 1 & 1 & 0 & 1 \\ 1 & 1 & 1 & 0 \\ 1 & 1 & 1 & 1 \end{bmatrix} \qquad\qquad [4.19]$$

The reader should consult the definitions of the explanatory variables (ORG, RACE, PLOC) and substitute the values of these explanatory variables for each sub-population g into matrix $\mathbf{X}$ of [4.14]. He should then confirm that the result of this operation is the matrix $\mathbf{X}$ of [4.19].

Readers unfamiliar with matrix expressions and the principles of matrix algebra should consult Davis (1973), Mather (1976), Rogers (1971), Wilson and Kirkby (1980) for discussion of the topic oriented to the needs of geographers and environmental scientists.

EXAMPLE 4.2   *Evaluation of military policemen by negro soldiers from different regions*

In the preceding example we have considered the case in which the linear logit model is dichotomous in form. In this example we will examine another data set from Stouffer's study of the American soldier (Theil 1970, 117); one for which a multiple-category linear logit model is appropriate. The data which are presented in Table 4.4 consist of

Table 4.4
Evaluation of military policemen by negro soldiers

Sub-populations		Response categories			
		$r = 1$	$r = 2$	$r = 3$	
Region of origin	Present camp	Fair most of the time	Half fair half unfair	Unfair most of the time	Total
$g = 1$: North	North	118 (0.258)	207 (0.450)	134 (0.292)	459 (1.0)
$g = 2$: North	South	181 (0.146)	514 (0.416)	542 (0.438)	1,237 (1.0)
$g = 3$: South	North	253 (0.330)	313 (0.409)	200 (0.261)	766 (1.0)
$g = 4$: South	South	653 (0.270)	1,006 (0.416)	760 (0.314)	2,419 (1.0)

evaluations of white military policemen by negro soliders. The evaluations of military policemen are classified into three response categories: (a) 'fair most of the time'; (b) 'about half fair, half unfair'; (c) 'unfair most of the time'. The sub-populations are defined on the basis of the cross-classification of two potential explanatory variables: (a) the soldier's region of origin; (b) the location of the soldier's present camp.

To the data of Table 4.4, a multiple-category linear logit model of the form [4.11] can be fitted. This model takes the form of a set of two linear logit equations.

$$E_A(\bar{L}_{13|g}) = E_A \left[\log_e \frac{f_{1|g}}{f_{3|g}} \right] = \beta_{11} + \beta_{21}X_{g2} + \beta_{31}X_{g3}$$

[4.20]

$$E_A(\bar{L}_{23|g}) = E_A \left[\log_e \frac{f_{2|g}}{f_{3|g}} \right] = \beta_{12} + \beta_{22}X_{g2} + \beta_{32}X_{g3}$$

where

$X_{g2} = \text{ORG}_g$ = 1 whenever sub-population g includes soldiers from the south, 0 from the north

$X_{g3} = \text{PLOC}_g$ = 1 whenever sub-population g includes soldiers whose present camp is in the south, 0 if present camp is in the north

$f_{1|g}/f_{3|g}$ = odds that negro soldiers of type g will regard white military policemen as 'fair most of the time' rather than 'unfair most of the time'

$f_{2|g}/f_{3|g}$ = odds that negro soldiers of type g will regard military policemen as 'half fair, half unfair' rather than 'unfair most of the time'

Theil (1970, 121) reports the following weighted least squares estimates of the parameters of [4.20].

$$\hat{\beta}_{11} = -0.39 \quad \hat{\beta}_{21} = 0.78 \quad \hat{\beta}_{31} = -0.59$$
$$\quad (0.09) \qquad\qquad (0.08) \qquad\qquad (0.09)$$

[4.21]

$$\hat{\beta}_{12} = \quad 0.28 \quad \hat{\beta}_{22} = 0.27 \quad \hat{\beta}_{32} = -0.30$$
$$\quad (0.08) \qquad\qquad (0.07) \qquad\qquad (0.08)$$

The parameter estimates for the two explanatory variables (ORG and PLOC) have the same signs in both equations in [4.20] and these conform with our expectations about locational loyalty, inertia and racial attitudes in the United States. The estimates ($\hat{\beta}_{21}$, $\hat{\beta}_{31}$) relating to the first equation in [4.20] tell us that ORG and PLOC have a positive and negative effect respectively on the odds that a negro soldier will regard white military policemen as 'fair most of the time' rather than 'unfair most of the time'. In other words, negro soldiers of southern origin will evaluate white military policemen more favourably than negro soldiers of northern origin, but negro soldiers currently located in a southern camp will evaluate white military policemen less favourably than negro soldiers located in northern camps. Expressed another way, we can state that a southern origin and a northern camp experience have the effect of making negro soldiers less dissatisfied with white military policemen. The parameter estimates ($\hat{\beta}_{22}$ and $\hat{\beta}_{32}$) relating to the second equation in [4.20] which compares 'half unfair, half fair' with 'unfair most of the time' corroborate this inference.

Using the parameter estimates [4.21] we can derive predicted logit values, for each sub-population g of the form

$$\hat{L}_{13|g} = x'_g\hat{\beta}_1 = \hat{\beta}_{11} + \hat{\beta}_{21}X_{g2} + \hat{\beta}_{31}X_{g3}$$

$$\hat{L}_{23|g} = x'_g\hat{\beta}_2 = \hat{\beta}_{12} + \hat{\beta}_{22}X_{g2} + \hat{\beta}_{32}X_{g3}$$

[4.22]

We can then substitute these values into a grouped data analogy to the individual-case, multiple-category logistic models of Chapters 2 and 3 (see [2.99] and [2.102])

$$\hat{P}_{1|g} = \frac{e^{x'_g\hat{\beta}_1}}{1 + \sum_{s=1}^{2} e^{x'_g\hat{\beta}_s}} \qquad \hat{P}_{2|g} = \frac{e^{x'_g\hat{\beta}_2}}{1 + \sum_{s=1}^{2} e^{x'_g\hat{\beta}_s}}$$

$$\hat{P}_{3|g} = \frac{1}{1 + \sum_{s=1}^{2} e^{x'_g\hat{\beta}_s}}$$

[4.23]

These equations allow us to obtain the predicted probabilities of the selection of each response category by members of each sub-population g. The predictions can then be compared with the original observed response category proportions ($f_{1|g}$, $f_{2|g}$ and $f_{3|g}$) in each sub-population and an assessment of the goodness-of-fit of the model can be obtained. (Alternatively the fit can be assessed by considering the differences between the observed logits $\hat{L}_{12|g}$, $\hat{L}_{23|g}$ and the predicted logits $\hat{L}_{12|g}$ and $\hat{L}_{23|g}$). Using the parameter estimates [4.21] and the equations [4.23] Theil (1970, 123) computed the predicted probabilities shown in Table 4.5. These should be compared with the observed proportions in Table 4.4.

As in Example 4.1, it will be useful at this point to consider the matrix expression of model [4.20]. This takes the form

$$
\begin{bmatrix}
E_A(\hat{L}_{13|1}) \\
E_A(\hat{L}_{23|1}) \\
E_A(\hat{L}_{13|2}) \\
E_A(\hat{L}_{23|2}) \\
E_A(\hat{L}_{13|3}) \\
E_A(\hat{L}_{23|3}) \\
E_A(\hat{L}_{13|4}) \\
E_A(\hat{L}_{23|4})
\end{bmatrix}
=
\begin{bmatrix}
X_{11} & X_{12} & X_{13} & 0 & 0 & 0 \\
0 & 0 & 0 & X_{11} & X_{12} & X_{13} \\
X_{21} & X_{22} & X_{23} & 0 & 0 & 0 \\
0 & 0 & 0 & X_{21} & X_{22} & X_{23} \\
X_{31} & X_{32} & X_{33} & 0 & 0 & 0 \\
0 & 0 & 0 & X_{31} & X_{32} & X_{33} \\
X_{41} & X_{42} & X_{43} & 0 & 0 & 0 \\
0 & 0 & 0 & X_{41} & X_{42} & X_{43}
\end{bmatrix}
\begin{bmatrix}
\beta_{11} \\
\beta_{21} \\
\beta_{31} \\
\beta_{12} \\
\beta_{22} \\
\beta_{32}
\end{bmatrix}
$$

[4.24]

or $E_A(\hat{L}) = X\beta$

[4.25]

The logit vector on the left-hand side of this matrix expression has two

Sub-populations		Response categories		
		$r = 1$ Fair most of the time	$r = 2$ Half fair half unfair	$r = 3$ Unfair most of the time
Region of origin	Present camp			
$g = 1$: North	North	0.225	0.442	0.333
$g = 2$: North	South	0.159	0.418	0.423
$g = 3$: South	North	0.350	0.412	0.238
$g = 4$: South	South	0.264	0.415	0.322

logit values ($\bar{L}_{13|g}$ and $\bar{L}_{23|g}$) for each sub-population g. Once again, using the principles of matrix multiplication, any particular element in the logit vector is given by multiplying the elements of that particular row of the **X** matrix by elements in the **β** parameter vector. For example, the fifth element in the logit vector is given by:

$$E_A(\bar{L}_{13|3}) = [X_{31}\ X_{32}\ X_{33}\ 0\ 0\ 0] \begin{bmatrix} \beta_{11} \\ \beta_{21} \\ \beta_{31} \\ \beta_{12} \\ \beta_{22} \\ \beta_{32} \end{bmatrix} \qquad [4.26]$$

or more compactly,

$$E_A(\bar{L}_{13|3}) = X_{31}\beta_{11} + X_{32}\beta_{21} + X_{33}\beta_{31} = x'_3\boldsymbol{\beta}_1 \qquad [4.27]$$

The sixth element in the logit vector, written in the form of [4.27] is

$$E_A(\bar{L}_{23|3}) = X_{31}\beta_{12} + X_{32}\beta_{22} + X_{33}\beta_{32} = x'_3\boldsymbol{\beta}_2 \qquad [4.28]$$

The explanatory variables ($X_{g2} = \text{ORG}_g$ and $X_{g3} = \text{PLOC}_g$) in model [4.20] are both dummy variables and X_{g1}, the constant term, is represented by a column of ones. As a result, the design matrix X is composed entirely of 0 and 1's. Matrix **X** in [4.24] can therefore be written explicitly as

$$\mathbf{X} = \begin{bmatrix} 1 & 0 & 0 & 0 & 0 & 0 \\ 0 & 0 & 0 & 1 & 0 & 0 \\ 1 & 0 & 1 & 0 & 0 & 0 \\ 0 & 0 & 0 & 1 & 0 & 1 \\ 1 & 1 & 0 & 0 & 0 & 0 \\ 0 & 0 & 0 & 1 & 1 & 0 \\ 1 & 1 & 1 & 0 & 0 & 0 \\ 0 & 0 & 0 & 1 & 1 & 1 \end{bmatrix} \qquad [4.29]$$

The reader should consult the definitions of the explanatory variables (ORG and PLOC) and substitute the values of these explanatory variables for each sub-population g into matrix $\mathbf{X}$ of [4.24]. He should then confirm that the result of this operation is the matrix $\mathbf{X}$ of [4.29].

4.2 Weighted least squares estimation of cell (f) linear logit models

In Section 2.4 we discussed the principles of weighted least squares estimation, and we saw that WLS parameter estimates are obtained, in the dichotomous linear logit model case, by minimizing

$$\sum_g w_g \left(\bar{L}_{12|g} - \hat{L}_{12|g}\right)^2 \tag{4.30}$$

where the weights are defined as:

$$w_g = n_{+|g} f_{1|g} \left(1 - f_{1|g}\right) \tag{4.31}$$

We saw that weights of this form imply that, given $n_{+|g}$, less weight is allocated as $f_{1|g}$ approaches 0 or 1, and that a zero weight is given when $f_{1|g}$ equals 0 or 1. To overcome this difficulty we considered Berkson's suggested replacement working values. Using elementary calculus we then considered the principles involved in the minimization of [4.30]. We saw that this involves taking the first partial derivatives of [4.30] with respect to the parameter estimates, setting each of these to zero, and then solving the resulting set of simultaneous equations which are known as the 'normal equations'. In this way we derived the required WLS estimators.

In our discussion of WLS estimation in Chapter 2 we were concerned with a dichotomous logit model which contained just one explanatory variable and two parameters. In this chapter we are concerned with the more general case in which we have dichotomous logit models which contain several explanatory variables and parameters. It is necessary, therefore, to generalize the procedure outlined in Chapter 2, and to do this we must employ the matrix expressions introduced in Example 4.1.

In matrix terms, equation [4.30] becomes:

$$(\bar{\mathbf{L}} - \hat{\mathbf{L}})' \mathbf{V}_{\bar{L}}^{-1} (\bar{\mathbf{L}} - \hat{\mathbf{L}}) \tag{4.32}$$

In other words, given that $\hat{\mathbf{L}} = \mathbf{X}\hat{\boldsymbol{\beta}}$, we can say that the weighted least squares principle implies that parameter estimates are obtained by minimizing the quadratic form:

$$(\bar{\mathbf{L}} - \mathbf{X}\hat{\boldsymbol{\beta}})' \mathbf{V}_{\bar{L}}^{-1} (\bar{\mathbf{L}} - \mathbf{X}\hat{\boldsymbol{\beta}}) \tag{4.33}$$

With the exception of $\mathbf{V}_{\bar{L}}^{-1}$ we have met all the terms in this expression before in [4.15]. This new matrix, $\mathbf{V}_{\bar{L}}^{-1}$, is a matrix of weights which has terms of the form [4.31] down its principal diagonal and zeros elsewhere.

$$\mathbf{V_L^{-1}} = \begin{bmatrix} w_1 & 0 & . & . & . & 0 \\ 0 & w_2 & . & . & . & 0 \\ . & . & . & & . & \\ . & . & . & . & . & \\ . & . & & . & . & \\ 0 & 0 & . & . & . & w_G \end{bmatrix}$$ [4.34]

This matrix of weights is the inverse of $\mathbf{V_L}$ which is a consistent estimator of the variance-covariance matrix of observed logits $\mathbf{\bar{L}}$.

To obtain the required estimates we take the first partial derivatives of [4.33] with respect to $\hat{\boldsymbol{\beta}}$ and set these equal to zero. The simultaneous equations obtained in this manner are known as the normal equations and take the form:

$$(\mathbf{X'V_L^{-1}X})\hat{\boldsymbol{\beta}} = \mathbf{X'V_L^{-1}\bar{L}}$$ [4.35]

These equations represent the matrix equivalent of the normal algebra equations [2.30] and [2.31] and their solution[†] is:

$$\hat{\boldsymbol{\beta}} = (\mathbf{X'V_L^{-1}X})^{-1}\mathbf{X'V_L^{-1}\bar{L}}$$ [4.36]

which is known as the generalized least squares estimator of $\boldsymbol{\beta}$. Standard errors of $\hat{\boldsymbol{\beta}}$ can be obtained by taking the square roots of the values along the principal diagonal of the matrix:

$$\mathbf{V_{\boldsymbol{\beta}}} = (\mathbf{X'V_L^{-1}X})^{-1}$$ [4.37]

where $\mathbf{V_{\boldsymbol{\beta}}}$ is a consistent estimator of the variance-covariance matrix of $\hat{\boldsymbol{\beta}}$.

The advantage of writing the weighted least squares procedure of Chapter 2 in matrix terms can be seen in the compact forms of expressions [4.35], [4.36] and [4.37]. Given that the dichotomous logit models in this chapter contain several explanatory variables and parameters, the normal algebra equivalents to [4.35], [4.36] and [4.37] would become extremely cumbersome to write out and work with. To confirm this, the reader should consider the single explanatory variable normal algebra equivalents to [4.35] and [4.36] which are given in [2.30], [2.31] and [2.32]. He should then consider what form these would take if there were five, ten or fifteen explanatory variables.

Another advantage of using matrix expressions is that [4.35], [4.36] and [4.37] remain the appropriate equations when our linear ligit models are multiple-category rather than dichotomous.[‡] The only

---

[†] Assuming that there are no linearly dependent columns in $\mathbf{X}$ (i.e. that it is of full rank) and that the matrix $(\mathbf{X'V_L^{-1}X})$ is non-singular (i.e. has a non-zero determinant).

[‡] In developing this generalization to the multiple category case we normally make assumptions equivalent to those outlined in Section 4.1 for the dichotomous case. We normally assume that the observed proportions $f_{r|g}$ are the results of independent random drawings from a multinomial population with probabilities $P_{r|g}$, and that the drawings from different sub-populations are independent. Under these assumptions, it follows that the observed proportions $(f_{1|g}, \dots f_{R|g})$ for any sub-population g have, asymptotically, a multivariate normal distribution with means $(P_{1|g}, \dots, P_{R|g})$

difference is that the form of the $\mathbf{V_L}^{-1}$ matrix changes. In the multiple response category case the weight [4.31] used in the dichotomous case has to be modified to the form:

$$\mathbf{W}_g = n_{+|g} \begin{bmatrix} f_{1|g}(1-f_{1|g}) & -f_{1|g}f_{2|g} & \cdots & -f_{1|g}f_{R-1|g} \\ -f_{2|g}f_{1|g} & f_{2|g}(1-f_{2|g}) & \cdots & -f_{2|g}f_{R-1|g} \\ \cdot & \cdot & & \cdot \\ \cdot & \cdot & & \cdot \\ \cdot & \cdot & & \cdot \\ -f_{R-1|g}f_{1|g} & -f_{R-1|g}f_{2|g} & \cdots & f_{R-1|g}(1-f_{R-1|g}) \end{bmatrix}$$

$$[4.38]$$

In other words, in the R-category case, the weight for sub-population g becomes a $(R-1) \times (R-1)$ matrix with diagonal elements which are similar to the weight [4.31], and with non-zero off-diagonal elements which reflect the non-zero correlations between pairs of error terms, e.g. between $(\bar{L}_{2R|g} - L_{2R|g})$ and $(\bar{L}_{4R|g} - L_{4R|g})$, or between $(\bar{L}_{3R|g} - L_{3R|g})$ and $(\bar{L}_{5R|g} - L_{5R|g})$ for the same sub-population g. The error terms in question are the multiple-category equivalents to those on the right-hand side of [4.2]. They are correlated because for each

---

and variances and covariances of the form (see Grizzle, Starmer and Koch 1969, 490; Lehnen and Koch 1974a, 288–9, Parks 1980, 297).

$$\mathbf{V}_{\mathbf{P}_g} = \frac{1}{n_{+|g}} \begin{bmatrix} P_{1|g}(1-P_{1|g}) & -P_{1|g}P_{2|g} & \cdots & -P_{1|g}P_{R|g} \\ -P_{2|g}P_{1|g} & P_{2|g}(1-P_{2|g}) & \cdots & -P_{2|g}P_{R|g} \\ \cdot & \cdot & & \cdot \\ \cdot & \cdot & & \cdot \\ \cdot & \cdot & & \cdot \\ -P_{R|g}P_{1|g} & -P_{R|g}P_{2|g} & \cdots & P_{R|g}(1-P_{R|g}) \end{bmatrix}$$

i.e. a $R \times R$ matrix with variances along the principal diagonal and covariances in the off-diagonal positions. Furthermore, it follows that the observed logits $\bar{L}_{rR|g}$ for any sub-population g also have, asymptotically, a multivariate normal distribution with means $L_{rR|g}$ and variances and covariances in the form of a $(R-1) \times (R-1)$ matrix (see Theil 1970, 146; Parks 1980, 297).

$$\mathbf{V}_{\mathbf{L}_g} = \frac{1}{n_{+|g}} \begin{bmatrix} 1/P_{R|g}+1/P_{1|g} & 1/P_{R|g} & \cdots & 1/P_{R|g} \\ 1/P_{R|g} & 1/P_{R|g}+1/P_{2|g} & \cdots & 1/P_{R|g} \\ \cdot & \cdot & & \cdot \\ \cdot & \cdot & & \cdot \\ \cdot & \cdot & & \cdot \\ 1/P_{R|g} & 1/P_{R|g} & \cdots & 1/P_{R|g}+1/P_{R-1|g} \end{bmatrix}$$

In addition, it follows that $(\bar{L}_{rR|g} - L_{rR|g})$, the error terms (the multiple-category equivalents to those on the right-hand side of the model [4.2]) for any sub-population g also have, asymptotically, a multivariate normal distribution with zero means and variances and covariances identical to those in $\mathbf{V}_{\mathbf{L}_g}$.

From the above, it follows that $f_{r|g}$ and $\bar{L}_{rR|g}$ are consistent estimators of $P_{r|g}$ and $L_{rR|g}$ respectively, and that if we replace $P_{r|g}$ in the two matrices above by $f_{r|g}$, then $\mathbf{V}_{\mathbf{f}_g}$ and $\mathbf{V}_{\mathbf{L}_g}$ are consistent estimators of $\mathbf{V}_{\mathbf{P}_g}$ and $\mathbf{V}_{\mathbf{L}_g}$.

sub-population g both the true probabilities and the observed proportions must sum to one (i.e. $\Sigma_{r=1}^{R} P_{r|g} = 1$ and $\Sigma_{r=1}^{R} f_{r|g} = 1$). As a result, if, by some sampling error, $f_{1|g}$ for example exceeds the true $P_{1|g}$, then one or all of the remaining $f_{r|g}$, $r = 2 \ldots R$, *must* understate the true $P_{r|g}$. Conversely, if one of the observed proportions understates the true probability, then one or all of the remaining observed proportions must exceed the true probability. These inter-relationships caused by the 'sum-to-one' constraint create non-zero covariances between the pairs of error terms within each sub-population. As a result the $\mathbf{V}_{\dot{L}}^{-1}$ matrix cannot be a simple diagonal matrix, as in [4.34], but must be a 'block-diagonal' matrix with sub-matrices of the form [4.38] along the principal diagonal. For example, in the three category logit model case of Example 4.2, the $\mathbf{V}_{\dot{L}}^{-1}$ matrix takes the form:

$$
\mathbf{V}_{\dot{L}}^{-1} =
\begin{bmatrix}
 & & 0 & 0 & 0 & 0 & 0 & 0 \\
\mathbf{W}_1 & & & & & & & \\
 & & 0 & 0 & 0 & 0 & 0 & 0 \\
0 & 0 & & & 0 & 0 & 0 & 0 \\
 & & \mathbf{W}_2 & & & & & \\
0 & 0 & & & 0 & 0 & 0 & 0 \\
0 & 0 & 0 & 0 & & & 0 & 0 \\
 & & & & \mathbf{W}_3 & & & \\
0 & 0 & 0 & 0 & & & 0 & 0 \\
0 & 0 & 0 & 0 & 0 & 0 & & \\
 & & & & & & \mathbf{W}_4 & \\
0 & 0 & 0 & 0 & 0 & 0 & &
\end{bmatrix}
\qquad [4.39]
$$

where

$$
\mathbf{W}_g = n_{+|g}
\begin{bmatrix}
f_{1|g}\,(1 - f_{1|g}) & -f_{1|g}\,f_{2|g} \\
-f_{2|g}\,f_{1|g} & f_{2|g}\,(1 - f_{2|g})
\end{bmatrix}
\qquad [4.40]
$$

and where all the other matrices are as shown in [4.24].

It should be noted that the elements within the matrix of weights [4.38] will be zero whenever an observed proportion $f_{r|g}$ equals 0 or 1. To overcome this difficulty, Grizzle, Starmer and Koch (1969) have suggested using replacement values similar to those suggested by Berkson (1953, 1955) for the dichotomous case (see [2.27] and footnote, p. 32). When there are R possible response categories they suggest replacing any occasional zero observed frequency ($n_{r|g} = 0$) in Table 4.2 by $1/R$. This has the effect of making the estimate of $f_{r|g}$ equal to $1/Rn_{+|g}$.

The weighted least squares procedure provides, therefore, a computationally convenient, non-iterative method of deriving the parameters of our cell (f) linear logit models, and the parameter estimators can be shown to have very desirable asymptotic properties. Moreover, the principles of the procedure will be familiar to those readers who have experience of least squares estimation of the regression models in cells (a) to (c) of Table 1.1. However, in adopting the procedure we must not lose sight of two of its basic assumptions.

1. The procedure is oriented towards moderate and large size samples. In particular, the size $n_{+|g}$ of each sub-population must be sufficiently

large[†] to justify the approximation of the unknown probability values $P_{r|g}$ in the linear logit model by the observed proportions $f_{r|g}$, and also to justify the estimation of the variance–covariance matrix of observed logits and/or error terms by $\mathbf{V}_{\bar{L}}$, the inverse of the matrix of weights defined in [4.34] for the dichotomous case and in [4.38] and [4.39] for the multiple-category case. (See Grizzle, Starmer and Koch 1969; Theil 1970, 137–46; Koch *et al.* 1977, 154–8 for further discussion of these assumptions.)

2. We must be prepared to argue that the data in our cell (f) type contingency tables (e.g. Tables 4.2, 4.3, 4.4) could conceivably have arisen from an underlying product multinomial sampling model (see the discussion of this type of sampling model in Ch. 1). Data from an explicit stratified random sampling design, or data which we can assume to be equivalent to a stratified random sample, will meet this condition if the strata are incorporated into the definition of the sub-populations.[‡] On the other hand, data which have arisen from more complex sample designs must be treated with care. In such cases, the procedure outlined above represents only an approximation for use in the preliminary stages of analysis. Brief comments on this issue are provided by Lehnen and Koch (1974a, 286–9) and a detailed consideration is provided by Koch *et al.* (1975).

4.3 Goodness-of-fit and test statistics

In the previous section we saw that parameter estimates can be derived using the weighted least squares procedure by finding the values of $\boldsymbol{\beta}$ which minimize the quadratic form [4.33]. For any particular data set and any particular logit model, the value at which [4.33] is minimized can be denoted Q, and this value can be used as a measure of the goodness-of-fit of the model in question to the original data. The goodness-of-fit statistic Q is thus:

$$Q = (\bar{\mathbf{L}} - \mathbf{X}\hat{\boldsymbol{\beta}})' \, \mathbf{V}_{\bar{L}}^{-1} \, (\bar{\mathbf{L}} - \mathbf{X}\hat{\boldsymbol{\beta}}) \qquad [4.41]$$

which can be written alternatively as:

$$Q = \bar{\mathbf{L}}'\mathbf{V}_{\bar{L}}^{-1}\bar{\mathbf{L}} - \hat{\boldsymbol{\beta}}'(\mathbf{X}'\mathbf{V}_{\bar{L}}^{-1}\mathbf{X})\hat{\boldsymbol{\beta}} \qquad [4.42]$$

Under the hypothesis that the model fits, Q has approximately a chi-square distribution with degrees of freedom equal to the difference in the first dimensions of $\bar{\mathbf{L}}$ and $\hat{\boldsymbol{\beta}}$ as long as the sample sizes $(n_{+|g})$ of each sub-population g are sufficiently large. Small values of Q for a given number of degrees of freedom indicate that the model provides an adequate fit to the data.[§]

[†] What represents 'sufficiently large' is a controversial issue, but it should be noted that it may not be as large as some critics of the procedure might suggest. Koch *et al.* (1977, 157) and Imrey *et al.* (1976, 622) consider the issue. They advise that the majority of the proportions $f_{r|g}$ used in forming the observed logits $\bar{L}_{rR|g}$ should be based on five or more observed responses (i.e. $n_{r|g} \geqslant 5$).

[‡] In addition, most simple random samples can be assumed to meet this conditon.

[§] Test statistics such as Q are known as generalized Wald (1943) statistics.

If the goodness-of-fit statistic Q indicates that the model adequately describes the data, tests of linear hypotheses about the parameters in the model can then be undertaken. In particular, for a general null hypothesis (H_0) of the form:

$$H_0 : C\hat{\beta} = 0 \qquad\qquad [4.43]$$

where C is a matrix or vector which is used to specify the hypothesis to be tested[†], and 0 is a vector of zeros a suitable test statistic is:

$$Q_C = (C\hat{\beta})' \left[C(X'V_L^{-1}X)^{-1}C' \right] (C\hat{\beta}) \qquad\qquad [4.44]$$

Under the null hypothesis, and in large samples, this test statistic has approximately a chi-square distribution with degrees of freedom equal to the number of rows in C.

Although test statistic Q_C looks rather complex, it is much easier to apply than it may seem and it can be used, together with suitable forms of the C matrix or vector, to specify the joint tests, partial joint tests, and separate tests which we considered in Section 2.6.

4.3.1   Joint tests

In this case we test the null hypothesis that all the parameters in the model associated with the explanatory variables are equal to zero. For example, in the case of our dichotomous American soldiers linear logit model in Example 4.1 this would imply:

$$H_0 : \beta_2 = \beta_3 = \beta_4 = 0 \qquad\qquad [4.45]$$

To convert this null hypothesis into the general matrix form [4.43] we must specify an appropriate form of matrix C. In this case [4.43] would take the form:

$$H_0 : \begin{bmatrix} 0 & 1 & 0 & 0 \\ 0 & 0 & 1 & 0 \\ 0 & 0 & 0 & 1 \end{bmatrix} \begin{bmatrix} \beta_1 \\ \beta_2 \\ \beta_3 \\ \beta_4 \end{bmatrix} = \begin{bmatrix} 0 \\ 0 \\ 0 \end{bmatrix} \qquad\qquad [4.46]$$

$$H_0 : \qquad C \qquad\qquad \beta \quad = \quad 0$$

The reader should use the principles of matrix multiplication to satisfy himself that the matrix expression in [4.46] is equivalent to the expression in [4.45].

We now substitute the matrix C from [4.46] into the general test statistic [4.44], and also substitute into this statistic the appropriate explanatory variable matrix X, matrix of weights V_L^{-1}, and vector of parameter estimates $\hat{\beta}$. In this case these come from Example 4.1. If the value of the test statistic Q_C we derive from this substitution is greater than the tabulated value of chi-square at, say, the conventional 5 per cent level of significance with degrees of freedom equal to the number

---

[†] More formally we can state that C is a known matrix of full rank.

of rows in $\mathbf{C}$ (in our case 3) we can reject the null hypothesis $\beta_2 = \beta_3 = \beta_4 = 0$ with a 5 per cent chance of error.

4.3.2 *Partial joint tests*

Closely related to the overall joint test are tests in which we consider jointly not *all* the explanatory variable parameters in the model but only a subset. Such tests are very useful, for they enable us to test whether a model can be simplified without a significant loss of explanatory power.

Once again, in the case of Example 4.1, we might wish to test the null hypothesis that the subset of parameters β_3 and β_4 are jointly equal to zero

$$H_0 : \beta_3 = \beta_4 = 0 \qquad [4.47]$$

In this case the general null hypothesis expression [4.43] would take the form:

$$H_0 : \begin{bmatrix} 0 & 0 & 1 & 0 \\ 0 & 0 & 0 & 1 \end{bmatrix} \begin{bmatrix} \beta_1 \\ \beta_2 \\ \beta_3 \\ \beta_4 \end{bmatrix} = \begin{bmatrix} 0 \\ 0 \end{bmatrix} \qquad [4.48]$$

$$H_0 : \qquad \mathbf{C} \qquad\quad \boldsymbol{\beta} \quad = \quad \mathbf{0}$$

We now substitute the matrix $\mathbf{C}$ from [4.48] and the matrices and vectors $\mathbf{X}$, $\mathbf{V_L}^{-1}$ and $\hat{\boldsymbol{\beta}}$ from Example 4.1 into the the general test statistic [4.44]. If the value of the statistic Q_C we derive from this substitution is greater than the tabulated value of chi-square at, say, the conventional 5 per cent level of significance with d.f. = dim $\mathbf{C}$ (i.e. in our case 2), then we can reject the null hypothesis $H_0 : \beta_3 = \beta_4 = 0$ with a 5 per cent chance of error. The implication would then be that the fitted model in Example 4.1 could not be simplified by the removal of the variables X_3 and X_4 (i.e. by the constraining to zero of β_3 and β_4) without a significant loss of explanatory power.

A special form of partial joint test occurs when we wish to test whether various subsets of parameters or pairs of parameters are equal in value. For example, in the case of Example 4.1 we might wish to test the null hypothesis

$$H_0 : \beta_2 = \beta_4 \qquad [4.49]$$

In this case the general expression [4.43] would take the form:

$$H_0 : [0 \quad 1 \quad 0 \quad -1] \begin{bmatrix} \beta_1 \\ \beta_2 \\ \beta_3 \\ \beta_4 \end{bmatrix} = 0 \qquad [4.50]$$

$$H_0 : \qquad \mathbf{C} \qquad\quad \boldsymbol{\beta} \quad = 0$$

where $\mathbf{C}$ is now a row vector and 0 is a scalar (a single element or value). We now substitute $\mathbf{C}$ from [4.50] and $\mathbf{X}$, $\mathbf{V_L}^{-1}$ and $\hat{\boldsymbol{\beta}}$ from Example 4.1 into the general test statistic [4.44]. If the value of Q_C we derive from this substitution is greater than the tabulated value of chi-square at the conventional 5 per cent level of significance with d·f· = dim $\mathbf{C}$ (i.e. our case 1), then we can reject the null hypothesis $H_0: \beta_2 = \beta_4$ with a 5 per cent chance of error. On the other hand, if the value of Q_C is smaller than the tabulated value of chi-square we can suggest that β_3 equals β_4, and this could be an important inference in substantive terms or may be of value in future model refinement and estimation.

4.3.3 Separate tests

These are tests of the value of a single parameter. For example, in the case of the model of Example 4.1 we might wish to test the null hypothesis

$$H_0 : \beta_3 = 0 \qquad\qquad\qquad [4.51]$$

In this case the general expression [4.43] would take the form:

$$H_0: \begin{bmatrix} 0 & 0 & 1 & 0 \end{bmatrix} \begin{bmatrix} \beta_1 \\ \beta_2 \\ \beta_3 \\ \beta_4 \end{bmatrix} = 0 \qquad\qquad [4.52]$$

$$H_0: \qquad \mathbf{C} \qquad\qquad \boldsymbol{\beta} \qquad = 0$$

and we would substitute $\mathbf{C}$ from [4.52] and $\mathbf{X}$, $\mathbf{V_L}^{-1}$ and $\hat{\boldsymbol{\beta}}$ from Example 4.1 into the general test statistic [4.44]. Once again, if the value of Q_C we derive from this substitution is greater than the tabulated value of chi-square at the conventional 5 per cent level of significance with d.f. = dim $\mathbf{C}$ (i.e. in our case 1), then we can reject the null hypothesis $H_0: \beta_3 = 0$ with a 5 per cent chance of error.

An alternative separate test would involve a comparison of the parameter estimate with its estimated standard error derived from $\mathbf{V_{\hat{\beta}}}$, the consistent estimator of the variance–covariance matrix of $\hat{\boldsymbol{\beta}}$ given in equation [4.37]. For example, in the case of null hypothesis [4.51] we would consider the ratio

$$\frac{\hat{\beta}_3}{\sqrt{\mathrm{Var}(\hat{\beta}_3)}} \qquad\qquad\qquad [4.53]$$

This would provide an informal 't statistic' of the type used in Chapters 2 and 3 and values greater than $+2$ or less than -2 would be taken to provide an informal indication of parameters which were 'significantly' different from zero.

EXAMPLE 4.3   *Automobile accidents and the accident environment in North Carolina*

In this example we will illustrate the use of a number of the test statistics discussed above. The example is taken from work reported by Landis *et al.* (1976) and it concerns the relationship between the severity of injury in car accidents in North Carolina and certain potential explanatory variables which characterize the accident environment; variables such as location, time of day, weather conditions, the year and model of the car, and characteristics of the driver. The model we will consider relates to just one particular group of car accidents, those for the years 1966 or 1968–72 which relate to single vehicles travelling at medium speed in open country whose driver was male and who had not been drinking prior to the accident. For this particular group, Table 4.6 presents the proportion of drivers suffering non-severe or severe injury within twelve sub-populations defined on the basis of the cross-classification of three potential explanatory variables: (a) weather conditions (good or bad); (b) time of day (day or night); (c) model year of car (1966 and before, 1967–69 or 1970–73).

To the data of Table 4.6 a dichotomous linear logit model of the form [4.6] can be fitted using the weighted least squares procedure outlined above. This model takes the form:

$$E_A(\bar{L}_{12|g}) = E_A\left[\log_e \frac{f_{1|g}}{f_{2|g}}\right] = \beta_1 + \beta_2 X_{g2}$$

$$+ \beta_3 X_{g3} + \beta_4 X_{g4} + \beta_5 X_{g5}$$

[4.54]

where:

$f_{1|g}/f_{2|g}$ = the odds that injury will be 'non-severe' for car accidents with characteristics g.

X_{g2} = 1 whenever accident occurred in good weather conditions, -1 in bad weather conditions.

Table 4.6
Severity of driver injury in one particular type of North Carolina automobile accident

	Sub-populations			Response categories		
	Weather	Time	Model Year	$r = 1$ Not severe	$r = 2$ Severe	Total
$g = 1$:	Good	Day	–1966	5,633 (0.863)	898 (0.137)	6,531 (1.0)
$g = 2$:	Good	Day	1967–69	2,371 (0.902)	259 (0.098)	2,630 (1.0)
$g = 3$:	Good	Day	1970–73	1,022 (0.911)	100 (0.089)	1,122 (1.0)
$g = 4$:	Good	Night	–1966	7,583 (0.833)	1,526 (0.167)	9,109 (1.0)
$g = 5$:	Good	Night	1967–69	3,314 (0.880)	451 (0.120)	3,765 (1.0)
$g = 6$:	Good	Night	1970–73	1,308 (0.886)	168 (0.114)	1,476 (1.0)
$g = 7$:	Bad	Day	–1966	3,915 (0.902)	428 (0.098)	4,343 (1.0)
$g = 8$:	Bad	Day	1967–69	2,006 (0.931)	149 (0.069)	2,155 (1.0)
$g = 9$:	Bad	Day	1970–73	700 (0.942)	43 (0.058)	743 (1.0)
$g = 10$:	Bad	Night	–1966	3,793 (0.883)	504 (0.117)	4,297 (1.0)
$g = 11$:	Bad	Night	1967–69	1,924 (0.921)	166 (0.079)	2,090 (1.0)
$g = 12$:	Bad	Night	1970–73	718 (0.934)	51 (0.066)	769 (1.0)

X_{g3} = 1 whenever accident occurred during the day, -1 during the night.

X_{g4} = 1 whenever accident involved a car of '1966 or before' model type, -1 whenever accident involved a car of 1970–73 model type, 0 whenever accident involved a car of 1967–69 model type.

X_{g5} = 1 whenever accident involved a car of 1967–69 model type, -1 whenever accident involved a car of 1970–73 model type, 0 whenever accident involved a car of '1966 or before' model type.

In matrix terms the model [4.54] takes the form:

$$E_A(\mathbf{\bar{L}}) = \mathbf{X\beta} \qquad \qquad [4.55]$$

which written out in detail looks very similar to model [4.14] except that in this case there are now twelve sub-populations (g = 1, 2, . . . 12), five parameters, and the matrix $\mathbf{X}$ (using the definitions of the explanatory variables, above) looks as follows:

$$\mathbf{X} = \begin{bmatrix} 1 & 1 & 1 & 1 & 0 \\ 1 & 1 & 1 & 0 & 1 \\ 1 & 1 & 1 & -1 & -1 \\ 1 & 1 & -1 & 1 & 0 \\ 1 & 1 & -1 & 0 & 1 \\ 1 & 1 & -1 & -1 & -1 \\ 1 & -1 & 1 & 1 & 0 \\ 1 & -1 & 1 & 0 & 1 \\ 1 & -1 & 1 & -1 & -1 \\ 1 & -1 & -1 & 1 & 0 \\ 1 & -1 & -1 & 0 & 1 \\ 1 & -1 & -1 & -1 & -1 \end{bmatrix} \qquad [4.56]$$

The reader will clearly have noticed that the definition of the dummy variables in model [4.55] and their representation in the design matrix [4.56] differs from that in Examples 4.1 and 4.2. In this case, it can be seen that the dichotomous explanatory variables (X_2 = weather conditions and X_3 = time of day) are coded 1 and -1 instead of 1 and 0 as in Examples 4.1, 4.2 and Chapter 3. Furthermore, although the three-category explanatory variable 'model year of car' is entered in the usual way as $J-1 = 2$ dummy variables, these variables are coded, 1, 0, -1 and 0, 1, -1 instead of in the 1, 0, 0 and 0, 1, 0 pattern discussed in Chapter 3. These differences result from the fact that the categorical explanatory variables in this example are coded using a method known as 'centred effect' coding, and it should be noted that this new type of coding has an impact on the values of the parameter estimates and how we interpret them. In the following section of this chapter we will discuss this issue in detail. At this stage, however, given that the issue is of only secondary importance to our main objectives in presenting this example, no further details will be provided.

Landis *et at.* (1976, 211) report the following weighted least squares estimates of the parameters of model [4.54] or [4.55]

$$\hat{\beta}_1 = 2.2190 \qquad \hat{\beta}_2 = -0.2075 \qquad \hat{\beta}_3 = 0.1086$$
$$\quad (0.0236) \qquad\qquad (0.0172) \qquad\qquad (0.0159)$$

$$\hat{\beta}_4 = -0.2983 \qquad \hat{\beta}_5 = 0.0949$$
$$\quad (0.0248) \qquad\qquad (0.0293)$$

[4.57]

For this particular type of car accident in North Carolina (single vehicle accidents involving non-drinking males and occurring at medium speed in open country locations) the parameter estimates indicate the following:

1. $\hat{\beta}_1$ indicates the general or overall log-odds that injury will be non-severe rather than severe. (This overall mean term is a product of the 'centred effect coding' employed and will be discussed in the following section);

2. $\hat{\beta}_2$ indicates the differential effect of weather conditions at the time of the accident on the odds that injury will be non-severe rather than severe. The reported value of $\hat{\beta}_2$ is the effect of good weather and its negative is the effect of bad weather. It perhaps comes as something of a surprise to the layman in this field of study that, other factors being held constant, good weather reduces the odds of injury being non-severe (i.e. $\hat{\beta}_2$ has a negative sign) whilst bad weather increases the odds of injury being non-severe (i.e. the negative of $\hat{\beta}_2$ has a positive sign). The reasons for this result are beyond the scope of this brief example and the reader must refer to the original sources for further discussion.

3. $\hat{\beta}_3$ indicates the differential effect of time of day of accident on the odds that injury will be non-severe rather than severe. The reported value of $\hat{\beta}_3$ is the effect of day-time accidents. Other factors being held constant, day-time accidents are shown to increase the odds that injury will be non-severe (i.e. $\hat{\beta}_3$ has a positive sign) and night-time accidents to reduce the odds that accident will be non-severe;

4. $\hat{\beta}_4$ and $\hat{\beta}_5$ indicate the differential effect of '1966 or before' and '1967–69' model types on the odds that injury will be non-severe rather than severe. The effect ($\hat{\beta}_6$) of '1970–73' model types, which is not shown, can be calculated as the negative of the sum of $\hat{\beta}_4$ and $\hat{\beta}_5$ (i.e. $\hat{\beta}_6 = -(\hat{\beta}_4 + \hat{\beta}_5) = -(-0.2983 + 0.0949) = 0.2034$). Other factors being held constant, it appears that older types of car (i.e. older models) reduce the odds that injury will be non-severe whereas more recent models increase the odds that injury will be non-severe. The reasons for this relate to improvements in car design with respect to driver safety which were introduced during the 1960s and these issues are considered in more detail by Koch and Reinfurt (1974).

As stated above, one of the major purposes of this example is to illustrate the application of some of the test statistics discussed in the previous section. We will begin, therefore, by considering the goodness-of-fit of model [4.54] to the data of Table 4.6. We can determine this by substituting the **X** matrix [4.56], the parameter estimates [4.57] and the appropriate $V_{\bar{L}}^{-1}$ matrix of weights into equation [4.41], and computing the value of Q. In our case $Q = 1.9836$. Under the hypothesis that the model fits, Q has approximately a chi-square distribution with d.f. = dim $(\bar{L})$ − dim $(\hat{\beta})$, in our case 12−5 = 7. The tabulated value of chi-square for 7 d.f. at the conventional 5 per cent level of significance is 14.067, whereas $Q = 1.9836$. This indicates that model [4.54] fits the data of Table 4.6 satisfactorily.

Having accepted that model [4.54] provides a satisfactory fit to the data of Table 4.6, we can then consider separate, partial joint or joint tests of the type discussed above. In this case we will consider tests of the following null hypotheses:

(a) $H_0 : \beta_2 = 0$ (i.e. there is no significant weather condition effect)

(b) $H_0 : \beta_3 = 0$ (i.e. there is no significant time of day effect)

(c) $H_0 : \beta_4 = \beta_5 = 0$ (i.e. there is no significant car type effect)

(d) $H_0 : \beta_4 = \beta_5$ (i.e. the two car type dummy variable parameters are equal).

To test these null hypotheses using our general test statistic [4.44] we must specify appropriate C matrices or vectors. In our case these are:

(a) $C = \begin{bmatrix} 0 & 1 & 0 & 0 & 0 \end{bmatrix}$

(b) $C = \begin{bmatrix} 0 & 0 & 1 & 0 & 0 \end{bmatrix}$

(c) $C = \begin{bmatrix} 0 & 0 & 0 & 1 & 0 \\ 0 & 0 & 0 & 0 & 1 \end{bmatrix}$

(d) $C = \begin{bmatrix} 0 & 0 & 0 & 1 & -1 \end{bmatrix}$

and the associated Q_C test statistics are respectively:

(a) $Q_C = 146.10$;   (c) $Q_C = 157.79$;
(b) $Q_C = 46.52$;   (d) $Q_C = 107.91$

In all cases the value of Q_C is greater than the tabulated value of chi-square at the conventional 5 per cent level of significance with 1, 1, 2 and 1 degrees of freedom respectively; and we can reject each of the null hypotheses (a) to (d) with a 5 per cent chance of error.

4.4 Coding systems for categorical explanatory variables

In Example 4.4 we came into contact for the first time with an alternative method of coding categorical explanatory variables. The new method is termed 'centred effect' or simply 'effect' coding, and in the dichotomous case it involves coding one category +1 (or simply 1 for convenience) and the other category −1. This type of coding has strong links with analysis of variance models and it is the coding method most widely employed in the many important contributions to be found in the biometrics literature on the theory and application of cell (f) linear logit models (see for example Grizzle, Starmer and Koch 1969, and the work of Koch and his associates).

In contrast, the coding system we discussed in Chapter 3 (which in the dichotomous case involves coding one category 1 and the other category 0) is termed 'dummy' coding or 'cornered effect' coding. This type of coding is dominant in the econometrics literature, and in the discussion of dummy variable regression models. It is also the coding system most widely employed in geographical applications of categorical explanatory variables.

The names 'centred effect' and 'cornered effect' coding are related particularly to the interpretation of the $\hat{\beta}_1$ parameter estimate in our models (see Kritzer 1978; Reynolds 1977). In the case of 'centred effect' coding, the value of $\hat{\beta}_1$ is an estimate of the grand or overall mean of the log odds across all the sub-populations. On the other hand, in the case of 'cornered effect' or 'dummy' coding, the value of $\hat{\beta}_1$ is an estimate of the log odds for a single sub-population: the so-called 'anchor' or 'base' group (i.e. the 'corner' group) coded zero on all dummy variables. The other 'main effect' (i.e. non-interaction) parameter estimates in a 'centred effect' coded model are interpreted (as in analysis of variance models) as differential effects or deviations around the overall mean, whereas in a 'cornered effect' model they are interpreted (as we saw in Ch. 3 and in Example 4.1) as differences from the 'anchor' or 'base' group.

To illustrate some of these comments, let us now reconsider model [4.12] from Example 4.1. First, we will write out the design matrix **X** in both 'centred effect' and 'cornered effect' coding:

$$
\mathbf{X} = \begin{bmatrix} 1 & -1 & -1 & -1 \\ 1 & -1 & -1 & 1 \\ 1 & -1 & 1 & -1 \\ 1 & -1 & 1 & 1 \\ 1 & 1 & -1 & -1 \\ 1 & 1 & -1 & 1 \\ 1 & 1 & 1 & -1 \\ 1 & 1 & 1 & 1 \end{bmatrix} \quad \mathbf{X} = \begin{bmatrix} 1 & 0 & 0 & 0 \\ 1 & 0 & 0 & 1 \\ 1 & 0 & 1 & 0 \\ 1 & 0 & 1 & 1 \\ 1 & 1 & 0 & 0 \\ 1 & 1 & 0 & 1 \\ 1 & 1 & 1 & 0 \\ 1 & 1 & 1 & 1 \end{bmatrix} \quad \text{[4.58]}
$$

where columns 2, 3 and 4 in both matrices are ORG, RACE and PLOC respectively. We can then re-estimate model [4.12] using the same weighted least squares procedure as in Example 4.1, but now with the 'centred effect' coded design matrix. The parameter estimates derived from this estimation using the 'centred effect' design matrix can be compared to those derived in Example 4.1 using the 'cornered effect' design matrix as follows:

Cornered effect $\hat{\beta}_1 = -2.74$ $\hat{\beta}_2 = 2.60$ $\hat{\beta}_3 = 0.76$ $\hat{\beta}_4 = 1.54$
                  (0.08)   (0.06)   (0.06)   (0.06)

Centred effect $\hat{\beta}_1 = -0.29$ $\hat{\beta}_2 = 1.30$ $\hat{\beta}_3 = 0.38$ $\hat{\beta}_4 = 0.77$
                  (0.03)   (0.03)   (0.03)   (0.03)

[4.59]

As we noted in Example 4.1, $\hat{\beta}_1$ in the 'cornered effect' model can be taken to represent our estimate of the log odds of expressing a preference for a southern army camp amongst negroes of northern origin who are presently located in a northern camp. $\hat{\beta}_2$, $\hat{\beta}_3$ and $\hat{\beta}_4$ in the 'cornered effect' model provide us with estimates of the effect on these 'anchor category' log odds if soldiers are of southern rather than northern origin, are white rather than negro, and are presently located in a southern rather than a northern camp. In the 'centred effect' model, $\hat{\beta}_1$ represents an estimate of the overall or grand mean of the log odds of expressing a preference for a southern army camp, where the overall mean is computed across all eight sub-populations. $\hat{\beta}_2$ then represents the differential effect of region of origin on the overall mean log odds of expressing a preference for a southern camp. The reported value of $\hat{\beta}_2$ (1.30) is the differential effect of southern origin, and its negative (-1.30) is the differential effect of northern origin. Similarly, $\hat{\beta}_3$ and $\hat{\beta}_4$ represent the differential effects of race and location of present camp on the overall mean log odds of expressing a preference for a southern army camp. The reported values of $\hat{\beta}_3$ and $\hat{\beta}_4$ (0.38 and 0.77) represent the differential effects of being white and being currently located in a southern camp, and their negatives (-0.38 and -0.77) represent the differential effects of being negro and being currently located in a northern camp. In this simple example, where all the explanatory variables are dichotomous, it can be seen that $\hat{\beta}_2$, $\hat{\beta}_3$ and $\hat{\beta}_4$ in the 'centred effect' model are exactly half those in the 'cornered effect' model. However, the reader must note that this simple relationship between the two sets of parameters does not extend to the case in which the explanatory variables are multiple-category rather than dichotomous.

Despite the fact that different sets of parameter estimates are produced by the different coding methods, the reader should note that both methods produce identical predictions of the log odds (i.e. predicted logits) from the equation:

$$\hat{\mathbf{L}} = \mathbf{X}\hat{\boldsymbol{\beta}} \qquad\qquad [4.60]$$

To see this, we simply need to multiply the matrices in [4.58] by the associated parameter estimates in [4.59]. In both cases the results are as shown in Table 4.7.

Table 4.7
Predicted logits (log odds) and predicted probabilities from model [4.12]: both coding methods

Sub-populations			Predicted logit $\hat{L}_{12|g}$	Predicted probability $\hat{P}_{1|g}$
Region of origin	Race	Present camp		
g = 1: North	Negro	North	−2.737	0.061
g = 2: North	Negro	South	−1.202	0.232
g = 3: North	White	North	−1.979	0.121
g = 4: North	White	South	−0.445	0.391
g = 5: South	Negro	North	−0.135	0.467
g = 6: South	Negro	South	1.400	0.803
g = 7: South	White	North	0.623	0.651
g = 8: South	White	South	2.158	0.897

Furthermore, it follows that if the predicted logits from both coding methods are identical, then the predicted probabilities derived by substituting into the equations

$$\hat{P}_{1|g} = \frac{e^{x'_g\beta}}{1+e^{x'_g\beta}} \quad \text{and} \quad \hat{P}_{2|g} = \frac{1}{1+e^{x'_g\beta}} \qquad [4.61]$$

must also be identical. Table 4.7 shows the $\hat{P}_{1|g}$ values, the predicted probabilities of preferring a southern army camp. In addition, the goodness-of-fit statistic Q [4.41] for both 'cornered effect' and 'centred effect' models will be identical (in both cases in Example 4.1, $Q = 25.55$) and all hypothesis tests will produce identical Q_C statistics [4.44].

We saw in Chapter 3 that multiple-category explanatory variables can be handled by introducing one less dummy variable than the number of categories. In the case of a three-category explanatory variable (categories A, B, C) we would therefore introduce two dummy variables. Assuming X_{g1} is the constant term, in 'cornered effect' coding these would be defined as:

$X_{g2} = \quad$ 1 if sub-population g includes individuals of category A
$\qquad\qquad$ 0 otherwise

$X_{g3} = \quad$ 1 if sub-population g includes individuals of category B
$\qquad\qquad$ 0 otherwise

In 'centred effect' coding the variables would be defined as:

$X_{g2} = \quad$ 1 if sub-population g includes individuals of category A,
$\qquad\qquad$ 0 if sub-population g includes individuals of category B,
$\qquad\quad$ -1 if sub-population g includes individuals of category C.

$X_{g3} = \quad$ 1 if sub-population g includes individuals of category B,
$\qquad\qquad$ 0 if sub-population g includes individuals of category A,
$\qquad\quad$ -1 if sub-population g includes individuals of category C.

If our sub-populations $g = 1, \ldots 6$ included individuals of categories A, B, C, A, B, C respectively then the 'centred effect' and 'cornered effect' coded design matrices would look as follows:

$$\mathbf{X} = \begin{bmatrix} 1 & 1 & 0 \\ 1 & 0 & 1 \\ 1 & -1 & -1 \\ 1 & 1 & 0 \\ 1 & 0 & 1 \\ 1 & -1 & -1 \end{bmatrix} \qquad \mathbf{X} = \begin{bmatrix} 1 & 1 & 0 \\ 1 & 0 & 1 \\ 1 & 0 & 0 \\ 1 & 1 & 0 \\ 1 & 0 & 1 \\ 1 & 0 & 0 \end{bmatrix} \qquad [4.62]$$

In models in which there are no interaction effects (for further discussion of interaction see next section) the choice between the two coding methods is simply a matter of personal taste. In models which include interaction terms, it is suggested by some writers (e.g. Kritzer 1978) that 'centred effect' coding is preferable as it is not as sensitive as 'cornered effect' coding to arbitrary reversals of coding (i.e. +1 becoming −1 and vice versa, or 1 becoming 0 and vice versa). In addition, many of the published examples of cell (f) linear logit models which include interaction terms come from those subject areas which by tradition use 'centred effect' coding. As a result, from this point onwards, we will tend to adopt 'centred effect' coding in our discussion. However, the reader will encounter examples of both types of coding in his wider reading and he should also be aware that some of the most widely available computer programs will adopt 'centred effect' coding; whilst others (e.g. GLIM) will adopt 'cornered effect' coding. Clearly, the reader must attempt to become familiar with both coding systems.

4.5 Interactions, saturation, hierarchies and parsimony

In all the examples discussed so far in this chapter, we have considered models in which only main effect parameters are included. In other words, explanatory variables have entered our models in the simplest additive fashion and we have considered none of the interactions which might possibly exist amongst the explanatory variables. In the case of model [4.12], for example, although we considered only the main effects of the variables ORG, RACE and PLOC there may also be potentially important interactions between the variables ORG and RACE or between variables RACE and PLOC, or between ORG and PLOC. The discovery that any one of these interactions was important would indicate, in the context of model [4.12], that the form or magnitude of the relationship between a particular explanatory variable and the log odds of expressing a preference for a southern army camp was dependent upon the categories of another explanatory variable. For example, a significant interaction between ORG and RACE would indicate that the nature of the relationship between a soldier's region of origin and the odds of expressing a preference for a southern army camp was dependent upon the racial group (white or negro) to which the soldier belonged. Furthermore, in addition to these possible pairwise interactions, a complex or higher-order interaction might possibly exist between all three explanatory variables, ORG, RACE and PLOC.

The 'centred effect' coded design matrix for model [4.12] has been described in expression [4.58]. Let us now consider what the form of the design matrix will be should we extend model [4.12] to include: (a) an interaction effect between ORG and RACE; (b) an interaction effect between ORG and PLOC; (c) an interaction effect between RACE and PLOC. These extended design matrices are shown as (a), (b) and (c) on page 137.

It can clearly be seen that the interaction term (the final column in each matrix) is obtained by multiplying appropriate pairs of explanatory variables. Should we wish to include more than one interaction term into our model we would simply add extra columns to the appropriate matrix in [4.63]. For example, should we wish to include the interaction

(a)   (b)   (c)

$$
\begin{bmatrix}
1 & -1 & -1 & -1 & 1 \\
1 & -1 & -1 & 1 & 1 \\
1 & -1 & 1 & -1 & -1 \\
1 & -1 & 1 & 1 & -1 \\
1 & 1 & -1 & -1 & -1 \\
1 & 1 & -1 & 1 & -1 \\
1 & 1 & 1 & -1 & 1 \\
1 & 1 & 1 & 1 & 1
\end{bmatrix}
\begin{bmatrix}
1 & -1 & -1 & -1 & 1 \\
1 & -1 & -1 & 1 & -1 \\
1 & -1 & 1 & -1 & 1 \\
1 & -1 & 1 & 1 & -1 \\
1 & 1 & -1 & -1 & -1 \\
1 & 1 & -1 & 1 & 1 \\
1 & 1 & 1 & -1 & -1 \\
1 & 1 & 1 & 1 & 1
\end{bmatrix}
\begin{bmatrix}
1 & -1 & -1 & -1 & 1 \\
1 & -1 & -1 & 1 & -1 \\
1 & -1 & 1 & -1 & -1 \\
1 & -1 & 1 & 1 & 1 \\
1 & 1 & -1 & -1 & 1 \\
1 & 1 & -1 & 1 & -1 \\
1 & 1 & 1 & -1 & -1 \\
1 & 1 & 1 & 1 & 1
\end{bmatrix}
$$

[4.63]

effect between RACE and PLOC within matrix (a) we would simply add the final column of matrix (c) to matrix (a).

If we were to add all three possible pairwise interaction effects into our basic model [4.12] and, in addition, were to add the higher-order interaction effect between ORG, RACE and PLOC we would specify a special type of model whose design matrix would take the form:

$$
\mathbf{X} =
\begin{bmatrix}
1 & -1 & -1 & -1 & 1 & 1 & 1 & -1 \\
1 & -1 & -1 & 1 & 1 & -1 & -1 & 1 \\
1 & -1 & 1 & -1 & -1 & 1 & -1 & 1 \\
1 & -1 & 1 & 1 & -1 & -1 & 1 & -1 \\
1 & 1 & -1 & -1 & -1 & -1 & 1 & 1 \\
1 & 1 & -1 & 1 & -1 & 1 & -1 & -1 \\
1 & 1 & 1 & -1 & 1 & -1 & -1 & -1 \\
1 & 1 & 1 & 1 & 1 & 1 & 1 & 1
\end{bmatrix}
$$

[4.64]

(The reader should notice that the final column, the higher-order interaction term, is derived by multiplying together columns 2, 3 and 4, i.e. the variables ORG, RACE and PLOC. He should also be aware of the fact that associated with each column in this matrix will be a parameter, β_1, β_2, ... β_8 respectively). This special type of model is known as the *saturated* model, and the name *saturated* implies that the number of parameters in the model equals the number of logit terms $\bar{L}_{12|g}$, in the logit vector L in the matrix expression [4.14] of model [4.12]. In Example 4.1 we have eight sub-populations and eight observed logit values $\bar{L}_{12|g}$ in [4.14]. Associated with the design matrix [4.64] are eight parameters, one for each column. The number of parameters equals the number of observed logit values and thus the extended model defined by design matrix [4.64] is a saturated model. Clearly, as the number of parameters equals the number of observed logit values, the dimension of $\bar{L}$ equals the dimension of $\boldsymbol{\beta}$. Thus, in the goodness-of-fit statistic [4.41] there are *no* degrees of freedom and the value of Q will be zero. In other words, the saturated model will fit the observed data perfectly.

It appears, therefore, that given contingency tables such as Tables 4.1 and 4.2, we can fit linear logit models which include none, some, or all possible interactions between the explanatory variables and, unlike the situation in Chapters 2 and 3, we can readily specify the exact form of a model which must fit our original data perfectly. However, a question must be posed as to the value of a model which fits perfectly but has no degrees of freedom. A characteristic feature of all modelling in social and environmental science is a quest for the insight to be gained in the process of simplifying complex reality. What is the value of a model which replicates perfectly but does not simplify, which has as many parameters as original pieces of information (in our case as many parameters as observed log odds)? The answer to this question is that the saturated model is not an end in itself, it is a point at which we can *begin* the process of reduction to a parsimonious form.

When we have evidence, or suspect, that interactions between our explanatory variables are present, we would normally begin by first fitting the saturated model. We would then attempt, on the basis of the results of hypothesis tests using statistic Q_C given in equation [4.44], to reduce the number of parameters in that model and derive a second, more parsimonious, 'reduced-form' model. Usually, the way we would attempt to carry out this process is by employing a stepwise 'hierarchical' elimination procedure. In this procedure we would first consider whether it is possible to eliminate the highest order interaction effect (i.e. whether we are justified in constraining to zero its associated parameter). If so, we would eliminate it and consider whether it is possible to eliminate the interaction effects of the next highest order, and so on, until we find a level of interaction effects which we are not justified in eliminating. The actual details of how we do this need not concern us at this stage (we will consider them later in this chapter). It is sufficient at this stage to appreciate that the saturated model is not used as an end in itself but as a starting point in a process of creating a parsimonious representation of the interactions (if any) between the explanatory variables. In addition, it is important to note the hierarchical nature of the elimination procedure which is usually adopted.

The previous point is an important one, for in categorical data analysis in general, and in our discussion of log-linear models in Chapters 5 and 6 in particular, the reader will often encounter the concept of a hierarchical set of models. In rather formal terms, a hierarchical set of models can be defined as being composed of models in which higher-order interaction terms are only included if their lower-order relatives are also included. For example, in the case of our 'preference for army camp location amongst American soldiers' model, the interaction effect ORG × RACE × PLOC would, by this definition, only be included if all its lower-order relatives ORG, RACE, PLOC, ORG × RACE, ORG × PLOC and RACE × PLOC were also included: it could not be included if, say, ORG × RACE or PLOC were excluded. Conversely, if PLOC is excluded from the model, its higher-order relatives ORG × PLOC, RACE × PLOC and ORG × RACE × PLOC must also be excluded. Non-hierarchical models, on the other hand, are models which break this higher-lower-order relative principle (e.g.

models which include combinations of terms such as ORG+
RACE+RACE×PLOC   or   ORG+RACE+PLOC+ORG×PLOC+
ORG×RACE×PLOC). Considerable, and often bitter, controversy
surrounds the issue of the appropriateness of non-hierarchical models (see
Evers and Namboodri 1978; Fienberg 1978; Magidson et al. 1981, for
further discussion), and the controversy involves broader questions such as
the multiple-inference problem, the construction of simultaneous confidence intervals and the design of rigorous simultaneous testing procedures.
For our purposes at this stage it is sufficient to note this, and to by-pass the
issue by stressing the hierarchical nature of the elimination procedure which
is usually adopted.

Finally, the reader should note that although we normally represent
the hierarchy of potential interactions in the form shown in design
matrix [4.64], this is not the only possibility. In experimental design
terminology, design matrix [4.64] represents a crossed factorial design
but there is nothing sacrosanct in this. We can also consider using nested
designs, or combinations of the two and so on. A nested design matrix
is one in which by layering or nesting one effect inside another we can
create separate effects for different sub-groups. These separate effects
are then often referred to as 'conditional' effects because the design
matrix represents the conditional effect of one explanatory variable
within levels of another explanatory variable.

To illustrate this, let us reformulate design matrix [4.64] by
suggesting that the soldier's region of origin variable ORG is the
dominant factor or controlling explanatory variable, and nesting the
other explanatory variables RACE and PLOC within the categories of
variable ORG. The design matrix would then take the following form:

$$\mathbf{X} = \begin{bmatrix} 1 & -1 & -1 & -1 & 1 & 0 & 0 & 0 \\ 1 & -1 & -1 & 1 & -1 & 0 & 0 & 0 \\ 1 & -1 & 1 & -1 & -1 & 0 & 0 & 0 \\ 1 & -1 & 1 & 1 & 1 & 0 & 0 & 0 \\ 1 & 1 & 0 & 0 & 0 & -1 & -1 & 1 \\ 1 & 1 & 0 & 0 & 0 & -1 & 1 & -1 \\ 1 & 1 & 0 & 0 & 0 & 1 & -1 & -1 \\ 1 & 1 & 0 & 0 & 0 & 1 & 1 & 1 \end{bmatrix} \qquad [4.65]$$

It can be seen that design matrix [4.65] consists of a main effect for
ORG (column 2) and two sets of effects for RACE, PLOC and
RACE×PLOC; one set for soldiers from a northern region of origin
(column 3, 4 and 5), one set for soldiers of a southern region of origin
(columns 6, 7 and 8). The two sets of effects for RACE, PLOC and
RACE×PLOC are conditional upon the category of ORG. Further
illustrations and discussion of such nested design matrices and conditional effects are provided by Evers and Namboodri (1978), Kritzer
(1978) and Lehnen and Koch (1974a).

<u>*EXAMPLE 4.4*</u>   *Determinants of home ownership in Boston and Baltimore*

As an illustration of a cell (f) linear logit model in which interactions between explanatory variables are expected *a priori*, we will consider a study by Li (1977) of the determinants of home ownership in the Boston and Baltimore SMSAs. In this study, Li began by fitting an additive, main-effects-only model of the form:

$$E(\bar{L}_{12|g}) = E_A\left[\log_e \frac{f_{1|g}}{f_{2|g}}\right] = \beta_1 + \sum_{k=2}^{5}\beta_k X_{gk} + \sum_{k=6}^{8}\beta_k X_{gk} \\ + \sum_{k=9}^{11}\beta_k X_{gk} + \beta_{12}X_{g12}$$

[4.66]

where

$f_{1|g}/f_{2|g}$ = the odds that households of type g will be home owners rather than renters.

X_{g2} to X_{g5} = a set of four dummy variables representing five age of head of household categories.

X_{g6} to X_{g8} = a set of three dummy variables representing four income classes.

X_{g9} to X_{g11} = a set of three dummy variables representing four family size classes.

X_{g12} = a dummy variable denoting race of head of household (1 if black, 0 if non-black).

Using data for husband–wife families in the Boston and Baltimore SMSAs taken from the tables of metropolitan housing characteristics in the 1970 Census of Housing, Li fitted model [4.66] using the weighted least squares procedure. From this he derived the parameter estimates shown in Table 4.8. In this table the parameter estimate for the 'anchor' category in each variable is explicitly stated as zero and the estimated standard errors of the parameter estimates are shown in brackets to the right of the parameter estimate.

The parameter estimates in Table 4.8 indicate that:
(a) the odds of home ownership rather than renting increase progressively as the age of the head of household increases;
(b) the odds of home ownership rather than renting increase progressively as income increases;
(c) the odds of home ownership rather than renting first increase as family size increases but then decline as family size exceeds five persons (the household budget constraint would appear to compel the largest families to sacrifice housing consumption for non-housing consumption);
(d) the odds of home ownership rather than renting are significantly reduced if the head of the household is black.

On the basis of existing theories of housing consumption, Li anticipated that the main-effects-only model [4.66] would not be adequate and that significant interactions between the explanatory variables were to be expected. He anticipated that four of the interactions likely to be particularly important were: (a) between age of head of household and family size because of life-cycle relationships; (b)

Table 4.8
Parameter estimates for model [4.66], after Li (1977)

Variable	Boston SMSA	Baltimore SMSA
Constant (under 25 years, less than $5,000, two-person, white)	−2.699 (0.025)	−2.157 (0.022)
Age of head:		
Under 25	0.0	0.0
25–34	1.318 (0.024)	1.176 (0.020)
35–44	2.278 (0.024)	2,149 (0.020)
45–65	2.506 (0.023)	2.601 (0.019)
65 or older	2.699 (0.024)	2.803 (0.022)
Income:		
Less than $5,000	0.0	0.0
$5,000–9,999	0.233 (0.012)	0.324 (0.014)
$10,000–14,999	0.858 (0.012)	0.894 (0.014)
$15,000 or more	1.365 (0.013)	1.211 (0.015)
Family size		
Two persons	0.0	0.0
Three and four persons	0.520 (0.008)	0.590 (0.010)
Five persons	1.024 (0.011)	0.796 (0.014)
Six or more persons	0.972 (0.011)	0.469 (0.013)
Race of head:		
White (Non-black)	0.0	0.0
Black	−1.129 (0.019)	−1.155 (0.010)
Total Number of Families	544,781	420,264

Note: Numbers in parentheses are asymptotic standard errors.

between income and family size because of household budget constraints; (c) between age of head of household and income because their joint effect might serve as a proxy for wealth; (d) between race and income because of both the income effect and the substitution effect on the consumption of black households who face a higher relative price for housing as a result of racial discrimination.

To test the significance of these potential interactions, Li estimated a set of additional logit models which included a wide range of the possible interaction terms. It should be noted, however, that he did *not* do this in the manner suggested in the previous section. He did not first fit a saturated model and then eliminate unnecessary interactions using some form of hierarchical procedure. Nevertheless, it is useful for our purposes to consider just one of these models as an illustration of the nature of interaction between explanatory variables.

The model we will consider is the one in which Li allowed an interaction to exist between the age of the head of the household variable and the family size variable. In terms of our discussion in the previous section of the chapter, the model would be expected to take the form CONSTANT (1) + AGE (2–5) + INCOME (6–8) + SIZE (9–11) + RACE (12) + AGE × SIZE (13–24), where the columns of the associated design matrix are listed in brackets following

Table 4.9
Parameter estimates for
AGE × SIZE interaction

Age of head of household	Family size							
	Boston				**Baltimore**			
	2	3–4	5	6+	2	3–4	5	6+
Under 25	0.0	0.566	1.445	1.835	0.0	0.605	1.285	1.287
25–34	0.801	1.934	2.679	2.650	0.799	1.845	2.172	2.030
35–44	1.962	2.946	3.369	3.367	2.039	2.826	3.066	2.660
45–64	2.746	3.060	3.364	3.295	2.802	3.223	3.162	2.919
65 or older	2.826	3.238	2.824	3.106	2.936	3.383	3.171	2.741

Fig. 4.1
Parameter estimates for
Boston from Table 4.9 (after Li 1977)

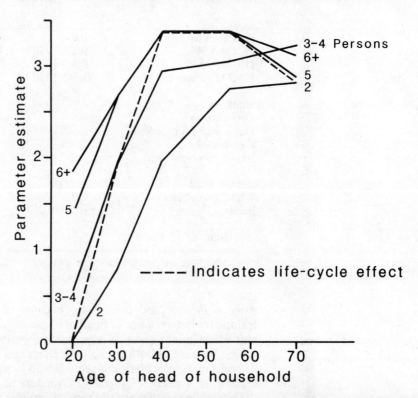

the name of the variable. In other words, the associated design matrix would be expected to have columns 13 to 24 which were the products of columns 2 to 5 and 9 to 11. Rather confusingly for our purposes, however, Li chose to adopt a different (but equivalent and perfectly acceptable) method of specifying the model and the design matrix. The specification of Li's model took the form CONSTANT (1)+INCOME (2–4)+RACE (5)+AGE×SIZE (6–24). In this case, instead of having 4+3 main effect terms for AGE and SIZE plus twelve interaction terms for AGE×SIZE, he simply used nineteen terms [†] to represent the combined effects of main and interaction terms.

[†] Nineteen terms because there are five age categories and four size categories in the original data, thus there are twenty categories of AGE × SIZE and one of these must be set to zero because a constant term is included in the model and perfect linear dependence with this would be created if all twenty AGE × SIZE terms were to be included. (See Ch. 3 for further discussion of this issue.)

The nineteen AGE × SIZE parameter estimates which Li derived from fitting this model are shown in Table 4.9 (the INCOME and RACE parameter estimates are not shown), and these estimates are graphed for the case of Boston in Figure 4.1. The nature of the interaction between AGE and SIZE stands out clearly in both table and diagram. The nature of the relationship between age of the head of the household and the odds of being a home-owner rather than renter is clearly dependent upon the family size class to which the household belongs. The life-cycle effect which causes the interaction is indicated by the dashed line in Fig. 4.1, and it can be observed in the parameter estimates of Table 4.9 by considering the progression across typical family sizes under the life-cycle hypothesis at various ages. In the case of Boston, this would give: (a) under 25 years old: two persons (parameter estimate = 0.0); (b) 25–34 years old: three to four persons (parameter estimate = 1.934); (c) 35–44 years old: five persons (3.369); (d) 45–64 years old: three to four persons (3.060); (e) 65 years or older: two persons (2.826).

4.6 Model selection

Earlier in this chapter we considered the issue of how to reduce a saturated cell (f) linear logit model to a more parsimonious 'reduced' form. It was suggested that this would involve a parameter elimination procedure which was hierarchical in nature, but no details of such a procedure were provided at that point. In the case of Example 4.1 or similar cases in which we have only two or three explanatory variables, it is a relatively simple matter to consider specifying all possible members of the hierarchical set of models and to consider devising an *ad hoc* elimination procedure which is hierarchical in nature. In cases in which there are only two explanatory variables there are only five possible models in the hierarchical set, and in cases in which there are only three explanatory variables there are only nineteen possible models in the hierarchical set. However, in the case in which there are four explanatory variables there are 167 possible models in the hierarchical set, and this figure rises dramatically for models with more than four explanatory variables. In these circumstances something rather more sophisticated and systematic than an *ad hoc* hierarchical elimination procedure is required to derive the desired parsimonious 'reduced' form model.

In Chapters 5 and 6 we will see that there are now some sophisticated and highly-regarded procedures which we can adopt to select the most appropriate log-linear model from a hierarchical set of such models. However, in the case of cell (f) linear logit models the choice of model selection procedures is less rich. Nevertheless, some systematic model selection procedures have been suggested, and we will concentrate in this chapter on a method suggested by Higgins and Koch (1977).

The procedure suggested by Higgins and Koch has two stages.

Stage one. At this stage, a range of explanatory variables, which on the basis of theory or the results of previous investigations are considered to have a relationship with the response variable, are screened. The aim of this screening is to select those explanatory variables responsible for the greatest amount of variation in response proportions amongst the sub-populations. The overall purpose is to reduce the size of the analysis to more manageable proportions by identifying a 'reasonable' subset of explanatory variables and focussing attention on the relationship between just this chosen subset of explanatory variables and the response variable. The chosen subset is said to be 'reasonable' in the sense that, once its constituent explanatory variables have been taken into account, none of the remaining original explanatory variables has significant effects. In addition, the subset is selected so as to maximize the explanatory capability of its constituent variables relative to the associated degrees of freedom.

The screening procedure relies on appropriately constructed Pearson chi-square statistics divided by their degrees of freedom. These are used as measures of the relative importance of certain combinations of variables in a multivariate relationship and, for those readers familiar with the technique, they can be thought of as being somewhat similar to the 'F–to–enter' statistics in stepwise multiple regression. The first explanatory variable selected as a member of the 'reasonable' subset is the one which has the largest chi-square per degree of freedom with regard to its first-order relationship to the response variable. Other explanatory variables are then selected as members of the subset by applying similar selection rules, using chi-square per degree of freedom for successively higher-order relationships.

The screening procedure also includes a method for terminating the selection process when the remaining explanatory variables are not statistically important enough to be included as members of the 'reasonable' subset. Two types of statistics are used to decide when to terminate selection: (i) Pearson chi-square statistics for the relationship of a specific explanatory variable with the response variable, summed over all possible combinations of explanatory variables that have already been selected; (ii) a statistic developed by Cochran (1954) and Mantel and Haenszel (1959), subsequently modified by Koch and Reinfurt (1974), (see also Koch 1978; Landis et al. 1978), possessing a chi-square distribution with one degree of freedom. This combines information with respect to the effect of a specific explanatory variable on the response variable over all possible combinations of the explanatory variables previously selected as members of the 'reasonable' subset. A termination statistic of type (i) reflects both the main effects of a specific explanatory variable and its interaction with previously selected explanatory variables. After the first few steps of the screening procedure, however, statistic (i) tends to lose its usefulness since the degrees of freedom increase rapidly making the selection process excessively stringent, and the data become so thinned that this statistic tends to lose its validity as a chi-square statistic. At this point, termination statistic (ii) is then used because it combines information across all combinations of previously selected explanatory variables and is not affected as much by the data thinning. It is also a highly sensitive tool for detecting the weak but consistent relationships which those explanatory variables not yet selected as members of the 'reasonable' subset may have with the response variable. In the case of both termination statistics, the criteria for terminating the selection process is *failure* to meet the conventional 5 per cent (0.05) or 10 per cent (0.10) levels of significance (i.e. termination X^2 statistics which fall below the critical values).

As a brief illustration of this first stage of the Higgins–Koch procedure, let us imagine a case in which we have a dichotomous response variable (A) and seven potential explanatory variables (B, C, D, E, F, G, H) with (2, 3, 2, 2, 2, 2, 2) categories respectively. Table 4.10 then shows hypothetical Pearson chi-square statistics for first, second, third and fourth order relationships between the response variable (A) and the explanatory variables ($B - H$), also degrees of freedom, termination statistics, etc.

The first-order relationships indicate that explanatory variable C has the largest chi-square per degree of freedom, and thus C becomes the first explanatory variable selected as part of the 'reasonable' subset. The next step is to determine which of the remaining six variables is to be selected for the subset once variable C has been adjusted for. The second-order relationships indicate that it is explanatory variable E which has the largest chi-square per degree of freedom and so, given that the termination statistic (which is of type i – see above) is significant at the conventional 5 per cent level, E becomes the second explanatory variable selected as part of the 'reasonable' subset. The next step is to

Table 4.10
Test statistics for an hypothetical example of stage one explanatory variable selection process

	Inclusion Statistics			Termination Statistics		
	X^2	d.f.	$X^2/d.f.$	X^2	Type	d.f.
First-order relationships						
$A \times B$	15.23	1	15.23			
$A \times C$	48.35	2	24.18			
$A \times D$	4.43	1	4.43			
$A \times E$	19.45	1	19.45			
$A \times F$	11.08	1	11.08			
$A \times G$	0.34	1	0.34			
$A \times H$	11.56	1	11.56			
Second-order relationships						
$A \times C \times B$	59.61	5	11.32			
$A \times C \times D$	56.34	5	11.27			
$A \times C \times E$	66.57	5	13.31	17.70	(i)	3
$A \times C \times F$	56.42	5	11.28			
$A \times C \times G$	48.24	5	9.65			
$A \times C \times H$	58.49	5	11.70			
Third-order relationships						
$A \times C \times E \times B$	80.44	11	7.31	12.78	(i)	6
$A \times C \times E \times D$	75.14	11	6.83			
$A \times C \times E \times F$	70.04	11	6.37			
$A \times C \times E \times G$	69.48	11	6.32			
$A \times C \times E \times H$	73.34	11	6.67			
Fourth-order relationships						
$A \times C \times E \times B \times D$	85.43	23	3.71	2.11	(ii)	1
$A \times C \times E \times B \times F$	84.47	23	3.67			
$A \times C \times E \times B \times G$	83.57	23	3.63			
$A \times C \times E \times B \times H$	82.77	23	3.60			

determine the relative importance of the five remaining variables, after adjustment for C and E. The third-order relationships indicate that it is explanatory variable B which has the largest chi-square per degree of freedom of the remaining explanatory variables. Given that the associated termination statistic is significant at the conventional 5 per cent level, B thus becomes the third explanatory variable selected as part of the 'reasonable' subset. Finally, the fourth-order relationships indicate that it is explanatory variable D which has the largest chi-square per degree of freedom of the remaining four explanatory variables. However, in this case, the termination statistic (which at this stage needs to be of type ii for the reasons discussed above) is not significant after adjustment for variables C, E and B. For this reason, explanatory variable D is not included in the 'reasonable' subset. Similarly (though not shown in the table) none of the other remaining variables F, G, H has a significant termination statistic at this stage, and so F, G, H are also not included in the 'reasonable' subset.

Table 4.10 indicates, therefore, that attention should be focussed on a model of the relationships between the explanatory variables C, E and B and the response variable A. To check that no more preferable three-explanatory variable model could be formed, we could at this point retain the first-choice variable C, but replace E and B with any two of the remaining variables D, F, G or H and check that our three-variable model C, E, B still possesses the largest chi-square per degree of freedom.

Stage two.   The second stage of the Higgins–Koch procedure involves fitting a series of cell (f) type linear logit models which relate the selected subset of explanatory variables to the response variable. This will usually involve the initial fitting of a saturated model in which all interactions between members of the selected subset of explanatory variables are included, and the subsequent reduction of this saturated model to a parsimonious form using some form of hierarchical parameter elimination procedure. As we will see in Example 4.5 below, in the available published illustrations of the procedure, this hierarchical elimination tends to remain rather *ad hoc* in nature, and it perhaps lacks sufficient concern with the thorny issue of simultaneous confidence intervals and testing procedures. Nevertheless, such is the reduction in the size of the overall model fitting task which is usually achieved by the Stage 1 explanatory variable selection procedure that the researcher will often be left with a manageable number of possible models in the hierarchical set. In such circumstances, it may often be feasible for the researcher to consider all possible members of the hierarchical set directly, or failing this, to adopt a hierarchical elimination procedure which may not be, statistically, the most elegant solution, but which in most circumstances will produce adequate results.

Although the Higgins–Koch procedure has been used successfully in a number of empirical applications, it should be applied cautiously and sympathetically, and the reader should remain aware that it represents a rather simplistic way of dealing with a complex analytical problem. The reader should also note that there is no claim that the 'reasonable' subset of explanatory variables identified in Stage 1 is the 'best' subset, since clearcut criteria for the 'best' subset are not usually available. As such, other 'reasonable' subsets may also exist and be worthy of investigation. Higgins and Koch suggest that if such alternative subsets can be identified (perhaps on the basis of judgement) they should be evaluated both with respect to the chi-square per degree of freedom criteria and the termination statistics. If they then appear to be of equal interest on these criteria or in their own right, then they too can be analysed further in Stage 2 of the procedure. Further discussion of the procedure and details of its constituent elements is to be found in Clarke and Koch (1976), Koch (1978), and Landis *et al*. (1978).

EXAMPLE 4.5 *Byssinosis amongst U.S. cotton textile workers*

As an illustration of the model selection procedure discussed above, we will consider Higgins and Koch's (1977) own analysis of a large cross-sectional occupational health survey. The objective of this analysis was to investigate the relationship between a respiratory ailment of cotton textile workers known as byssinosis, and certain characteristics of the workplace environment and the textile workers themselves. Survey information on 5,419 employees was available, and the fundamental question of interest was the relationship between employee complaints of byssinosis and the potential explanatory variables: sex, race, length of employment, smoking habits and dustiness of the workplace. The response and explanatory variables were classified into 2, 2, 2, 3, 2 and 3 categories respectively and the general profile of the 5,419 workers in the survey took the form shown in Table 4.11.

Table 4.11 Characteristics of 5,419 cotton textile workers in Higgins–Koch study

Byssinosis		Length of employment	
Yes	165 (0.03)	Less than 10 years	2,729 (0.50)
No	5,254 (0.97)	10 to 20 years	712 (0.13)
		20 or more years	1,978 (0.37)
Sex		Smoking habit	
Male	2,916 (0.54)	Smoker	3,189 (0.59)
Female	2,503 (0.46)	Non-Smoker	2,230 (0.41)
Race		Workplace	
White	3,516 (0.65)	Workplace 1 (most dusty)	669 (0.12)
Other Races	1,903 (0.35)	Workplace 2 (less dusty)	1,300 (0.24)
		Workplace 3 (least dusty)	3,450 (0.64)

Stage 1 of Higgins and Koch's analysis involved the selection of a 'reasonable' subset of these potential explanatory variables using the procedure discussed above. Table 4.12 shows the relevant inclusion and termination statistics.

On the basis of largest chi-square per degree of freedom the first variable to enter the 'reasonable' subset is workplace. Then, after adjustment for workplace, smoking has the largest chi-square per degree of freedom, and a significant termination statistic. It thus becomes the second member of the subset. At the third step, sex has the largest chi-square per degree of freedom of the remaining explanatory variables, but neither of the termination statistics (of type i or ii) is significant. Similarly, neither of the termination statistics of race (the

	Inclusion Statistics			Termination Statistics		
	X^2	d.f.	X^2/d.f.	X^2	Type	d.f.
First-order relationships						
Byssinosis × Sex	37.67	1	37.67			
Byssinosis × Race	6.22	1	6.22			
Byssinosis × Length of Employ	10.16	2	5.08			
Byssinosis × Smoking	20.09	1	20.09			
Byssinosis × Workplace	413.82	2	206.91			
Second-order relationships						
Byssinosis × Workplace × Sex	442.05	5	88.41			
Byssinosis × Workplace × Race	424.60	5	84.92			
Byssinosis × Workplace × Length of Employ	491.25	8	61.41			
Byssinosis × Workplace × Smoking	475.42	5	95.08	15.93	(i)	3
Third-order relationships						
Byssinosis × Workplace × Smoking × Sex	498.28	11	45.30	Not Sign.		Not Sign.
Byssinosis × Workplace × Smoking × Race	490.38	11	44.58	Not Sign.		Not Sign.
Byssinosis × Workplace × Smoking × Length of Employ	554.84	17	32.64	21.45	(i)	12

Table 4.12
Test statistics for stage one explanatory variable selection process – 5,419 cotton textile workers

variable with the next largest chi-square per degree of freedom) is significant. However, the remaining variable, length of employment, does have a significant termination statistic at the 5 per cent level, and so length of employment is included in the 'reasonable' subset. The fourth-order relationships are not shown in Table 4.12 as neither race nor sex has a significant termination statistic (type ii at this stage) after adjustment for workplace, smoking and length of employment. Table 4.12 indicates, therefore, that attention should be focussed on a model of the relationships between the explanatory variables, workplace, smoking and length of employment, and the response variable prevalence of byssinosis.

Stage 2 of Higgins and Koch's analysis involved them in fitting a series of cell (f) type linear logit models to byssinosis prevalence using the three explanatory variables chosen in Stage 1. At this stage, to reduce a problem of data thinning, the information for workplaces 2 and 3 was combined. The effect of this on the variable selection process of Stage 2 was checked, but it was found to cause no major changes.

The first model to be fitted by Higgins and Koch was a saturated model in which all the possible interactions between the three selected explanatory variables were included. The design matrix of this model took the form X_1 shown in [4.67] below. In this matrix, the second column represents the main effect of workplace (now reduced to a dichotomy), the third column represents the main effect of smoking, and the fourth and fifth columns represent the main effect of length of

$$\mathbf{X}_1 = \\ 12 \times 12 \quad \begin{bmatrix} 1 & 1 & 1 & 1 & 0 & 1 & 1 & 0 & 1 & 0 & 1 & 0 \\ 1 & 1 & -1 & 1 & 0 & -1 & 1 & 0 & -1 & 0 & -1 & 0 \\ 1 & 1 & 1 & 0 & 1 & 1 & 0 & 1 & 0 & 1 & 0 & 1 \\ 1 & 1 & -1 & 0 & 1 & -1 & 0 & 1 & 0 & -1 & 0 & -1 \\ 1 & 1 & 1 & -1 & -1 & 1 & -1 & -1 & -1 & -1 & -1 & -1 \\ 1 & 1 & -1 & -1 & -1 & -1 & -1 & -1 & 1 & 1 & 1 & 1 \\ 1 & -1 & 1 & 1 & 0 & -1 & -1 & 0 & 1 & 0 & -1 & 0 \\ 1 & -1 & -1 & 1 & 0 & 1 & -1 & 0 & -1 & 0 & 1 & 0 \\ 1 & -1 & 1 & 0 & 1 & -1 & 0 & -1 & 0 & 1 & 0 & -1 \\ 1 & -1 & -1 & 0 & 1 & 1 & 0 & -1 & 0 & -1 & 0 & 1 \\ 1 & -1 & 1 & -1 & -1 & -1 & 1 & 1 & -1 & -1 & 1 & 1 \\ 1 & -1 & -1 & -1 & -1 & 1 & 1 & 1 & 1 & 1 & -1 & -1 \end{bmatrix} \quad [4.67]$$

employment. The remaining columns represent interaction effects: column 6 – workplace × smoking; columns 7 and 8 – workplace × length of employment; columns 9 and 10 smoking × length of employment, columns 11 and 12 – workplace × smoking × length of employment. The top part of Table 4.13 shows the results of tests of the significance of these interaction and main effects using the general test statistic Q_C given in equation [4.44].[†] These results suggest that the three-way interaction (workplace × smoking × length of employment) and the two-way interaction (smoking × length of employment) are insignificant (i.e. we cannot reject the null hypotheses $H_0 : \beta_{11} = \beta_{12} = 0$ and $H_0 : \beta_9 = \beta_{10} = 0$).

On the basis of these results, Higgins and Koch eliminated columns 9–12 of design matrix $\mathbf{X}_1$ and estimated and tested a second model (design matrix $\mathbf{X}_2$) which was simply the saturated model, minus columns 9–12 of matrix $\mathbf{X}_1$. At this stage, Table 4.13 shows that Q, the overall goodness-of-fit statistic [4.41], is less than the tabulated value of chi-square for 4 d.f. at the conventional 5 per cent level of significance, thus indicating that model ($\mathbf{X}_2$) fits the byssinosis data adequately. In addition, the test statistics, Q_C, appear to indicate that no further terms can be eliminated from the model. However, inspection of the original survey data indicates that byssinosis prevalences are uniformly much larger at workplace 1 than at the combined workplaces 2 and 3. As a result, Higgins and Koch proposed a third model based upon an alternative, nested design, version of matrix $\mathbf{X}_2$.

The nested design matrix which Higgins and Koch suggested is shown

[†] Some statisticians would query the validity of testing all the main effects and interaction effects together in this way. Strictly, a hierarchical elimination procedure would demand an initial test of just the highest order interaction terms (workplace × smoking × length of employment). If these could be eliminated, a test of just the two-way interactions should then be performed, and so on, and some adjustment might be made for the repeated nature of the significance testing. Higgins and Koch's elimination procedure is more *ad hoc* and intuitive, and it does not explicitly address the issue of repeated significance testing, simultaneous confidence intervals, etc.

in [4.68] below as matrix $\mathbf{X}_3$.

$$\mathbf{X}_3 = \begin{bmatrix} 1 & 1 & 1 & 0 & & & & \\ 1 & -1 & 1 & 0 & & & & \\ 1 & 1 & 0 & 1 & & \mathbf{0} & & \\ 1 & -1 & 0 & 1 & & & & \\ 1 & 1 & -1 & -1 & & & & \\ 1 & -1 & -1 & -1 & & & & \\ & & & & 1 & 1 & 1 & 0 \\ & & & & 1 & -1 & 1 & 0 \\ & \mathbf{0} & & & 1 & 1 & 0 & 1 \\ & & & & 1 & -1 & 0 & 1 \\ & & & & 1 & 1 & -1 & -1 \\ & & & & 1 & -1 & -1 & -1 \end{bmatrix}$$

12×8

[4.68]

In this matrix, the workplaces are used as the dominant factor, or controlling explanatory variable, and the other explanatory variables (smoking and length of employment) are nested within the categories of the workplace variable. There are now two sets of effects for smoking and length of employment; one set for workplace 1 (columns 2–4), and one set for workplace 2 (columns 6–8). The two sets of effects for smoking and length of employment are conditional upon the workplace category. This new model is simply a reformulation of $\mathbf{X}_2$ and, consequently, it has the same goodness-of-fit statistic. However, the Q_C test statistics now indicate that the smoking and length of employment effects are not significant in the less dusty workplaces (2 and 3).

On the basis of the Q_C test statistic results for $\mathbf{X}_3$, Higgins and Koch estimated a fourth model based on design matrix $\mathbf{X}_4$ shown in [4.69].

$$\mathbf{X}_4 = \begin{bmatrix} 1 & 1 & 1 & 0 & 0 \\ 1 & -1 & 1 & 0 & 0 \\ 1 & 1 & 0 & 1 & 0 \\ 1 & -1 & 0 & 1 & 0 \\ 1 & 1 & -1 & -1 & 0 \\ 1 & -1 & -1 & -1 & 0 \\ & & & & 1 \\ & & & & 1 \\ & & \mathbf{0} & & 1 \\ & & & & 1 \\ & & & & 1 \\ & & & & 1 \end{bmatrix}$$

12×5

[4.69]

This allows smoking and length of employment effects for the most dusty workplace 1, but allows only a different overall average byssinosis prevalence level at the less dusty workplaces 2 and 3 (this is represented by column 5). The goodness-of-fit test statistic, Q, suggests that this more parsimonious model is also an adequate representation of the byssinosis data.

Finally, Higgins and Koch suggested that a further reduction in the model to just four parameters is possible. They based this suggestion upon the observation that, within workplace 1, byssinosis prevalences are similar for the two longer employment levels. As a result, they reduced the two columns (3 and 4) which represent length of employment in X_4, to just a single column. The goodness-of-fit statistic of this new model X_5 in Table 4.13 ($Q = 5.798$), indicates that this final

Table 4.13
Test statistics for design matrices X_1 – X_5 in Higgins and Koch byssinosis analysis

Source of variation	d.f.	Q_C	Q	d.f.
Design X_1				
Workplace: W	1	60.85*		
Smoking: S	1	8.67*		
Employment: E	2	14.02*		
$W \times S$	1	7.60*		
$W \times E$	2	12.52*		
$S \times E$	2	2.65		
$W \times S \times E$	2	1.61		
Design X_2			4.414	4
Workplace: W	1	68.84*		
Smoking: S	1	14.84*		
Employment: E	2	14.18*		
$W \times S$	1	13.32*		
$W \times E$	2	12.76*		
Design X_3			4.414	4
Workplace 1: Smoking	1	14.30*		
Employment	2	13.64*		
Workplace 2+3: Smoking	1	0.63		
Employment	2	1.02		
Difference in overall means of workplaces	1	68.84*		
Design X_4			5.797	7
Workplace 1: Smoking	1	14.30*		
Employment	2	13.64*		
Difference in overall means of workplaces	1	69.01*		
Design X_5			5.798	8
Workplace 1: Smoking	1	14.44*		
Employment	1	13.64*		
Difference in workplace 1 reference cell and workplace 2 and 3 overall mean	1	3.16		

Note: * indicates effect significant at conventional 5 per cent level

reduction in the number of parameters in the model is accompanied by only a very small (and insignificant) increase in the size of the goodness-of-fit statistic.

Both models X_4 and X_5 indicate that within workplace 1 (the dustiest of the workplaces), smoking and length of employment are important explanatory variables for byssinosis prevalence, whilst in the less dusty workplaces (2 and 3) they are not important. Model X_5 indicates (as can be seen in the final Q_C test statistic in Table 4.13) that non-smokers with less than ten years of employment at workplace 1 do not react in a significantly different way from the employees at workplaces 2 and 3. However, the model implies that smoking and increased length of employment interact with the dustiness at workplace 1 to significantly increase the prevalence of byssinosis.

4.7   A more general matrix formulation

Earlier in the chapter we expressed our cell (f) linear logit models in the general matrix form

$$E_A(\bar{L}) = \mathbf{X}\boldsymbol{\beta} \qquad [4.70]$$

On the left-hand side of this equation, $\bar{L}$ is a column vector of observed logit values; one value for each sub-population. For example, in the case of model [4.14] of Example 4.1, $\bar{L}$ takes the form

$$\bar{L} \atop 8 \times 1 = \begin{bmatrix} \bar{L}_{12|1} \\ \bar{L}_{12|2} \\ . \\ . \\ . \\ \bar{L}_{12|8} \end{bmatrix} = \begin{bmatrix} \log_e(f_{1|1}/f_{2|1}) \\ \log_e(f_{1|2}/f_{2|2}) \\ . \\ . \\ . \\ \log_e(f_{1|8}/f_{2|8}) \end{bmatrix} = \begin{bmatrix} \log_e f_{1|1} - \log_e f_{2|1} \\ \log_e f_{1|2} - \log_e f_{2|2} \\ . \\ . \\ . \\ \log_e f_{1|8} - \log_e f_{2|8} \end{bmatrix} \qquad [4.71]$$

Grizzle, Starmer and Koch (1969) suggest that logit vectors of this type can be generated from the observed response proportions in contingency tables such as Tables 4.2, 4.3 and 4.4 by using the computationally simple sequence of matrix operations

$$\bar{L} = \mathbf{K}[\log_e(\mathbf{A}f)] \qquad [4.72]$$

where f is a vector of observed proportions and $\mathbf{A}$ and $\mathbf{K}$ are matrices whose forms are determined by the structure of the data under investigation and the model to be fitted. In the case of [4.71], for example, the column vector f consists of the elements in successive rows of Table 4.3, and it takes the form

$$f = \begin{bmatrix} f_{1|1} \\ f_{2|1} \\ f_{1|2} \\ f_{2|2} \\ \cdot \\ \cdot \\ \cdot \\ f_{1|8} \\ f_{2|8} \end{bmatrix}$$

16×1

[4.73]

The matrices **A** and **K** take the form

$$\mathbf{A} = \begin{bmatrix} 1 & 0 & 0 & . & . & . & 0 & 0 \\ 0 & 1 & 0 & . & . & . & 0 & 0 \\ 0 & 0 & 1 & . & . & . & 0 & 0 \\ . & . & . & & & & . & . \\ . & . & . & & & & . & . \\ . & . & . & & & & . & . \\ 0 & 0 & 0 & . & . & . & 0 & 1 \end{bmatrix}$$

16×16

[4.74]

$$\mathbf{K} = \begin{bmatrix} 1 & -1 & 0 & 0 & 0 & 0 & . & . & . & 0 & 0 \\ 0 & 0 & 1 & -1 & 0 & 0 & . & . & . & 0 & 0 \\ 0 & 0 & 0 & 0 & 1 & -1 & . & . & . & 0 & 0 \\ . & & & . & & . & & & & . & \\ . & & & . & & . & & & & . & \\ . & & & . & & . & & & & . & \\ 0 & 0 & 0 & 0 & 0 & 0 & . & . & . & 1 & -1 \end{bmatrix}$$

8×16

Using the rules of matrix multiplication, the reader should satisfy himself that f, **A** and **K** of [4.73] and [4.74] produce $\bar{\mathbf{L}}$ of [4.71] when substituted into [4.72].

In this simple case, taken from Example 4.1, there may seem to the reader to be little advantage in using the sequence of matrix operations [4.72] to generate the logit vector. However, not all cases of cell (f) linear logit models are so simple. The advantage of the sequence of matrix operations [4.72] is that by appropriate choices of the operator matrices **A** and **K** we can define, in a systematic way, the necessary logit vectors for multiple response category models (e.g. that in Example 4.2), for multivariate response models (i.e. cases in which there are two or more, possibly multiple-category, response variables), and for a wide variety of much more complex contingency tables (e.g. those in which the response categories represent complex response profiles or those which are structually incomplete, etc.). Moreover, not only can this sequence be made a routine part of a standard computer program for fitting cell (f) linear logit models (e.g. as in the program GENCAT discussed in Ch.7), but the sequence can readily be generalized using additional operator matrices and additional logarithmic operations or exponentiations to define a wide range of other functions of the observed response proportions which are non-logit in form (see Forthofer and Koch 1973; Koch *et al.* 1977, 155). In this way, our linear logit models can be seen as merely one member (arguably the most important and widely used) of a rather broader class of linear models for cell (f) categorical data problems. This broader class of linear models and the weighted least squares approach to the estimation of all members of the class was first developed in a systematic fashion by Grizzle, Starmer and Koch (1969), and it has, subsequently, been widely used and extended by Koch and his associates (e.g. Koch and Reinfurt 1971; Koch *et al.* 1972, 1976, 1977; Forthofer and Koch 1973; Imrey *et al.* 1976; Landis and Koch 1977; Lehnen and Koch 1974a, b). The approach is often referred to in the literature as the GSK (Grizzle–Starmer–Koch) method (see Reynolds 1977; Kritzer, 1978 for further introductory accounts).

In Chapter 8 we will return to some of the issues raised by this more general matrix formulation of our cell (f) linear logit models. We will consider cases in which we have more complex contingency tables, and in which the sequence of matrix operations [4.72] plays a much more important role.

CHAPTER 5

All variables categorical but no division into response and explanatory

In the introduction to the previous chapter we noted that the problems of cell (f) of Table 1.1 lie in a zone of transition between the domains of two major elements of the unified approach to categorical data analysis. That is to say, the problems of cell (f) lie between the domains of the logistic, linear logit and probit models which we have considered in Chapters 2, 3 and 4 on the one hand, and log-linear models, a class of models which we have not yet encountered, on the other. As a result, we noted that two types of models are now widely used to handle the problems of cell (f): (a) linear logit models of the type we discussed in Chapter 4; (b) log-linear models whose parameters are estimated by an iterative proportional fitting procedure, iterative weighted least squares procedure or a Newton–Raphson procedure. Logically, having discussed the first of these type of models in Chapter 4, we should now turn to the second type. However, it is possible to regard log-linear models for cell (f) problems as a special case of the general class of log-linear models. For pedagogic purposes, it is therefore much more convenient to first introduce the reader to log-linear models in the context of the problems of cell (g) of Table 1.1 and then, having done that, to return to a discussion of the special characteristics of log-linear models for cell (f) problems.

This then is the pattern we will adopt. In this chapter we will turn first to cell (g) of Table 1.1; to the case in which all variables are categorical but in which there is no division of variables into response and explanatory. This can occur when all variables are 'natural' response variables, or, more likely, when the researcher chooses to treat the variables as such and when he chooses to make no use of any response/explanatory distinction which might exist. In such cases, our aim is not, as in previous chapters, to assess the effects of explanatory variables on a response variable. Instead, our aim is to describe the structural relationships among the response variables (i.e. to describe the structural relationships between the dimensions of the contingency table). In the case of cell (g) problems our contingency tables take on the classical contingency table form shown in Tables 1.2 and 1.4 (repeated for convenience as Tables 5.1 and 5.2). These tables and their

Table 5.1
An $I \times J$ two-dimensional
contingency table

Table 5.1
An $I \times J$ two-dimensional
contingency table

		Variable B				
		$j = 1$	2	$\ldots$	J	Total
Variable A	$i = 1$	n_{11}	n_{12}	$\ldots$	n_{1J}	n_{1+}
	$i = 2$	n_{21}	n_{22}	$\ldots$	n_{2J}	n_{2+}
	$\cdot$	$\cdot$	$\cdot$		$\cdot$	$\cdot$
	$\cdot$	$\cdot$	$\cdot$		$\cdot$	$\cdot$
	$i = I$	n_{I1}	n_{I2}	$\ldots$	n_{IJ}	n_{I+}
	Total	n_{+1}	n_{+2}	$\ldots$	n_{+J}	$n_{++} = N$

Table 5.2
A $3 \times 2 \times 2$ three-dimensional contingency table

		Variable C				
		$k = 1$		$k = 2$		
		Variable B		Variable B		
		$j = 1$	$j = 2$	$j = 1$	$j = 2$	Total
Variable A	$i = 1$	n_{111}	n_{121}	n_{112}	n_{122}	n_{1++}
	$i = 2$	n_{211}	n_{221}	n_{212}	n_{222}	n_{2++}
	$i = 3$	n_{311}	n_{321}	n_{312}	n_{322}	n_{3++}
	Total	n_{+11}	n_{+21}	n_{+12}	n_{+22}	$n_{+++} = N$

notation can readily be generalized to any appropriate higher-dimensional form.

5.1 The hypothesis of independence and the chi-square test

Faced with a two-dimensional contingency table of the form shown in Table 5.1, most geographers and environmental scientists will have been trained, in their introductory statistics courses, to consider using a chi-square test to assess whether the two variables A and B (e.g. soil type and vegetation cover, region and average industrial wage rate, etc.) are *independent* of each other. In other words, they will attempt to assess the validity of a hypothesis which states that

$$P_{ij} = \text{Prob}(A = i) \times \text{Prob}(B = j)$$
$$P_{ij} = P_{i+}P_{+j} \qquad \qquad \qquad \textbf{[5.1]}$$

where P_{ij} is the probability of an observation belonging to the ith category of variable A *and* the jth category of variable B, P_{i+} is the marginal probability of belonging to the ith category of variable A, and P_{+j} is the marginal probability of belonging to the jth category of variable B. The chi-square test, first suggested by Pearson (1900, 1904), involves estimating the values of the frequencies to be expected in each cell of the contingency table if the two variables (A and B) are independent (these estimated values are denoted $\hat{m}_{ij}$); and computing the sum of the differences between these expected frequency estimates and the observed cell frequencies n_{ij}. The Pearson chi-square statistic thus takes the form

$$X^2 = \sum_{i=1}^{I} \sum_{j=1}^{J} \frac{(n_{ij} - \hat{m}_{ij})^2}{\hat{m}_{ij}}$$

[5.2]

The value of the statistic depends upon the size of the differences $(n_{ij} - \hat{m}_{ij})$. If the two variables (A and B) are independent, the differences will be less than otherwise would be the case. Consequently, the value of the X^2 statistic will be smaller when the two variables are independent than otherwise.

By assuming that the observed cell frequencies (n_{ij}) are observations from a multinomial distribution (see Ch. 1 for discussion), and by further assuming that the expected frequencies are not too small, the statistic X^2 can be shown to have approximately a chi-square χ^2 distribution (where we use the Greek symbol χ^2 to distinguish the distribution from the Pearson statistic X^2). We can then test the hypothesis that variables A and B are independent by comparing the value of X^2 we obtain from [5.1] with the tabulated chi-square value for $(I-1) \times (J-1)$ degrees of freedom at the conventional 1, 5 or 10 per cent significance levels. Values of X^2 greater than the tabulated chi-square value for a given number of degrees of freedom and a given significance level would result in the rejection of the hypothesis of independence between variables A and B. (For further introduction to the chi-square test and distribution, see Everitt 1977).

5.2 Towards a log-linear model of independence

In the previous chapters of this book we have used linear models to represent categorical contingency table data. We will now consider how we might reformulate the traditional chi-square test of independence as a similar linear type of model, and we will consider what are the advantages which accrue from such a reformulation.

In general, if we assume that the observed cell frequencies (the n_{ij}'s) in our two-dimensional contingency table are observations from a multinomial distribution with sample size N and cell probabilities (P_{ij}), the frequency m_{ij}, to be expected in the ijth cell of the table is

$$m_{ij} = E(n_{ij}) = NP_{ij}$$

[5.3]

That is to say, the expected frequency, m_{ij}, is the expected value of n_{ij} (where n_{ij} is being viewed in this case as a random variable[†]), and it is

---

[†] Strictly, if we were to follow the usual convention in statistics we should signify that n_{ij} is a random variable by writing it in capitals as N_{ij}. However, in the log-linear modelling literature this is never done and we will thus follow the usual practice.

given by multiplying the total number of observations in the table by the probability that these observations will fall into the ijth cell. Under the hypothesis of independence between the variables (A and B) in our contingency table, however, we can substitute for P_{ij} from [5.1] and the general expression for the expected cell frequencies becomes in this case

$$m_{ij} = NP_{i+}P_{+j} \qquad\qquad [5.4]$$

If we now take the natural logarithm of both sides of [5.4] we get a linear model of the form

$$\log_e m_{ij} = \log_e N + \log_e P_{i+} + \log_e P_{+j} \qquad\qquad [5.5]$$

a form which is reminiscent of the models used in the analysis of variance. To see this a little more clearly, we can first re-express [5.4] in the equivalent form (see Bishop, Fienberg and Holland 1975, 28)

$$m_{ij} = \frac{m_{i+}m_{+j}}{N} \qquad\qquad [5.6]$$

and, by taking the natural logarithm of both sides of [5.6], re-express [5.5] as

$$\log_e m_{ij} = \log_e m_{i+} + \log_e m_{+j} - \log_e N \qquad\qquad [5.7]$$

If we next sum [5.7] over i, over j, and over i *and* j we get

$$\sum_{i=1}^{I} \log_e m_{ij} = \sum_{i=1}^{I} \log_e m_{i+} + I \log_e m_{+j} - I \log_e N \qquad\qquad [5.8]$$

$$\sum_{j=1}^{J} \log_e m_{ij} = J \log_e m_{i+} + \sum_{j=1}^{J} \log_e m_{+j} - J \log_e N \qquad\qquad [5.9]$$

$$\sum_{i=1}^{I}\sum_{j=1}^{J} \log_e m_{ij} = J \sum_{i=1}^{I} \log_e m_{ij} + I \sum_{j=1}^{J} \log_e m_{ij} - IJ \log_e N \qquad\qquad [5.10]$$

From this it is a matter of simple algebra[†] to show that the linear model

---

[†] To see this we can substitute the terms from [5.12] to [5.14] into [5.11] to give

$$\log_e m_{ij} = \frac{1}{J}\sum_{j=1}^{J} \log_e m_{ij} + \frac{1}{I}\sum_{i=1}^{I} \log_e m_{ij} - \frac{1}{IJ}\sum_{i=1}^{I}\sum_{j=1}^{J} \log_e m_{ij} .$$

We then substitute the expressions [5.8] to [5.10] into the above. This gives

$$\log_e m_{ij} = \frac{1}{J}\left(J\log_e m_{i+} + \sum_{j=1}^{J} \log_e m_{+j} - J\log_e N \right) + \frac{1}{I}\left(\sum_{i=1}^{I} \log_e m_{i+} \right.$$

$$\left. + I\log_e m_{+j} - I\log_e N \right) - \frac{1}{IJ}\left(J\sum_{i=1}^{I} \log_e m_{i+} + I\sum_{j=1}^{J} \log_e m_{+j} - IJ\log_e N \right)$$

$$\log_e m_{ij} = \log_e m_{i+} + \frac{1}{J}\sum_{j=1}^{J} \log_e m_{+j} - \log_e N + \frac{1}{I}\sum_{i=1}^{I} \log_e m_{i+}$$

$$+ \log_e m_{+j} - \log_e N - \frac{1}{I}\sum_{i=1}^{I} \log_e m_{i+} - \frac{1}{J}\sum_{j=1}^{J} \log_e m_{+j} + \log_e N$$

$$= \log_e m_{i+} + \log_e m_{+j} - \log_e N.$$

That is to say, beginning with [5.11] and [5.12] to [5.14] we can produce [5.7] and vice-versa.

[5.7], (and implicitly model [5.5]) can be written in the form:[†]

$$\log_e m_{ij} = \lambda + \lambda_i^A + \lambda_j^B \qquad\qquad [5.11]$$

where

$$\lambda = \frac{1}{IJ}\sum_{i=1}^{I}\sum_{j=1}^{J}\log_e m_{ij} \qquad\qquad [5.12]$$

$$\lambda_i^A = \frac{1}{J}\sum_{j=1}^{J}\log_e m_{ij} - \lambda \qquad\qquad [5.13]$$

$$\lambda_j^B = \frac{1}{I}\sum_{i=1}^{I}\log_e m_{ij} - \lambda \qquad\qquad [5.14]$$

The model which we have just derived is a model which is linear in the logarithms of the expected cell frequencies. As a result, it is known as a *log-linear* model. In this case, as it is derived from [5.4] it is known as the *log-linear model of independence*. Its parameters,[‡] λ, λ_i^A, λ_j^B, show marked similarities to those encountered in analysis-of-variance models, consequently analysis-of-variance type terminology is commonly used to describe them. In this context, the first parameter, λ, is seen to be similar to the overall or grand mean term of an analysis-of-variance model, and it represents the overall mean of the logarithms of expected frequencies. The λ_i^A and λ_j^B parameters are seen to be similar to the 'main effects' terms of an analysis-of-variance model. The parameters λ_i^A ($i = 1, \ldots, I$) represent the difference between the overall mean and the mean of the logarithms of expected frequencies in the J cells at level i of variable A. They represent, therefore, the main effects of being at level i ($i = 1, \ldots I$) of variable A. Similarly, the parameters λ_j^B ($j = 1, \ldots, J$) represent the main effects of being at level j of variable B. Since λ_i^A and λ_j^B in this interpretation represent deviations from the overall mean, the following analysis-of-variance type constraints are assumed to hold:

$$\sum_{i=1}^{I}\lambda_i^A = \sum_{j=1}^{J}\lambda_j^B = 0 \qquad\qquad [5.15]$$

The reader should note, however, that although this analysis of variance type terminology and constraint system is the most widely adopted interpretation of the log-linear model [5.11], it is not the only one. The reader must view this interpretation of the parameters as being consistent with an implicit[§] 'centred effect' coding system of the type

---

[†] For completeness we should add ($i = 1, \ldots, I$ and $J = 1, \ldots, J$) to model [5.11]. We should also add ($i = 1, \ldots, I$) to [5.13] and ($j = 1, \ldots, J$) to [5.14]. For convenience, many accounts of log-linear models (e.g. Fienberg 1977, 1980; Upton 1978a; Everitt 1977) omit such terms. We will adopt a similar shorthand in this book.

[‡] Other systems of notation are used for such models. In particular, the parameters are often denoted u, $u_{1(i)}$, $u_{2(j)}$ instead of λ, λ_i^A, λ_j^B (see Bishop, Fienberg and Holland 1975; Fienberg 1977, 1980). The issue is discussed by Wrigley (1980c) and the system of notation adopted in this book is that suggested in that discussion.

[§] By the term 'implicit' we mean that the log-linear model [5.11] could alternatively be written in matrix form with a 'centred effect' or 'cornered effect' design matrix made explicit. For example, assuming for convenience that $I = 2$ and $J = 2$, i.e. $i = 1, 2$ and $j = 1, 2$, and noting that in this case

discussed in Chapter 4. An alternative implicit coding system would be a 'cornered effect' coding system. If a 'cornered effect' system was adopted, the first parameter, λ, of the log-linear model [5.11] would be interpreted as the 'anchor' or base cell parameter and would represent the logarithm of the expected frequency in the 'anchor' or base cell. The parameters λ_i^A and λ_j^B in this alternative system would then be interpreted as differences from the 'anchor' cell, i.e. they would represent effects on the 'anchor' cell expected frequency of being in cell i of variable A and cell j of variable B rather than the 'anchor' cell. In this system, the constraints [5.15] would take the form

$$\lambda_I^A = \lambda_J^B = 0 \qquad\qquad\qquad [5.16]$$

where we have followed the convention (see Plackett 1974; Holt 1979; Fingleton 1981) of constraining the *last* category of each variable to zero, i.e. we have assumed that the 'anchor' or base cell is the IJth cell. Alternatively, we could adopt the convention of constraining the *first* category of each variable to zero, i.e. we could assume that the 'anchor' cell is the cell $i = 1$ and $j = 1$. The constraints would in this case be

$$\lambda_1^A = \lambda_1^B = 0 \qquad\qquad\qquad [5.17]$$

The choice between [5.16] and [5.17] is purely arbitrary.[†]

The reader should note that the majority of descriptions of log-linear models which he will encounter, particularly those in papers or books by North American authors, will adopt the 'centred effect' analysis-of-

---

the constraints [5.15] imply that $\lambda_2^A = -\lambda_1^A$ and $\lambda_2^B = -\lambda_1^B$ and the constraints [5.16] imply that $\lambda_2^A = 0$ and $\lambda_2^B = 0$, then in explicit matrix form the log-linear model would become

$$
\begin{bmatrix} \log_e m_{11} \\ \log_e m_{21} \\ \log_e m_{12} \\ \log_e m_{22} \end{bmatrix}
=
\begin{bmatrix} 1 & 1 & 1 \\ 1 & -1 & 1 \\ 1 & 1 & -1 \\ 1 & -1 & -1 \end{bmatrix}
\begin{bmatrix} \lambda \\ \lambda_1^A \\ \lambda_1^B \end{bmatrix}
\qquad
\begin{bmatrix} \log_e m_{11} \\ \log_e m_{21} \\ \log_e m_{12} \\ \log_e m_{22} \end{bmatrix}
=
\begin{bmatrix} 1 & 1 & 1 \\ 1 & 0 & 1 \\ 1 & 1 & 0 \\ 1 & 0 & 0 \end{bmatrix}
\begin{bmatrix} \lambda \\ \lambda_1^A \\ \lambda_1^B \end{bmatrix}
$$

in 'centred effect' and 'cornered effect' coding respectively. Expanded out, these matrix expressions can be seen to imply

$$\log_e m_{11} = \lambda + \lambda_1^A + \lambda_1^B \qquad\qquad \log_e m_{11} = \lambda + \lambda_1^A + {}_1^B$$
$$\log_e m_{21} = \lambda - \lambda_1^A + \lambda_1^B \qquad\qquad \log_e m_{21} = \lambda + \lambda_1^B$$
$$\log_e m_{12} = \lambda + \lambda_1^A - \lambda_1^B \qquad\qquad \log_e m_{12} = \lambda + \lambda_1^A$$
$$\log_e m_{22} = \lambda - \lambda_1^A - \lambda_1^B \qquad\qquad \log_e m_{22} = \lambda$$

In the 'centred effect' coding: λ is the overall mean term: λ_1^A is the differential effect on the overall mean of being at level 1 of variable A; $(-\lambda_1^A = \lambda_2^A)$ is the differential effect on the overall mean of being at level 2 of variable A; λ_1^B is the differential effect on the overall mean of being at level 1 of variable B; $(-\lambda_1^B = \lambda_2^B)$ is the differential effect on the overall mean of being at level 2 of variable B. In the 'cornered effect' coding: λ is the logarithm of the expected frequency in the 'anchor' cell IJ (i.e. cell 22); λ_1^A is the effect on the 'anchor' cell expected frequency of being at level 1 of variable A rather than level 2; λ_1^B is the effect on the 'anchor' cell expected frequency of being at level 1 of variable B rather than level 2.

[†] The reader should note that the initial releases of the GLIM package (Nelder 1975) used constraints of the form [5.16] and it is this initial GLIM system which was used to produce Holt's (1979) examples. However, more recent releases of the GLIM package, beginning with Release 3 (Baker and Nelder 1978) use constraints of the form [5.17]. Some commentators have become slightly confused by this change.

variance type interpretation of the parameters, and the constraints [5.15]. Many widely used computer programs for fitting log-linear models, for example the BMDP, ECTA and C-TAB programs which will be discussed in Chapter 7, also assume this system and use constraints [5.15]. However, the reader should be aware that other programs, particularly the important British computer package GLIM (see Ch. 7 for details; also Bowlby and Silk 1982), use contraints of the form [5.16] or [5.17] when fitting log-linear models (GLIM now uses the [5.17] system). Consequently, the parameter estimates derived from GLIM and similar programs must be interpreted using the 'cornered effect' coding interpretation described above. A number of British commentators on log-linear models (e.g. Holt 1979; Fingleton 1981) draw attention to this issue, and stress the importance of being aware that the parameter estimates of log-inear models and the interpretation of these estimates differ according to the constraint system [5.15] or [5.16] to [5.17] used.

In the remainder of this chapter, unless stated otherwise, we will follow the majority of current textbook accounts of log-linear models (e.g. Fienberg 1977, 1980; Upton 1978a; Everitt 1977; Reynolds 1977; Payne 1977) and adopt the 'centred effect' interpretation of the parameters. However, the reader must remain aware of this issue, and must take care when attempting to interpret the parameters of log-linear models.

5.3   A hierarchical set of log-linear models for two-dimensional contingency tables

I fear that the first act of most social scientists upon seeing a contingency table is to compute chi-square for it. Sometimes this approach is enlightening, sometimes wasteful, but sometimes it does not go quite far enough. (Mosteller 1968 p.1).

The question which remains unanswered in the previous section concerns the issue of what advantages accrue from the reformulation of the traditional chi-square test as a log-linear model of independence. To answer this question we must first note that the log-linear model of independence is merely a special case of the most general log-linear model for a two-dimensional contingency table. This most general model has the form

$$\log_e m_{ij} = \lambda + \lambda_i^A + \lambda_i^B + \lambda_{ij}^{AB} \qquad\qquad [5.18]$$

and it is known as the *saturated* log-linear model for the two-dimensional table. As its name suggests, the saturated model has as many parameters as there are cells in the contingency table. Consequently, as we saw when we first encountered saturated models in Section 4.5, the model will fit the observed cell frequencies perfectly and there will be no degrees of freedom. The additional parameters λ_{ij}^{AB} ($i = 1, \ldots, I$; $j = 1, \ldots, J$) which have been.added to those in the model of independence [5.11], are known as 'interaction' parameters or two-variable or two-way effect terms. As in the case of the 'main effect' parameters λ_i^A and λ_j^B, we assume that these new interaction parameters are subject to constraints. In the 'centred effect' interpretation these constraints take the form:

$$\sum_{i=1}^{I} \lambda_{ij}^{AB} = \sum_{j=1}^{J} \lambda_{ij}^{AB} = 0 \qquad\qquad [5.19]$$

and are added to the constraints [5.15]. In the 'cornered effect' interpretation they take either of the forms:

$$\lambda_{Ij}^{AB} = \lambda_{iJ}^{AB} = 0 \quad\text{or}\quad \lambda_{1j}^{AB} = \lambda_{i1}^{AB} = 0 \qquad [5.20]$$

and are added to the constraints [5.16] or [5.17] respectively.

In terms of the new interaction parameters, λ_{ij}^{AB}, a test of the hypothesis of independence is equivalent to testing whether $\lambda_{ij}^{AB} = 0$ for $i = 1, \ldots, I$ and $j = 1, \ldots, J$, and the model of independence can be seen as a special case of the saturated log-linear model in which the interaction parameters are set to zero. However, the model of independence is clearly not the *only* special case of the saturated model. By setting other terms in the saturated model to zero, we can specify a range of log-linear models. Each of these has a totally different interpretation, and is associated with a particular hypothesis about the nature of the structural relationships between the two variables A and B in our two-dimensional contingency table. For example, if we set the interaction parameters λ_{ij}^{AB} and the main effect parameters λ_j^B to zero, we are left with the model

$$\log_e m_{ij} = \lambda + \lambda_i^A \qquad\qquad [5.21]$$

which is a representation of the hypothesis that the categories of the B variable are equally probable. Similarly, if we set the interaction parameters and the main effect parameters λ_i^A to zero, we derive the model

$$\log_e m_{ij} = \lambda + \lambda_j^B \qquad\qquad [5.22]$$

which is a representation of the hypothesis that the categories of the A variable are equally probable. In addition, if we set the interaction parameters and all the main effect parameters to zero, we derive the model

$$\log_e m_{ij} = \lambda \qquad\qquad [5.23]$$

which is a representation of the hypothesis that all the categories of both variables are equally probable. What this implies, therefore, is that reformulation of the traditional chi-square test as a log-linear model of independence, allows us to see it (and the hypothesis of independence) as merely one of the possible models (hypotheses) which can be derived from the most general log-linear model for a two-dimensional contingency table. This opens up to the geographer and environmental scientist a wider and richer range of hypotheses than he would traditionally consider when evaluating the structure of a contingency table. It also forces him to consider the analysis of contingency tables as part of the broader linear model framework (many aspects of which he is likely to be familiar with and to use in his research), and as part of the unified approach to categorical data analysis which we are in the process of outlining in this book.

Taken together, models [5.18], [5.11], [5.21], [5.22] and [5.23] can be seen to define the type of hierarchical set of models which we introduced in Section 4.5. That is to say, as we move from model [5.23] through [5.22], [5.21] and [5.11] to the saturated model [5.18], and vice versa, parameters are added and deleted in a progressive fashion. The interaction parameters λ_{ij}^{AB} are only included if all their lower order relatives are also included, and we do not consider non-hierarchical models of the type

$$\log_e m_{ij} = \lambda + \lambda_{ij}^{AB} \qquad\qquad [5.24]$$

or

$$\log_e m_{ij} = \lambda + \lambda_j^B + \lambda_{ij}^{AB} \qquad\qquad [5.25]$$

which break this higher/lower-order relative principle. In practice, most standard accounts of log-linear models which the reader will encounter, and virtually all applications of log-linear models in the social and environmental sciences, confine their attention to hierarchical sets of models. There are both technical and interpretative reasons for this. The technical reasons concern the simplification of estimation techniques possible in the case of hierarchical models, and the attractive properties of goodness-of-fit statistics for such models. The interpretative reasons are more complex. They relate to a feature which Nelder (1976) refers to as *marginality*, and are commented upon by Fienberg (1977, 39, 1978; Knoke and Burke 1980, 72). In simple terms, however, we can state that if interaction terms are non-negligible (i.e. non-zero), then it is of little interest that the main effects in the same model may be negligible. For simplicity, we will likewise confine our discussion to hierarchical log-linear models in this book.

Stated in rather crude terms, what we try to do in the log-linear model approach to the analysis of contingency tables is to eliminate, in a hierarchical fashion, the parameters of the saturated model which are *not* essential to a description of the structural relationships between the variables in the contingency table. By reducing the number of parameters, we thus try to reduce the model to the most parsimonious form consistent with the relationships between the variables revealed in the contingency table data. The log-linear model approach to the analysis of contingency tables such as Table 5.1 involves, therefore, the fitting of a hierarchical set of models, and the selection of one of these models as the most 'acceptable' representation of the structural relationships between the variables in the contingency table on the basis of its goodness-of-fit, parsimony and substantive meaning.

EXAMPLE 5.1   *Some simple two-dimensional contingency tables*

The log-linear modelling approach to the analysis of two-dimensional tables can now be illustrated using three simple examples.

1. **Pebbles in glacial till**. Anderson (1955) classified a large number of pebbles in glacial till as faceted or unfaceted, and striated or unstriated in terms of the lithologic composition of the pebbles. Krumbein and Graybill (1965, 187) report the subset of Anderson's data shown in Table 5.3. Here the pebbles are simply classified according to whether they are faceted or not, and are divided into two lithologic types, granite and metamorphic rocks.

Table 5.3
Pebbles in glacial till (after Anderson 1955, reported in Krumbein and Graybill 1965, 187).

Variable A		Variable B		
		Faceted	*Unfaceted*	*Total*
	Granite	41	170	211
	Metamorphic	14	42	56

To this 2×2 contingency table we can now fit a hierarchical set of log-linear models. The first model we fit is the saturated model

$$\log_e m_{ij} = \lambda + \lambda_i^A + \lambda_j^B + \lambda_{ij}^{AB} \qquad [5.26]$$

Adopting a 'centred effect' coding system and the constraint system [5.15] and [5.19], and using a computer program such as BMDP, ECTA or C-TAB, the parameter estimates for this model are:[†]

$$\hat{\lambda} = 3.807$$
$$\hat{\lambda}_1^A = 0.618 \qquad \hat{\lambda}_2^A = -0.618$$
$$\hat{\lambda}_1^B = -0.630 \qquad \hat{\lambda}_2^B = 0.630$$
$$\hat{\lambda}_{11}^{AB} = -0.081 \qquad \hat{\lambda}_{21}^{AB} = 0.081 \qquad \hat{\lambda}_{12}^{AB} = 0.081 \qquad \hat{\lambda}_{22}^{AB} = -0.081$$

In the case of the saturated model, the estimates of the expected cell frequencies are identical to the observed cell frequencies in Table 5.3.

The second model we fit is the model of independence

$$\log_e m_{ij} = \lambda + \lambda_i^A + \lambda_j^B \qquad [5.27]$$

Under the same type of constraints, the parameter estimates for this model are

$$\hat{\lambda} = 3.783$$
$$\hat{\lambda}_1^A = 0.663 \qquad \hat{\lambda}_2^A = -0.663$$
$$\hat{\lambda}_1^B = -0.675 \qquad \hat{\lambda}_2^B = 0.675$$

and estimates of the expected cell frequencies are

$$\hat{m}_{11} = 43.46 \qquad \hat{m}_{21} = 11.54 \qquad \hat{m}_{12} = 167.54 \qquad \hat{m}_{22} = 44.46$$

[†] A full discussion of how these estimates are derived is contained in Section 5.6. However, the reader can, at this stage, readily compute the parameter estimates of model [5.27] and its associated expected cell frequency estimates by hand. To do this, substitute marginal frequencies from Table 5.3 into equation [5.51] and then the $\hat{m}_{ij}$ values in place of the m_{ij} terms in equations [5.12] to [5.14].

The third model we fit is the no-B-effect model

$$\log_e m_{ij} = \lambda + \lambda_i^A \qquad\qquad [5.28]$$

which is a representation of the hypothesis that the categories of the B variable are equally probable within given levels of variable A. Under the same type of constraints, the parameter estimates for this model are

$$\hat{\lambda} \quad = 3.995$$

$$\hat{\lambda}_1^A \quad = 0.663 \qquad \hat{\lambda}_1^A = -0.663$$

and the estimates of the expected cell frequencies are

$$\hat{m}_{11} = 105.5 \quad \hat{m}_{21} = 28.0 \quad \hat{m}_{12} = 105.5 \quad \hat{m}_{22} = 28.0$$

The fourth model we fit is the no-A-effect model

$$\log_e m_{ij} = \lambda + \lambda_j^B \qquad\qquad [5.29]$$

which is a representation of the hypothesis that the categories of the A variable are equally probable within given levels of variable B. Under the same type of constraints the parameter estimates for this model are

$$\hat{\lambda} \quad = \quad 3.989$$

$$\hat{\lambda}_1^B = -0.675 \quad \hat{\lambda}_2^B = 0.675$$

and the estimates of the expected frequencies are

$$\hat{m}_{11} = 27.5 \quad \hat{m}_{21} = 27.5 \quad \hat{m}_{12} = 106.0 \quad \hat{m}_{22} = 106.0$$

Finally, the fifth model we fit is the equiprobability model

$$\log_e m_{ij} = \lambda \qquad\qquad [5.30]$$

The parameter estimate for this model is

$$\hat{\lambda} = 4.201$$

and the estimates of the expected cell frequencies are

$$\hat{m}_{11} = 66.75 \quad \hat{m}_{21} = 66.75 \quad \hat{m}_{12} = 66.75 \quad \hat{m}_{22} = 66.75$$

The question we now pose is which of these fitted models is the most 'acceptable' representation of the structural relationships between the variables in Table 5.3, and we choose the most 'acceptable' model on the basis of its goodness-of-fit, parsimony and substantive meaning. The goodness-of-fit of each model can be summarized using either the Pearson chi-square statistic [5.2] or the *likelihood-ratio* chi-square statistic

$$G^2 = 2 \sum_{i=1}^{I} \sum_{j=1}^{J} n_{ij}[\log_e(n_{ij}/\hat{m}_{ij})] \qquad\qquad [5.31]$$

If the fitted model provides an adequate representation of the observed data and the total sample is large,[†] both the Pearson chi-square statistic and G^2 have approximate χ^2 distributions with degrees of freedom given by the expression,

$$\text{d·f·} = \text{number of cells in table } -\text{number of parameters}$$
$$\text{in the model that require estimating} \qquad [5.32]$$

Although it is less well-known amongst geographers than the Pearson chi-square statistic, the likelihood-ratio chi-square statistic (G^2) has the advantage that it is divisible into additive portions. We will see later that this is a most useful property. As a result, G^2 is the goodness-of-fit statistic which is most widely used in practice, and we will adopt it from this point onwards.

Table 5.4 shows the goodness-of-fit of each member of the hierarchical set of models. Other than the saturated model, the only model with a non-significant G^2 value (i.e. the only model, other than the saturated, with an acceptable fit to the data) is the model of independence. On the grounds of parsimony and goodness-of-fit this is clearly the model we should choose as the most 'acceptable' representation of the structural relationships between the variables in Table 5.3. It implies that we have no reason for rejecting the hypothesis that granite and metamorphic rock pebbles are independent in their tendency towards faceting. This is in accordance with Anderson's (1955) original results. He showed that although faceting is not independent of rock type when a wide variety of rocks are considered, independence is found within more restricted lithologic types, as in this example.

For purposes of comparison, Table 5.5 shows the parameter estimates, expected cell frequency estimates, and G^2 statistics for the hierarchical set of log-linear models fitted to Table 5.3 using the alternative 'cornered effect' coding system, the constraint system [5.17] and [5.20], and the GLIM computer program. The reader should note that the expected cell frequency estimates and the G^2 statistics remain the same, but the parameter estimates take different values under this alternative 'cornered effect' coding and parameter constraint system. As an exercise, the reader should revise the discussion in Section 5.2 and interpret the parameter estimates under both 'centred effect' and 'cornered effect' coding systems.

2. Lifetime residential mobility and retirement migration. As part of a much wider study of the characteristics of retired migrants in Britain, Law and Warnes (1980) considered the relationship between previous residential mobility (since marriage) and retirement migration for a small matched sample of retired migrants and non-migrants. A sample of 201 migrants in two retirement areas (the North Wales coastal area and the County of Dorset) were interviewed, together with 100 non-migrants (matched to have approximately the same characteristics) in the London and Manchester metropolitan areas. Table 5.6 presents this data in the form of a 4×2 two-dimensional contingency table.

To this contingency table we can fit the hierarchical set of log-linear models shown in Table 5.7. On the basis of the G^2 goodness-of-fit

---

[†] An excellent (but advanced) discussion of the small-sample properties of these statistics, and of some possible modifications of G^2, is provided by Fienberg (1980, 172–76).

Table 5.4
Goodness-of-fit of the hierarchical set of log-linear models fitted to Table 5.3

Model		G^2	d.f.	Expression for d.f.
Saturated:	$\log_e m_{ij} = \lambda + \lambda_i^A + \lambda_j^B + \lambda_{ij}^{AB}$	0.0	0	$IJ - [1 + (I-1) + (J-1) + (I-1)(J-1)]$
Independence:	$\log_e m_{ij} = \lambda + \lambda_i^A + \lambda_j^B$	0.81	1	$IJ - [1 + (I-1) + (J-1)]$
No B-effect:	$\log_e m_{ij} = \lambda + \lambda_i^A$	99.36*	2	$IJ - [1 + (I-1)]$
No A-effect:	$\log_e m_{ij} = \lambda + \lambda_j^B$	96.68*	2	$IJ - [1 + (J-1)]$
Equiprobability:	$\log_e m_{ij} = \lambda$	195.20*	3	$IJ - 1$

* indicates G^2 value in upper 5 per cent tail of χ^2 distribution with d.f. as indicated.

Table 5.5
Results under 'cornered effect' coding system and using GLIM

Model	$\hat{\lambda}$	$\hat{\lambda}_2^A$	$\hat{\lambda}_2^B$	$\hat{\lambda}_{22}^{AB}$	$\hat{m}_{11}$	$\hat{m}_{21}$	$\hat{m}_{12}$	$\hat{m}_{22}$	G^2
Saturated	3.714	−1.075	1.422	−0.324	41.00	14.00	170.00	42.00	0.0
Independence	3.772	−1.327	1.349	0	43.46	11.54	167.54	44.46	0.81
No B-effect	4.659	−1.327	0	0	105.50	28.00	105.50	28.00	99.36
No A-effect	3.314	0	1.349	0	27.50	27.50	106.00	106.00	96.68
Equiprobability	4.201	0	0	0	66.75	66.75	66.75	66.75	195.20

Table 5.6
Post-marriage to retirement residential mobility for a small sample of retired households (adapted from Law and Warnes 1980, 199).

		Variable B		
		Retired migrants	*Retired non-migrants*	*Total*
Variable A	No moves	41	11	52
	Local moves only	47	61	108
	1–2 longer moves	59	19	78
	3+ longer moves	54	9	63
	Total	201	100	301

statistics, the only model with an acceptable fit to the data in Table 5.6 is seen to be the saturated model. In other words, in this case, we cannot remove the two-variable interaction parameters λ_{ij}^{AB} as there is a significant interaction between the two variables in Table 5.6. On the whole, more retired migrants have experienced longer distance moves since marriage than retired non-migrants.

Table 5.8 shows the parameter estimates for the saturated model

Table 5.7
Goodness-of-fit of the hierarchical set of log-linear models fitted to Table 5.6

Model		G^2	d.f.
Saturated:	$\log_e m_{ij} = \lambda + \lambda_i^A + \lambda_j^B + \lambda_{ij}^{AB}$	0.0	0
Independence:	$\log_e m_{ij} = \lambda + \lambda_i^A + \lambda_j^B$	42.87*	3
No B-effect:	$\log_e m_{ij} = \lambda + \lambda_i^A$	77.43*	4
No A-effect:	$\log_e m_{ij} = \lambda + \lambda_j^A$	65.69*	6
Equiprobability:	$\log_e m_{ij} = \lambda$	100.20*	7

* indicates G^2 value in upper 5 per cent tail of χ^2 distribution with d.f. as indicated.

Table 5.8
Parameter estimates for saturated model under alternative coding and constraint systems

	Centred effect coding constraints [5.15], [5.19]	Cornered effect coding constraints [5.17], [5.20]
$\hat{\lambda}$	3.410	3.714
$\hat{\lambda}_1^A$	−0.354	0.0
$\hat{\lambda}_2^A$	0.570	0.137
$\hat{\lambda}_3^A$	0.101	0.364
$\hat{\lambda}_4^A$	−0.317	0.275
$\hat{\lambda}_1^B$	0.497	0.0
$\hat{\lambda}_2^B$	−0.497	−1.316
$\hat{\lambda}_{11}^{AB}$	0.160	0.0
$\hat{\lambda}_{21}^{AB}$	−0.628	0.0
$\hat{\lambda}_{31}^{AB}$	0.069	0.0
$\hat{\lambda}_{41}^{AB}$	0.398	0.0
$\hat{\lambda}_{12}^{AB}$	−0.160	0.0
$\hat{\lambda}_{22}^{AB}$	0.628	1.576
$\hat{\lambda}_{32}^{AB}$	−0.069	0.183
$\hat{\lambda}_{42}^{AB}$	−0.398	−0.476

computed using: (a) the 'centred effect' coding system, constraints [5.15], [5.19], and the BMDP program; (b) the 'cornered effect' coding system, constraints [5.17], [5.20], and the GLIM program. The estimates of the expected cell frequencies are identical for both systems and, in the case of saturated model, equal to the observed cell frequencies in Table 5.6. Once again, the reader should note how important it is to be aware that the parameter estimates and the interpretation of these estimates differ according to the coding and constraint system used. As an exercise, he should attempt to interpret these parameters under both systems using the discussion in Section 5.2 as a guide.

3. Farm acreage under woodland. Table 5.9 shows the number of farms in three counties of a single region in the U.S.A. with less than or more than 20 per cent of their acreage under woodland. Visual inspection of this table indicates that there is little difference between counties in terms of the numbers of farms with more than or less than 20 per cent of their acreage under woodland. Is this borne out by the log-linear model which best represents these data?

Table 5.9
Farm acreage under woodland in three countries in a single region

		Variable B		
		Number of farms with less than 20% of their acreage under woodland	Number of farms with more than 20% of their acreage under woodland	Total
Variable A	County 1	105	26	131
	County 2	125	31	156
	County 3	117	29	146
	Total	347	86	433

Table 5.10 shows the G^2 statistics for the hierarchical set of log-linear models fitted to Table 5.9. It can readily be seen that the most parsimonious model with an acceptable fit to the data is the no-A-effect model. This indicates that categories of the A variable are equally probable within given levels of variable B. Hence, the A variable in Table 5.9 is redundant, and we can consider reducing the dimensionality of the table by collapsing it over this variable.

Table 5.10
Goodness-of-fit of the hierarchical set of log-linear models fitted to Table 5.9

Model	G^2	d.f.
Saturated:	0.0	0
Independence:	0.00003	2
No B-effect:	168.60*	3
No A-effect:	2.21	4
Equiprobability:	170.80*	5

* indicates G^2 value in upper 5 per cent tail of χ^2 distribution with d.f. as indicated.

5.4 Log-linear models for multidimensional contingency tables

So far, we have confined our attention to log-linear models for two-dimensional contingency tables. These models represent a valuable alternative to the traditional tests for independence and measures of association in two-dimensional tables; tests and measures with which most geographers and environmental scientists are likely to be familiar (for reviews see Everitt 1977; Lewis 1977; Upton 1978a). The real advantages of log-linear models, however, are to be seen most clearly in the case of *multidimensional* tables. Such tables are currently treated in a rather inadequate fashion in the available quantitative methods texts for geographers and environmental scientists, yet such tables are the rule rather than the exception in research work in the social and environmental sciences.

The simplest type of multidimensional table is the type of three-dimensional shown in Table 5.2. In this case, the two-dimensional saturated log-linear model [5.18] extends to the form

$$\log_e m_{ijk} = \lambda + \lambda_i^A + \lambda_j^B + \lambda_k^C + \lambda_{ij}^{AB} + \lambda_{ik}^{AC} + \lambda_{jk}^{BC} + \lambda_{ijk}^{ABC} \qquad [5.33]$$

To take account of the new third dimension or variable in our table, we have included the parameters λ_k^C, λ_{ik}^{AC}, λ_{jk}^{BC} and λ_{ijk}^{ABC}. The final parameters λ_{ijk}^{ABC} ($i = 1, \ldots, I$, $j = 1, \ldots, J$, $k = 1, \ldots, K$) are known as the three-variable interaction effects, and they are subject to constraints of the same type as those in two-dimensional models. In the 'centred effect' interpretation these additional constraints take the form

$$\sum_{i=1}^{I} \lambda_{ijk}^{ABC} = \sum_{j=1}^{J} \lambda_{ijk}^{ABC} = \sum_{k=1}^{K} \lambda_{ijk}^{ABC} = 0 \qquad \text{[5.34]}$$

and are added to [5.15] and an extended version of [5.19]. In the 'cornered effect' interpretation they take either of the forms

$$\lambda_{Ijk}^{ABC} = \lambda_{iJk}^{ABC} = \lambda_{ijK}^{ABC} = 0 \quad \text{or} \quad \lambda_{1jk}^{ABC} = \lambda_{i1k}^{ABC} = \lambda_{ij1}^{ABC} = 0 \qquad \text{[5.35]}$$

and are added to [5.16] or [5.17] and an extended [5.20].

Once again, the saturated model [5.33] has as many parameters as there are cells in our three-dimensional contingency table. To achieve a parsimonious model which has a good fit to our observed data, what we must do is try to eliminate from the saturated model the parameters which are not necessary to describe the structural relationships which exist between the variables in our contingency table.

By setting different parameters in the saturated model to zero a range of models for the three-dimensional table can be specified. Each model implies a particular hypothesis about the relationships between the three variables. Clearly, there are many different combinations of parameters in the saturated model which can be set to zero. Table 5.11 shows nine distinct types of model formed by setting different combinations of parameters to zero. In some cases, there are a number of possible models of the same type. For example, there are three possible models of type 6 which can be formed by: (a) setting the three-variable interaction parameters λ_{ijk}^{ABC} to zero; (b) setting two of the three sets of two-variable interaction parameters (λ_{ij}^{AB} and λ_{jk}^{BC}, λ_{ij}^{AB} and λ_{ik}^{AC}, or λ_{ik}^{AC} and λ_{jk}^{BC}) to zero; (c) setting a single set of main effect parameters (λ_j^B, λ_i^A or λ_k^C) to zero. The reader should note that the particular ordering of the main effect parameters given in the parentheses above corresponds to that given for the two-variable interaction parameters for reasons of hierarchical ordering. For example, if λ_i^A is set to zero then the hierarchical principle implies that both λ_{ij}^{AB} and λ_{ik}^{AC} must be set to zero. In the case of model type 6, the three possible models are therefore:

$$\log_e m_{ijk} = \lambda + \lambda_i^A + \lambda_k^C + \lambda_{ik}^{AC}$$

$$\log_e m_{ijk} = \lambda + \lambda_j^B + \lambda_k^C + \lambda_{jk}^{BC} \qquad \text{[5.36]}$$

$$\log_e m_{ijk} = \lambda + \lambda_i^A + \lambda_j^B + \lambda_{ij}^{AB}$$

Taken together, the nineteen log-linear models in Table 5.11, and the nine distinct model types define what is termed the hierarchical set of models for a three-dimensional contingency table. As we noted in Chapter 4, the main feature of this hierarchical set of models is the fact that parameters are omitted only in descending order of dimensionality.

That is to say, higher-order parameters are only included in a model if *all* their lower-order relatives are also included. For example, if λ_{ik}^{AC} is included in the model then λ_i^A and λ_k^C must also be included. By the same token, models such as:

$$\log_e m_{ijk} = \lambda + \lambda_i^A + \lambda_j^B + \lambda_k^C + \lambda_{ijk}^{ABC}$$

$$\log_e m_{ijk} = \lambda + \lambda_i^A + \lambda_j^B + \lambda_{ik}^{AC} + \lambda_{jk}^{BC}$$

[5.37]

are not permissible hierarchical models, as some necessary lower-order parameters are missing from them. As noted in the case of two-dimensional models, we will confine our discussion to hierarchical log-linear models, and we will not consider non-hierarchical models such as [5.37].

Each of the model types in the hierarchical set in Table 5.11 implies a particular hypothesis about the relationships between the three variables in our contingency table. These hypotheses are discussed in detail by Bishop *et al.* (1975, 37), Payne (1977, 119) and Haberman (1978, 197) and can be summarised as follows.

Type 1 hypothesis. The saturated model. This implies that the association between every pair of variables varies with the level of the third.

Type 2 hypothesis. The no three-variable interaction or pairwise association model. This implies that each pair of variables is associated but each two-variable effect is unaffected by the level of the third variable.

Type 3 hypothesis. The conditional independence model. This implies that a pair of variables are independent given the third. For example, variable B may be independent of C given A. In this particular case, the cell probabilities in our contingency table can be expressed as:

$$P_{ijk} = \frac{\text{Prob}(B = j \text{ and } A = i)\ \text{Prob}(C = k \text{ and } A = i)}{\text{Prob}(A = i)}$$

[5.38]

Table 5.11
Hierarchical set of log-linear models for a three-dimensional contingency table

Type 4 hypothesis. The multiple independence model. This implies that two variables considered as a joint variable are independent of the third. For example, the joint variable AB may be independent of the third

Model type	Number of models of such type	λ terms set to zero	Model specification
1	1	None	$\log_e m_{ijk} = \lambda + \lambda_i^A + \lambda_j^B + \lambda_k^C + \lambda_{ij}^{AB} + \lambda_{ik}^{AC} + \lambda_{jk}^{BC} + \lambda_{ijk}^{ABC}$
2	1	λ^{ABC}	$\log_e m_{ijk} = \lambda + \lambda_i^A + \lambda_j^B + \lambda_k^C + \lambda_{ij}^{AB} + \lambda_{ik}^{AC} + \lambda_{jk}^{BC}$
3	3	$\lambda^{ABC}, \lambda^{AB}$	$\log_e m_{ijk} = \lambda + \lambda_i^A + \lambda_j^B + \lambda_k^C + \lambda_{ik}^{AC} + \lambda_{jk}^{BC}$
4	3	$\lambda^{ABC}, \lambda^{AB}, \lambda^{BC}$	$\log_e m_{ijk} = \lambda + \lambda_i^A + \lambda_j^B + \lambda_k^C + \lambda_{ik}^{AC}$
5	1	$\lambda^{ABC}, \lambda^{AB}, \lambda^{AC}, \lambda^{BC}$	$\log_e m_{ijk} = \lambda + \lambda_i^A + \lambda_j^B + \lambda_k^C$
6	3	$\lambda^{ABC}, \lambda^{AB}, \lambda^{BC}, \lambda^B$	$\log_e m_{ijk} = \lambda + \lambda_i^A + \lambda_k^C + \lambda_{ik}^{AC}$
7	3	$\lambda^{ABC}, \lambda^{AB}, \lambda^{AC}, \lambda^{BC}, \lambda^A$	$\log_e m_{ijk} = \lambda + \lambda_j^B + \lambda_k^C$
8	3	$\lambda^{ABC}, \lambda^{AB}, \lambda^{AC}, \lambda^{BC}, \lambda^A, \lambda^C$	$\log_e m_{ijk} = \lambda + \lambda_j^B$
9	1	$\lambda^{ABC}, \lambda^{AB}, \lambda^{AC}, \lambda^{BC}, \lambda^A, \lambda^B, \lambda^C$	$\log_e m_{ijk} = \lambda$

variable C. In this case, the cell probabilities can be expressed as:

$$P_{ijk} = \text{Prob}(A = i \text{ and } B = j) \times \text{Prob}(C = k) \qquad [5.39]$$

Type 5 hypothesis. The mutual independence model. This implies that the three variables are mutually independent. This is the three-dimensional equivalent to the model of independence for two-dimensional tables, and the cell probabilities in this case can be expressed as:

$$P_{ijk} = \text{Prob}(A = i) \times \text{Prob}(B = j) \times \text{Prob}(C = k) \qquad [5.40]$$

Type 6 hypothesis. This model implies that all categories of one variable are equiprobable given the other two. For example, the specific case of a model of this type shown in Table 5.11 implies that all categories of variable B are equiprobable given variables A and C.

Type 7 hypothesis. This model implies that all categories of one variable are equiprobable given the other two variables, and that the other two variables are independent. For example, the specific case of a model of this type shown in Table 5.11 implies that all categories of variable A are equiprobable given B and C and that variables B and C are independent.

Type 8 hypothesis. This model implies that, given one variable, all combinations of the categories of the other two variables are equally probable. For example, the specific case of a model of this type shown in Table 5.11 implies that all combinations of the categories of variables A and C are equally probable given variable B.

Type 9 hypothesis. This model implies that all combinations of the three variables are equally likely.

It can be seen that models of types 6-9 involve the exclusion of one or more of the one-variable or 'main' effect terms. Such models are referred to by Bishop, Fienberg and Holland (1975, 38) as 'non-comprehensive' models and they are much less commonly used than models of types 1–5. Non-comprehensive models indicate that one or more of the variables in the table are redundant, and the implication is that the dimensionality of the table can be reduced accordingly.

Extension of log-linear models from three-dimensional tables to four-dimensional and higher-dimensional tables is straightforward. We use the same principles as in the movement from two-dimensional tables to three-dimensional tables. In the case of a four-dimensional $I \times J \times K \times L$ table for example, the saturated log-linear model becomes

$$\log_e m_{ijkl} = \lambda + \lambda_i^A + \lambda_j^B + \lambda_k^C + \lambda_l^D + \lambda_{ij}^{AB} + \lambda_{ik}^{AC} + \lambda_{il}^{AD} + \lambda_{jk}^{BC}$$
$$+ \lambda_{jl}^{BD} + \lambda_{kl}^{CD} + \lambda_{ijk}^{ABC} + \lambda_{ijl}^{ABD} + \lambda_{ikl}^{ACD} + \lambda_{jkl}^{BCD} + \lambda_{ijkl}^{ABCD} \qquad [5.41]$$

where to take account of the new fourth dimension we have included the parameters λ_l^D, λ_{il}^{AD}, λ_{jl}^{BD}, λ_{kl}^{CD}, λ_{ijl}^{ABD}, λ_{ikl}^{ACD}, λ_{jkl}^{BCD}, λ_{ijkl}^{ABCD}. The λ_{ijkl}^{ABCD} parameters are known as four-variable interaction effects and are subject to constraints of the form

$$\sum_{i=1}^{I} \lambda_{ijkl}^{ABCD} = \sum_{j=1}^{J} \lambda_{ijkl}^{ABCD} = \sum_{k=1}^{K} \lambda_{ijkl}^{ABCD} = \sum_{l=1}^{L} \lambda_{ijkl}^{ABCD} = 0 \qquad [5.42]$$

in the 'centred effect' interpretation, or

$$\lambda_{Ijkl}^{ABCD} = \lambda_{iJkl}^{ABCD} = \lambda_{ijKl}^{ABCD} = \lambda_{ijkL}^{ABCD} = 0, \text{ or}$$

[5.43]

$$\lambda_{1jkl}^{ABCD} = \lambda_{i1kl}^{ABCD} = \lambda_{ij1l}^{ABCD} = \lambda_{ijk1}^{ABCD} = 0$$

in the 'cornered effect' interpretation.

Once again, this saturated model has as many parameters as there are cells in our four-dimensional contingency table. To achieve a parsimonious model which has a good fit to our observed data we must attempt to eliminate from the saturated model the parameters which are not necessary to describe the structural relationships which exist between the variables in our contingency table. Once again we will normally employ a hierarchical process of parameter elimination and consider only those log-linear models which are members of the hierarchical set. In the case of four-dimensional contingency tables, however, there are now 167 models in the hierarchical set rather than the nineteen models listed in Table 5.11 for the case of three-dimensional tables, and this figure rises rapidly to thousands of models in the hierarchical set for higher-dimensional tables. As a result, sophisticated methods of selecting the most 'acceptable' model from such hierarchical sets are necessary, and we will consider such methods later in this chapter.

Table 5.12 shows that the 167 log-linear models in the four-dimensional hierarchical set can be divided into twenty-eight model types. Each of the model types implies a particular hypothesis about the relationships between the four variables in our contingency table. Most of the model types can be interpreted in terms similar to those used to describe three-dimensional models. For example, model types 7, 9, 10, 13, 14, 16, 20, 22–28 can be described in terms of the concepts of independence, conditional independence, equiprobability, and/or conditional equiprobability (e.g. model type 20 implies that the categories of variable D are conditionally equiprobable given the levels of the joint variable ABC; and model type 14 implies that variables B and C are conditionally independent, given the level of variable A, and variable D and the joint variable AC are conditionally independent given the level of variable B). In addition: model type 2 is the no four-variable interaction model; model type 6 is the no three-or four-variable interaction model; model type 1 is the saturated model; model type 3 implies that the three-variable interaction parameters (λ_{jkl}^{BCD}) are zero in the conditional three-dimensional table that is obtained when the level of variable A is given, and so on (see Goodman 1970; Haberman 1978, 259 for further details). It can be seen that the last nine model types, 20–28, involve the exclusion of one or more of the one-variable or 'main' effects. These are, therefore, 'non-comprehensive' models which indicate that one or more of the variables in the four-dimensional table are redundant and that dimensionality of the tables can be reduced accordingly. As a result, they can be seen to be models of the three and two-dimensional types shown in Table 5.11 and equations [5.18], [5.11], [5.21] and [5.22].

Log-linear models for three-dimensional, four-dimensional and higher-dimensional tables can be seen, therefore, to be relatively simple extensions of log-linear models for two-dimensional tables. The advantages of using such models, and of considering multidimensional

Table 5.12 Hierarchical set of log-linear models for a four-dimensional contingency table

Model Type	Number of models of such type	Model specification
1	1	$\log_e m_{ijkl} = \lambda + \lambda_i^A + \lambda_j^B + \lambda_k^C + \lambda_l^D + \lambda_{ij}^{AB} + \lambda_{ik}^{AC} + \lambda_{il}^{AD} + \lambda_{jk}^{BC} + \lambda_{jl}^{BD} + \lambda_{kl}^{CD}$ $+ \lambda_{ijk}^{ABC} + \lambda_{ijl}^{ABD} + \lambda_{ikl}^{ACD} + \lambda_{jkl}^{BCD} + \lambda_{ijkl}^{ABCD}$
2	1	$\log_e m_{ijkl} = \lambda + \lambda_i^A + \lambda_j^B + \lambda_k^C + \lambda_l^D + \lambda_{ij}^{AB} + \lambda_{ik}^{AC} + \lambda_{il}^{AD} + \lambda_{jk}^{BC} + \lambda_{jl}^{BD} + \lambda_{kl}^{CD}$ $+ \lambda_{ijk}^{ABC} + \lambda_{ijl}^{ABD} + \lambda_{ikl}^{ACD} + \lambda_{jkl}^{BCD}$
3	4	$\log_e m_{ijkl} = \lambda + \lambda_i^A + \lambda_j^B + \lambda_k^C + \lambda_l^D + \lambda_{ij}^{AB} + \lambda_{ik}^{AC} + \lambda_{il}^{AD} + \lambda_{jk}^{BC} + \lambda_{jl}^{BD} + \lambda_{kl}^{CD}$ $+ \lambda_{ijk}^{ABC} + \lambda_{ijl}^{ABD} + \lambda_{ikl}^{ACD}$
4	6	$\log_e m_{ijkl} = \lambda + \lambda_i^A + \lambda_j^B + \lambda_k^C + \lambda_l^D + \lambda_{ij}^{AB} + \lambda_{ik}^{AC} + \lambda_{il}^{AD} + \lambda_{jk}^{BC} + \lambda_{jl}^{BD} + \lambda_{kl}^{CD}$ $+ \lambda_{ijk}^{ABC} + \lambda_{ijl}^{ABD}$
5	4	$\log_e m_{ijkl} = \lambda + \lambda_i^A + \lambda_j^B + \lambda_k^C + \lambda_l^D + \lambda_{ij}^{AB} + \lambda_{ik}^{AC} + \lambda_{il}^{AD} + \lambda_{jk}^{BC} + \lambda_{jl}^{BD} + \lambda_{kl}^{CD}$ $+ \lambda_{ijk}^{ABC}$
6	1	$\log_e m_{ijkl} = \lambda + \lambda_i^A + \lambda_j^B + \lambda_k^C + \lambda_l^D + \lambda_{ij}^{AB} + \lambda_{ik}^{AC} + \lambda_{il}^{AD} + \lambda_{jk}^{BC} + \lambda_{jl}^{BD} + \lambda_{kl}^{CD}$
7	6	$\log_e m_{ijkl} = \lambda + \lambda_i^A + \lambda_j^B + \lambda_k^C + \lambda_l^D + \lambda_{ij}^{AB} + \lambda_{ik}^{AC} + \lambda_{il}^{AD} + \lambda_{jk}^{BC} + \lambda_{jl}^{BD} + \lambda_{ijk}^{ABC}$ $+ \lambda_{ijl}^{ABD}$
8	12	$\log_e m_{ijkl} = \lambda + \lambda_i^A + \lambda_j^B + \lambda_k^C + \lambda_l^D + \lambda_{ij}^{AB} + \lambda_{ik}^{AC} + \lambda_{il}^{AD} + \lambda_{jk}^{BC} + \lambda_{jl}^{BD} + \lambda_{ijk}^{ABC}$
9	12	$\log_e m_{ijkl} = \lambda + \lambda_i^A + \lambda_j^B + \lambda_k^C + \lambda_l^D + \lambda_{ij}^{AB} + \lambda_{ik}^{AC} + \lambda_{il}^{AD} + \lambda_{jk}^{BC} + \lambda_{ijk}^{ABC}$
10	4	$\log_e m_{ijkl} = \lambda + \lambda_i^A + \lambda_j^B + \lambda_k^C + \lambda_l^D + \lambda_{ij}^{AB} + \lambda_{ik}^{AC} + \lambda_{jk}^{BC} + \lambda_{ijk}^{ABC}$
11	6	$\log_e m_{ijkl} = \lambda + \lambda_i^A + \lambda_j^B + \lambda_k^C + \lambda_l^D + \lambda_{ij}^{AB} + \lambda_{ik}^{AC} + \lambda_{il}^{AD} + \lambda_{jk}^{BC} + \lambda_{jl}^{BD}$
12	15	$\log_e m_{ijkl} = \lambda + \lambda_i^A + \lambda_j^B + \lambda_k^C + \lambda_l^D + \lambda_{ij}^{AB} + \lambda_{ik}^{AC} + \lambda_{il}^{AD} + \lambda_{jk}^{BC}$
13	4	$\log_e m_{ijkl} = \lambda + \lambda_i^A + \lambda_j^B + \lambda_k^C + \lambda_l^D + \lambda_{ij}^{AB} + \lambda_{ik}^{AC} + \lambda_{il}^{AD}$
14	12	$\log_e m_{ijkl} = \lambda + \lambda_i^A + \lambda_j^B + \lambda_k^C + \lambda_l^D + \lambda_{ij}^{AB} + \lambda_{ik}^{AC} + \lambda_{jl}^{BD}$
15	4	$\log_e m_{ijkl} = \lambda + \lambda_i^A + \lambda_j^B + \lambda_k^C + \lambda_l^D + \lambda_{ij}^{AB} + \lambda_{ik}^{AC} + \lambda_{jk}^{BC}$
16	3	$\log_e m_{ijkl} = \lambda + \lambda_i^A + \lambda_j^B + \lambda_k^C + \lambda_l^D + \lambda_{ij}^{AB} + \lambda_{kl}^{CD}$
17	12	$\log_e m_{ijkl} = \lambda + \lambda_i^A + \lambda_j^B + \lambda_k^C + \lambda_l^D + \lambda_{ij}^{AB} + \lambda_{ik}^{AC}$
18	6	$\log_e m_{ijkl} = \lambda + \lambda_i^A + \lambda_j^B + \lambda_k^C + \lambda_l^D + \lambda_{ij}^{AB}$
19	1	$\log_e m_{ijkl} = \lambda + \lambda_i^A + \lambda_j^B + \lambda_k^C + \lambda_l^D$
20	4	$\log_e m_{ijkl} = \lambda + \lambda_i^A + \lambda_j^B + \lambda_k^C + \lambda_{ij}^{AB} + \lambda_{ik}^{AC} + \lambda_{jk}^{BC} + \lambda_{ijk}^{ABC}$
21	4	$\log_e m_{ijkl} = \lambda + \lambda_i^A + \lambda_j^B + \lambda_k^C + \lambda_{ij}^{AB} + \lambda_{ik}^{AC} + \lambda_{jk}^{BC}$
22	12	$\log_e m_{ijkl} = \lambda + \lambda_i^A + \lambda_j^B + \lambda_k^C + \lambda_{ij}^{AB} + \lambda_{ik}^{AC}$
23	12	$\log_e m_{ijkl} = \lambda + \lambda_i^A + \lambda_j^B + \lambda_k^C + \lambda_{ij}^{AB}$
24	4	$\log_e m_{ijkl} = \lambda + \lambda_i^A + \lambda_j^B + \lambda_k^C$
25	6	$\log_e m_{ijkl} = \lambda + \lambda_i^A + \lambda_j^B + \lambda_{ij}^{AB}$
26	6	$\log_e m_{ijkl} = \lambda + \lambda_i^A + \lambda_j^B$
27	4	$\log_e m_{ijkl} = \lambda + \lambda_i^A$
28	1	$\log_e m_{ijkl} = \lambda$

tables directly rather than by splitting or collapsing them into a series of two-dimensional tables (as implied by many existing statistics texts for geographers and environmental scientists), are enormous. Fienberg (1977, 1) notes that the practice of splitting or collapsing an essentially multidimensional table into a series of two-dimensional tables and computing chi-square statistics for the two-dimensional tables (a practice which is widespread in geography and environmental science) has the following disadvantages.

1. It confuses the *marginal* relationships between the pair of categorical variables included in the two-dimensional table with the relationship when other variables are present.

2. It does not allow for the *simultaneous* examination of the various pairwise relationships.

3. It ignores the possibility of three-variable or higher-order interactions among the variables.

As a result, two-dimensional tables formed in this manner by collapsing an essentially multidimensional underlying table can give rise to fallacious or paradoxical results. Some extreme possibilities of this type were demonstrated by Simpson (1951) and have subsequently been referred to as *Simpson's paradox* (see Blyth 1972 for further discussion). Upton (1978a, 43) and Fingleton (1981) provide illustrations of Simpson's paradox in which relationships (positive or negative) between two variables (A and B) in a three-dimensional table become reversed (i.e. become negative and positive respectively) in the two-dimensional table of variables A and B formed by collapsing the three-dimensional table over the third variable C. It is clear, therefore, that multidimensional tables should be analysed directly, rather than as a series of two-dimensional tables, because of the possibility of fallacious or paradoxical results if interactions exist between variables but are ignored. However, this statement is not meant to imply that table collapsing should never be undertaken. In certain circumstances, a degree of table collapsing is possible and useful, and we will consider this issue further in Chapter 9.

EXAMPLE 5.2   *Age, decay and use of buildings in north-east London*

An example of the application of log-linear modelling to a three-dimensional, cell (g) type contingency table is to be found in the geographical literature in the work of Lewis (1977, 141–6). Lewis was concerned with investigating the relationships between building age (variable A), building decay (variable B) and building use (variable C) in part of north-east London. Information on 1,407 buildings in the study area was available and these buildings were classified into five age categories (post 1939; 1914–39; 1870–1914; 1830–70; pre 1830), five decay categories (very little or no decay; some slight decay; much decay but largely superficial; substantial decay; severe decay with major structural weakness), and three use categories (residential; manufacturing and warehousing; offices). To the three-dimensional $5 \times 5 \times 3$ contingency table formed from this information, Lewis fitted a hierarchical set of log-linear models, some members of which are shown in Table 5.13.

We can now consider the goodness-of-fit of the models in Table 5.13

Table 5.13
Some members of the hierarchical set of log-linear models fitted by Lewis to his three-dimensional contingency table

Model specification	G^2	d.f.
$\log_e m_{ijk} = \lambda + \lambda_i^A + \lambda_j^B + \lambda_k^C + \lambda_{ij}^{AB} + \lambda_{ik}^{AC} + \lambda_{jk}^{BC} + \lambda_{ijk}^{ABC}$	0.0	0
$\log_e m_{ijk} = \lambda + \lambda_i^A + \lambda_j^B + \lambda_k^C + \lambda_{ij}^{AB} + \lambda_{ik}^{AC} + \lambda_{jk}^{BC}$	39.6	32
$\log_e m_{ijk} = \lambda + \lambda_i^A + \lambda_j^B + \lambda_k^C + \lambda_{ij}^{AB} + \lambda_{ik}^{AC}$	435.5*	40
$\log_e m_{ijk} = \lambda + \lambda_i^A + \lambda_j^B + \lambda_k^C + \lambda_{ij}^{AB} + \lambda_{jk}^{BC}$	344.4*	40
$\log_e m_{ijk} = \lambda + \lambda_i^A + \lambda_j^B + \lambda_k^C + \lambda_{ik}^{AC} + \lambda_{jk}^{BC}$	1,140.1*	48
$\log_e m_{ijk} = \lambda + \lambda_i^A + \lambda_j^B + \lambda_k^C$	2,079.7*	66
$\log_e m_{ijk} = \lambda$	3,505.0*	74

*indicates G^2 value in upper 5 per cent tail of χ^2 distribution with d.f. as indicated.

on the basis of G^2, the likelihood ratio chi-square statistic introduced in Example 5.1. We find that, other than the saturated model, the only model with a non-significant G^2 value (i.e. the only model other than the saturated, with an acceptable fit to the data) is the model in which the three-variable interaction parameters are set to zero. The reader will recall that this model is the type-2 model of Table 5.11, i.e. the pairwise association model. It implies that each pair of variables is associated but that the association of each pair is unaffected by the level of the third variable. That is to say, building age and building decay are associated but their association is unaffected by building use; building decay and building use are associated but their association is unaffected by building age; and so on.

EXAMPLE 5.3   *Industrial location in Hull*

A second illustration of the log-linear modelling of three-dimensional contingency tables can also be taken from the work of Lewis (1977, 168). In this example, Lewis summarizes the industrial structure of the port city of Hull using the three-dimensional contingency table shown in Table 5.14. In this table, industrial plants are classified according to: industrial plant size (variable A) – divided into three categories;

Table 5.14
A classification of the industrial plants of Hull (adapted from Lewis 1977, 168)

		Location						
		Access to waterfront			No access to waterfront			
		Industry type			Industry type			
Industrial plant size		1	2	3	1	2	3	Total
	1	24	15	25	28	19	40	151
	2	19	11	20	28	27	41	146
	3	14	16	30	21	23	45	149
	Total	57	42	75	77	69	126	446

industry category/type of the plant (variable B) – divided into three categories; industrial plant location in terms of access to the water front (variable C) – divided into two categories. To the three-dimensional $3 \times 2 \times 2$ contingency table formed in this way, Lewis fitted a hierarchical set of log-linear models, some members of which are shown in Table 5.15.

On the basis of the G^2 statistics, the most parsimonious model with an acceptable fit to the data is the model with no three-variable interaction parameters, no two-variable interaction parameters and no A variable main effect parameters. The reader will recall that this model is one of the type-7 models of Table 5.11, i.e. a model which implies that all categories of variable A (industrial plant size) are equiprobable given variables B and C (industrial plant category and location) and that industrial plant category and location in terms of access to waterfront are independent. The reader will recall that this type of model is one of the types of model known as 'non-comprehensive'. It implies that one of the variables in the table (variable A) is redundant, and the implication is that the dimensionality of the table can be reduced from three to two. This implication can be confirmed from an inspection of Table 5.16; the expected cell frequency estimates obtained under this model.

Table 5.15
Some members of the hierarchical set of log-linear models fitted by Lewis to his Hull industry contingency table

Model specification	G^2	d.f.
$\log_e m_{ijk} = \lambda + \lambda_i^A + \lambda_j^B + \lambda_k^C + \lambda_{ij}^{AB} + \lambda_{ik}^{AC} + \lambda_{jk}^{BC} + \lambda_{ijk}^{ABC}$	0.0	0
$\log_e m_{ijk} = \lambda + \lambda_i^A + \lambda_j^B + \lambda_k^C + \lambda_{ij}^{AB} + \lambda_{ik}^{AC} + \lambda_{jk}^{BC}$	1.039	4
$\log_e m_{ijk} = \lambda + \lambda_i^A + \lambda_j^B + \lambda_k^C + \lambda_{ij}^{AB} + \lambda_{jk}^{BC}$	3.318	6
$\log_e m_{ijk} = \lambda + \lambda_i^A + \lambda_j^B + \lambda_k^C + \lambda_{jk}^{BC}$	8.674	10
$\log_e m_{ijk} = \lambda + \lambda_i^A + \lambda_j^B + \lambda_k^C$	9.677	12
$\log_e m_{ijk} = \lambda + \lambda_j^B + \lambda_k^C + \lambda_{jk}^{BC}$	8.759	12
$\log_e m_{ijk} = \lambda + \lambda_j^B + \lambda_k^C$	9.762	14
$\log_e m_{ijk} = \lambda + \lambda_j^B$	31.472*	15
$\log_e m_{ijk} = \lambda$	60.015*	17

* indicates G^2 value in upper 5 per cent tail of χ^2 distribution with d.f. as indicated.

Table 5.16
Estimated expected frequencies in Table 5.14 under model $\log_e m_{ijk} = \lambda + \lambda_j^B + \lambda_k^C$ (cell frequencies from Lewis 1977, 168)

		Location					
		Access to waterfront			No access to waterfront		
Industrial plant size		Industry type			Industry type		
		1	2	3	1	2	3
	1	17.4	14.4	26.1	27.2	22.6	40.9
	2	17.4	14.4	26.1	27.2	22.6	40.9
	3	17.4	14.4	26.1	27.2	22.6	40.9

One useful feature of the G^2 statistic is that it has the property of being additive in a nested hierarchy of models. As a result, in a hierarchical set of models, such as that in Table 5.15, we can subtract the G^2 statistic for one model from the G^2 statistic of another in the same nested sub-set in order to assess the 'change' in goodness-of-fit which results from constraining additional parameters to zero. More formally, we can say that if G^2 (x) and G^2 (y) are goodness-of-fit statistics with degrees of freedom d.f.$_{(x)}$ and d.f.$_{(y)}$ respectively, and model (y) is lower in the nested hierarchy than model (x), then the statistic

$$G^2(y-x) = G^2(y)-G^2(x) \qquad \textbf{[5.44]}$$

with degrees of freedom d.f.$_{(y)}$ − d.f.$_{(x)}$ is distributed asymptotically as χ^2, and provides a test of the difference between models (x) and (y). In other words [5.44] provides a test of whether the reduction in goodness-of-fit which results from constraining to zero certain parameters in model (x) to produce the more parsimonious model (y) is statistically justifiable or, alternatively expressed, a test of the null hypothesis that the constrained parameters are equal to zero.

As an example of the use of test statistic [5.44] let us compare the third and fourth models in Table 5.15.

Model	Parameters included	G^2	d.f.
x	λ, λ_i^A, λ_j^B, λ_k^C, λ_{ij}^{AB}, λ_{jk}^{BC}	3.318	6
y	λ, λ_i^A, λ_j^B, λ_k^C, λ_{jk}^{BC}	8.678	10
Difference: $(y-x)$	λ_{ij}^{AB} with λ, λ_i^A, λ_j^B, λ_k^C	5.360	4
	λ_{jk}^{BC} occurring in both models		

Referring the observed value, 5.360, of the test statistic [5.44] to a χ^2 distribution with four degrees of freedom, we find that at the conventional 5 per cent significance level the observed value of the test statistic falls below the tabulated χ^2 value of 9.488. We thus accept the hypothesis that the two-variable interaction parameters λ_{ij}^{AB} are zero, and suggest that it is statistically justifiable to choose model y, the more parsimonious of the two models. The reader should now utilize the same test to compare the seventh and eighth models in Table 5.15. He should satisfy himself that the difference in G^2 values (31.472–9.762) suggests that the main effect parameters λ_k^C are non-zero and that, consequently, we must choose model seven, the less parsimonious of the two models.

Finally, the reader should note that in Table 5.15 it is possible to find more than one pair of models which differ only by the same parameters. For example, the pair of models:

$$\log_e m_{ijk} = \lambda + \lambda_i^A + \lambda_j^B + \lambda_k^C + \lambda_{jk}^{BC}$$

$$\log_e m_{ijk} = \lambda + \lambda_i^A + \lambda_j^B + \lambda_k^C \qquad \textbf{[5.45]}$$

differ only by the parameters λ_{jk}^{BC}, and the pair of models:

$$\log_e m_{ijk} = \lambda + \lambda_j^B + \lambda_k^C + \lambda_{jk}^{BC}$$

$$\log_e m_{ijk} = \lambda + \lambda_j^B + \lambda_k^C$$

[5.46]

also differ only by the parameters λ_{jk}^{BC}. In this particular case, the G^2 statistic differences in the two pairs of models $(9.677 - 8.674)$ and $(9.762 - 8.759)$ are identical. However, in other cases (particularly in higher-dimensional models) such G^2 differences will not all be identical. As a result, it must be stressed that the test statistic [5.44] is a test for the significance of a particular set of parameters conditional upon, or *given* that a certain set of other parameters have been included; e.g. in [5.46] we would be concerned with testing the significance of λ_{jk}^{BC} given that λ, λ_j^B and λ_k^C were included in the model. Later in this chapter, when we consider a model selection procedure known as Brown's screening procedure, we will need to return to this point.

5.5 Abbreviated notation systems for log-linear models

Models of the type (5.41) demonstrate that log-linear models for higher-dimensional tables can rapidly become rather cumbersome to write out in the notation we have utilized up to this point. Consequently, there is a strong case for developing an abbreviated notation system which can be used to summarize log-linear models; particularly those for higher-dimensional tables. One suggested abbreviation system which has become widely used is known as the *fitted marginals* notation system. This system exploits the principles of parameter ordering implicit in a hierarchical set of log-linear models, and describes a model by means of the highest order parameters included in that model. For example, the saturated model [5.41] can be abbreviated using this notation to the form $[ABCD]$. By virtue of the hierarchical requirement, the highest order interaction term, λ_{ijkl}^{ABCD}, implies that all lower-order relatives must also be included in the model, i.e. A, B, C, D, AB, AC, AD, BC, BD, CD, ABC, ABD, ACD, BCD must also be included in the model if the four-variable interaction $ABCD$ is included. The essence of this abbreviated notation, therefore, is that certain λ parameters are automatically included; consequently there is no need to write out the implict lower-order parameters.

To get a feel for this notation system, let us write out some of the models shown in Table 5.11 in both standard and abbreviated fashion :

$[ABC]$ implies $\quad \log_e m_{ijk} = \lambda + \lambda_i^A + \lambda_j^B + \lambda_k^C + \lambda_{ij}^{AB} + \lambda_{ik}^{AC} + \lambda_{jk}^{BC} + \lambda_{ijk}^{ABC}$

$[AB]\,[AC]\,[BC]$

implies $\quad \log_e m_{ijk} = \lambda + \lambda_i^A + \lambda_j^B + \lambda_k^C + \lambda_{ij}^{AB} + \lambda_{ik}^{AC} + \lambda_{jk}^{BC}$

$[AC]\,[BC]$ implies $\log_e m_{ijk} = \lambda + \lambda_i^A + \lambda_j^B + \lambda_k^C + \lambda_{ik}^{AC} + \lambda_{jk}^{BC}$

$[B]\,[AC]$ implies $\quad \log_e m_{ijk} = \lambda + \lambda_i^A + \lambda_j^B + \lambda_k^C + \lambda_{ik}^{AC}$

$[A]\,[B]\,[C]$ implies $\log_e m_{ijk} = \lambda + \lambda_i^A + \lambda_j^B + \lambda_k^C$

$[AC]$ implies $\qquad \log_e m_{ijk} = \lambda + \lambda_i^A + \lambda_k^C + \lambda_{ik}^{AC}$

$[B]\,[C]$ implies $\qquad \log_e m_{ijk} = \lambda + \lambda_j^B + \lambda_k^C$

The final part $[AC]$ of the abbreviation $[B]\,[AC]$ can be seen, therefore, to imply λ_{ik}^{AC}, λ_i^A and λ_k^C by virtue of hierarchical requirements. The first part $[B]$ implies that we must add the λ_j^B term. In addition, we add, as always, the λ term.

Using this notation system[†], it can be seen that most models can be defined in an economical fashion, and the advantages of the system become greater as the number of dimensions in our contingency table increase. For example, in a four-dimensional table with variables A, B, C and D, the model

$$\log_e m_{ijkl} = \lambda + \lambda_i^A + \lambda_j^B + \lambda_k^C + \lambda_l^D + \lambda_{ij}^{AB} + \lambda_{il}^{AD} + \lambda_{jk}^{BC}$$
$$+ \lambda_{jl}^{BD} + \lambda_{ijl}^{ABD} \qquad\qquad [5.47]$$

can be written in abbreviated form as $[BC]\,[ABD]$, whereas the model

$$\log_e m_{ijkl} = \lambda + \lambda_i^A + \lambda_j^B + \lambda_k^C + \lambda_l^D + \lambda_{jk}^{BC} + \lambda_{il}^{AD} \qquad [5.48]$$

can be written in abbreviated form as $[BC]\,[AD]$. Likewise, in a five-dimensional table with variables A, B, C, D, E, the model

$$\log_e m_{ijklm} = \lambda + \lambda_i^A + \lambda_j^B + \lambda_k^C + \lambda_l^D + \lambda_m^E + \lambda_{ij}^{AB} + \lambda_{ik}^{AC} + \lambda_{il}^{AD}$$
$$+ \lambda_{im}^{AE} + \lambda_{jk}^{BC} + \lambda_{jl}^{BD} + \lambda_{jm}^{BE} + \lambda_{kl}^{CD} + \lambda_{ijk}^{ABC} + \lambda_{ijl}^{ABD} \qquad [5.49]$$
$$+ \lambda_{ijm}^{ABE} + \lambda_{ikl}^{ACD} + \lambda_{jkl}^{BCD} + \lambda_{ijkl}^{ABCD}$$

can be written in abbreviated form as $[ABE]\,[ABCD]$.

In addition to compactness of expression, this abbreviated notation for log-linear models also communicates another important feature of the analysis. The combination of variables listed in the abbreviation define sub-tables ('marginal' tables) formed from the fully cross-classified contingency table. When estimating the expected cell frequencies in the full contingency table using a particular log-linear model, the expected frequencies in the marginal tables specified by the abbreviation must exactly equal the corresponding observed frequencies in the same set of sub-tables. The estimation procedures which we will discuss later in this chapter will always ensure this and will produce a perfect fit between expected and observed frequencies for the specified marginals. As a result, this abbreviation system is called the fitted marginals notation system.

Although this fitted marginals notation system is the abbreviation system most commonly used, British readers should also be aware that another closely related system is sometimes preferred by British writers and is adopted in the widely used GLIM computer package (Baker and Nelder 1978). This alternative abbreviated notation system was de-

---

[†] In this system, we have followed Fienberg (1977) and written the model-defining marginals, e.g. $[AB]\,[AC]\,[BC]$, in such a way that they are distinguished by square brackets. Other writers, e.g. Payne (1977), Knoke and Burke (1980), Upton (1978a) distinguish the model-defining marginals using curly brackets or slashes, e.g. $\{AB\}\,\{AC\}\,\{BC\}$ or $AB/AC/BC$.

veloped by Wilkinson and Rogers (1973) and is discussed and used by
writers such as Williams (1976), Whittaker and Aitkin (1978).

In this alternative system, the four-dimensional saturated model
[5.41] is abbreviated to the form $A^*B^*C^*D$, and the three-dimensional
saturated model [5.33] to the form A^*B^*C. The symbol $*$ is the so-called
'crossing operator' and it implies the presence of all appropriate
lower-order parameters in the model. It should not be confused with
the multiplication operator which is denoted by the dot symbol. To
see this, the reader should note that the three-dimensional saturated
model can alternatively be written out more fully as
$A^*B^*C = A+B+C+A.B+A\cdot C+B\cdot C+A\cdot B\cdot C$. In this system, the
operators $*$, $\cdot$, and $+$ can be combined using brackets where necessary to produce compound model formulae. For example,
$A^*(B+C) = A+B+C+A\cdot B+A\cdot C$. Also, the minus sign operator can
be used to delete terms. For example, $A^*B-A\cdot B = A+B$, and A^*B^*C-
$A\cdot(B+C) = A+B+C+B\cdot C$.

To illustrate this system, let us reconsider those models from Table
5.11 which we expressed in 'fitted marginals' notation above. In this
alternative abbreviation they become

A^*B^*C $\qquad$ implies $\log_e m_{ijk} = \lambda + \lambda_i^A + \lambda_j^B + \lambda_k^C + \lambda_{ij}^{AB} + \lambda_{ik}^{AC}$
$$+ \lambda_{jk}^{BC} + \lambda_{ijk}^{ABC}$$

$A^*B^* + A^*C + B^*C$ $\quad$ implies $\log_e m_{ijk} = \lambda + \lambda_i^A + \lambda_j^B + \lambda_k^C + \lambda_{ij}^{AB} + \lambda_{ik}^{AC}$
or $A^*B^*C - A\cdot B\cdot C$. $\qquad\qquad\qquad\qquad$ $+ \lambda_{jk}^{BC}$

$A^*C + B^*C$ $\qquad$ implies $\log_e m_{ijk} = \lambda + \lambda_i^A + \lambda_j^B + \lambda_k^C + \lambda_{ik}^{AC} + \lambda_{jk}^{BC}$
or $C^*(A+B)$

$A^*C + B$ $\qquad$ implies $\log_e m_{ijk} = \lambda + \lambda_i^A + \lambda_j^B + \lambda_k^C + \lambda_{ik}^{AC}$

$A+B+C$ $\qquad$ implies $\log_e m_{ijk} = \lambda + \lambda_i^A + \lambda_j^B + \lambda_k^C$

A^*C $\qquad$ implies $\log_e m_{ijk} = \lambda + \lambda_i^A + \lambda_k^C + \lambda_{ik}^{AC}$

$B+C$ $\qquad$ implies $\log_e m_{ijk} = \lambda + \lambda_j^B + \lambda_k^C$

The reader will note the close similarity between this alternative
abbreviation system and the fitted marginals system. This can be
confirmed by considering the four-dimensional models [5.47] and [5.48],
which in fitted marginal notation were $[BC][ABD]$ and $[BC][AD]$. In
the alternative abbreviation system they become $B^*C + A^*B^*D$ and
$B^*C + A^*D$.

Clearly, there is little in principle to choose between the two
abbreviation systems, and the reader must expect to encounter both in
his wider reading. For the novice in the field of log-linear modelling
however, the Wilkinson–Rogers–GLIM notation does have the advantage that the elements implied by the 'crossing operator', $*$, can always
be written out in full (e.g. $A^*B + B^*C$ can be written as
$A+B+C+A\cdot B+B\cdot C$). The novice can, therefore, begin by writing out
his models in full and can move to progressively more abbreviated
notation as he gains in experience. The fitted marginals notation, on the
other hand, does not allow such graduation and the novice can often

find the fitted marginals abbreviations of higher-dimensional models to be rather daunting when he first encounters them. For this reason, most subsequent examples of log-linear models in this chapter and in Chapters 6 and 9, when not written in explicit $\log_e m_{ijk} = \lambda + \lambda_i^A + \lambda_j^B + \ldots$ form, will be written in *both* elemental Wilkinson–Rogers–GLIM notation (e.g. $A + B + A \cdot B$ instead of $A * B$) and fitted marginals notation. As an illustration of this, Table 5.17 shows model types 1–19 from Table 5.12 (the hierarchical set of log-linear models for four-dimensional contingency tables) written in both fitted marginals and elemental Wilkinson-Rogers-GLIM notation.

5.6 Estimation of the parameters and the expected cell frequencies

Table 5.17
Model types 1–19 from Table 5.12 in alternative notation system

Before an acceptable model can be chosen from a set of hierarchical log-linear models such as that shown in Tables 5.11 and 5.12, estimates of the cell frequencies to be expected under each model are required. These enable us to assess the competing models by testing the goodness-of-fit of their estimated expected cell frequencies to the observed cell frequencies. In addition, estimates of the parameters of the models and the standard errors of the parameters are required.

Model type	Fitted marginals	Elemental Wilkinson–Rogers–GLIM specification
1	[ABCD]	$A+B+C+D+A\cdot B+A\cdot C+A\cdot D+B\cdot C+B\cdot D+C\cdot D+A\cdot B\cdot C$ $+A\cdot B\cdot D+A\cdot C\cdot D+B\cdot C\cdot D+A\cdot B\cdot C\cdot D$
2	[ABC] [ABD] [ACD] [BCD]	$A+B+C+D+A\cdot B+A\cdot C+A\cdot D+B\cdot C+B\cdot D+C\cdot D+A\cdot B\cdot C$ $+A\cdot B\cdot D+A\cdot C\cdot D+B\cdot C\cdot D$
3	[ABC] [ABD] [ACD]	$A+B+C+D+A\cdot B+A\cdot C+A\cdot D+B\cdot C+B\cdot D+C\cdot D+A\cdot B\cdot C$ $+A\cdot B\cdot D+A\cdot C\cdot D$
4	[ABC] [ABD] [CD]	$A+B+C+D+A\cdot B+A\cdot C+A\cdot D+B\cdot C+B\cdot D+C\cdot D+A\cdot B\cdot C$ $+A\cdot B\cdot D$
5	[ABC] [AD] [BD] [CD]	$A+B+C+D+A\cdot B+A\cdot C+A\cdot D+B\cdot C+B\cdot D+C\cdot D+A\cdot B\cdot C$
6	[AB] [AC] [AD] [BC] [BD] [CD]	$A+B+C+D+A\cdot B+A\cdot C+A\cdot D+B\cdot C+B\cdot D+C\cdot D$
7	[ABC] [ABD]	$A+B+C+D+A\cdot B+A\cdot C+A\cdot D+B\cdot C+B\cdot D+A\cdot B\cdot C+A\cdot B\cdot D$
8	[ABC] [AD] [BD]	$A+B+C+D+A\cdot B+A\cdot C+A\cdot D+B\cdot C+B\cdot D+A\cdot B\cdot C$
9	[ABC] [AD]	$A+B+C+D+A\cdot B+A\cdot C+A\cdot D+B\cdot C+A\cdot B\cdot C$
10	[ABC] [D]	$A+B+C+D+A\cdot B+A\cdot C+B\cdot C+A\cdot B\cdot C$
11	[AB] [AC] [AD] [BC] [BD]	$A+B+C+D+A\cdot B+A\cdot C+A\cdot D+B\cdot C+B\cdot D$
12	[AB] [AC] [AD] [BC]	$A+B+C+D+A\cdot B+A\cdot C+A\cdot D+B\cdot C$
13	[AB] [AC] [AD]	$A+B+C+D+A\cdot B+A\cdot C+A\cdot D$
14	[AB] [AC] [BD]	$A+B+C+D+A\cdot B+A\cdot C+B\cdot D$
15	[AB] [AC] [BC] [D]	$A+B+C+D+A\cdot B+A\cdot C+B\cdot C$
16	[AB] [CD]	$A+B+C+D+A\cdot B+C\cdot D$
17	[AB] [AC] [D]	$A+B+C+D+A\cdot B+A\cdot C$
18	[AB] [C] [D]	$A+B+C+D+A\cdot B$
19	[A] [B] [C] [D]	$A+B+C+D$

We will now consider the various procedures which are available to us to derive such estimates.

Three estimation procedures dominate current practice. These are: (a) the iterative proportional fitting (Deming–Stephan) procedure; (b) the iterative weighted least squares procedure; (c) the Newton–Raphson procedure. All these procedures are *maximum likelihood* estimation methods, and all three produce maximum likelihood estimators (MLE's), e.g. $\hat{m}_{ijk}$ and $\hat{\lambda}_{ij}^{AB}$ which have the very desirable properties which we discussed in Section 2.4. Many of the important results concerning maximum likelihood estimation of log-linear models were provided by Birch (1963), (see Bishop, Fienberg and Holland 1975, 57–83 for further discussion). In particular, Birch showed that the MLE's of the expected cell frequencies are functions of the observed marginal totals corresponding to the highest order effects in the log-linear model, and that a variety of different sampling schemes (see discussion in Section 1.4) give rise to the same MLE's.

For certain models, the MLE's of the expected cell frequencies can be obtained directly from simple closed-form expressions. For example, in the case of the two-dimensional model of independence:

$$\log_e m_{ij} = \lambda + \lambda_i^A + \lambda_j^B \qquad\qquad [5.50]$$

it is well known that the MLE's of the expected cell frequencies take the form:

$$\hat{m}_{ij} = (n_{i+}n_{+j})/N \qquad\qquad [5.51]$$

Similarly, in the case of the three-dimensional model of mutual independence:

$$\log_e m_{ijk} = \lambda + \lambda_i^A + \lambda_j^B + \lambda_k^C \qquad\qquad [5.52]$$

the MLE's of the expected cell frequencies take the form

$$\hat{m}_{ijk} = (n_{i++}n_{+j+}n_{++k})/N^2 \qquad\qquad [5.53]$$

Most of the other two-and three-dimensional log-linear models also have closed-form MLE's. For example, the type-3 and type-4 three-dimensional conditional independence and multiple independence log-linear models

$$\log_e m_{ijk} = \lambda + \lambda_i^A + \lambda_j^B + \lambda_k^C + \lambda_{ik}^{AC} + \lambda_{jk}^{BC} \qquad\qquad [5.54]$$

and

$$\log_e m_{ijk} = \lambda + \lambda_i^A + \lambda_j^B + \lambda_k^C + \lambda_{ij}^{AB} \qquad\qquad [5.55]$$

have MLE's of the expected cell frequencies which take the form

$$\hat{m}_{ijk} = \frac{n_{i+k}n_{+jk}}{n_{++k}} \qquad \text{and} \quad \hat{m}_{ijk} = \frac{n_{ij+}n_{++k}}{N} \qquad\qquad [5.56]$$

respectively. However, the pairwise association three-dimensional log-linear model

$$\log_e m_{ijk} = \lambda + \lambda_i^A + \lambda_j^B + \lambda_k^C + \lambda_{ij}^{AB} + \lambda_{ik}^{AC} + \lambda_{jk}^{BC} \qquad [5.57]$$

and many models for higher-dimensional tables do *not* have closed-form MLE's. In such cases, some sort of iterative numerical optimization procedure which ultimately converges to the required estimates is needed. The three estimation procedures which dominate current practice are, in essence, merely different types of numerical optimization procedure.

Bishop, Fienberg and Holland (1975, 76) provide a general scheme for determining whether a particular log-linear model has closed-form MLE's, and Goodman (1971b) provides a procedure for deriving these closed-form MLE's (see also the introductory discussion by Reynolds 1977, 145–52). However, the now widespread availability and simplicity of standard computer programs which use iterative procedures to estimate all log-linear models (regardless of whether a model can be estimated using a closed-form expression or not) has rendered such schemes redundant in most applied research contexts. It is far more important for the geographer or environmental scientist to know something about the iterative estimation procedures which dominate current practice than to be able to identify which log-linear models have closed-form MLE's and which do not, and it is to these iterative procedures that we now turn our attention.

5.6.1   *The iterative proportional fitting procedure*

This procedure, which was first introduced for work with census data by Deming and Stephan (1940a, b), is the most commonly used of the iterative estimation procedures for log-linear models and is employed by the widely used BMDP, ECTA and C-TAB programs (see Ch. 7). The name 'iterative proportional fitting' derives from the fact that each iteration in the procedure involves a proportional adjustment of a row or column of the table of estimated expected cell frequencies. Preliminary estimates of the expected cell frequencies for a particular log-linear model (usually each cell is given the value 1 as an initial estimate) are successively adjusted to fit each of the marginal totals of observed frequencies listed in the 'fitted marginals' abbreviation of that model.

If we take the example of a three-dimensional log-linear model of the form [5.57], which can be abbreviated in 'fitted marginals' notation as [AB] [AC] [BC], the initial estimates are adjusted first to fit [AB], then to fit [AC], and finally to equal the [BC] observed frequencies. With each new fit, however, the previous adjustments (e.g. to the [AB] and [AC] observed frequencies) are lost. Hence, it is necessary to begin a new cycle of the iteration process, starting with the final estimates of the first cycle. At each step of this new cycle, the estimates of the expected frequencies are adjusted to fit one set of marginal totals, but they do not fit the other two. Fortunately, however, as the cycles increase, the estimates come closer and closer to satisfying simultaneously all specified observed marginal totals. Ultimately, an arbitrarily small difference between estimated marginal totals and the specified observed marginal totals is reached, and further iterations only change the estimates of the expected cell frequencies by a tiny amount. At this stage, the iteration procedure is concluded. Generally, convergence is very rapid. It seldom takes more than ten cycles, and for models which have closed-form MLE's it produces the maximum likelihood estimates at the end of the first cycle.[†]

As an illustration of this procedure we will consider the process of fitting the three-dimensional pairwise association model [5.57] to the contingency table shown in Table 5.18. This table gives data on voting for the Governorship of Georgia and has three variables; the candidates and the birthplace and education level of the voter. It has previously been analysed by Reynolds (1977, 154) and comes originally from Orum and McCranie (1970). Two sets of marginal totals [AC] and [BC] are reported in the table. The other set of marginal totals [AB] which are of interest to us are $n_{11+} = 61$, $n_{12+} = 61$, $n_{21+} = 37$, $n_{22+} = 17$.

Table 5.18
Vote for Governorship of Georgia (adapted from Reynolds 1977, 154 and Orum and McCranie 1970)

		Variable C					
		Low educational level			High educational level		
		Variable B			Variable B		
		Rural birth place	Urban birth place	Marginal total [AC]	Rural birth place	Urban birth place	Marginal total [AC]
Variable A	Candidate 1 Votes	17	16	$n_{1+1}= 33$	44	45	$n_{1+2}= 89$
	Candidate 2 Votes	28	8	$n_{2+1}= 36$	9	9	$n_{2+2}= 18$
	Marginal total [BC]	$n_{+11} = 45$	$n_{+21}= 24$		$n_{+12} = 53$	$n_{+22}= 54$	

The iterative proportional fitting procedure begins by assuming an initial estimate of 1 for each expected frequency, i.e. $\hat{m}_{ijk}^{(0)} = 1$ for all i,j and k, and it proceeds by adjusting these initial estimates proportionally to satisfy the first set of observed marginal totals [AB]. This is done by calculating

$$\hat{m}_{ijk}^{(1)} = \left(\hat{m}_{ijk}^{(0)} \times n_{ij+} \right)/\hat{m}_{ij+}^{(0)} \qquad [5.58]$$

and it results in the [AB] marginal totals of the estimates $\hat{m}_{ij+}^{(1)}$ being set equal to the observed marginal totals n_{ij+}. These revised estimated expected frequencies are now adjusted to satisfy the second set of observed marginal totals [AC] by calculating

$$\hat{m}_{ijk}^{(2)} = \left(\hat{m}_{ijk}^{(1)} \times n_{i+k} \right)/\hat{m}_{i+k}^{(1)} \qquad [5.59]$$

This results in the [AC] marginal totals of the estimates $\hat{m}_{i+k}^{(2)}$ being set

[†] This statement is true for tables of less than seven dimensions. In seven-dimensional or higher-dimensional tables, additional cycles may be necessary.

equal to the observed marginal totals n_{i+k}. Finally, the first cycle is completed by adjusting these revised estimated expected frequencies to satisfy the third set of observed marginal totals $[BC]$ using

$$\hat{m}_{ijk}^{(3)} = \left(\hat{m}_{ijk}^{(2)} \times n_{+jk} \right)/\hat{m}_{+jk}^{(2)}$$

[5.60]

and this results in the $[BC]$ marginal totals of the estimates $\hat{m}_{+jk}^{(3)}$ being set equal to the observed marginal totals n_{+jk}. The second cycle now begins using the estimates obtained from [5.60] in equation [5.58], and the process is continued until it converges. Figure 5.1 shows this process in graphical terms.

Fig. 5.1
Iterative proportional fitting of pairwise association log-linear model [5.57] to Table 5.18 (cell frequencies from Reynolds 1977, 155). Boxes represent tables of expected cell frequencies

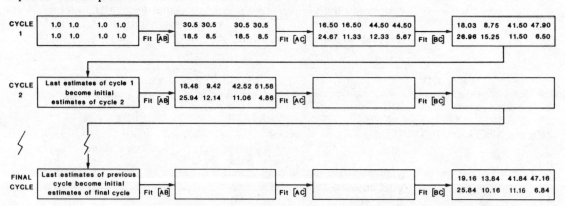

It can be seen from the preceding description that the iterative proportional fitting procedure does not provide estimates of the parameters, or of the asymptotic variances of the parameter estimates, as by-products of the computation. However, once the expected cell frequencies have been estimated, the required parameter estimates can be derived in a straightforward manner. They can be computed by simply substituting estimated expected cell frequencies in place of the m_{ij} terms in the two-dimensional expressions [5.12] to [5.14], and in place of the m_{ijk}, m_{ijkl}, etc. terms in expressions similar to [5.12] to [5.14] for the three-dimensional, four-dimensional and higher-dimensional cases. For example, in the case of the three-dimensional pairwise association model which we have just considered, the parameter estimators $\hat{\lambda}$, $\hat{\lambda}^A, \ldots, \hat{\lambda}^{BC}$ are given by substituting the $\hat{m}_{ijk}$ terms in place of the m_{ijk} terms in expressions of the form

$$\lambda = \frac{1}{IJK} \sum_{i=1}^{I} \sum_{j=1}^{J} \sum_{k=1}^{K} \log_e m_{ijk}$$

$$\lambda_i^A = \frac{1}{JK} \sum_{j=1}^{J} \sum_{k=1}^{K} \log_e m_{ijk} - \lambda$$

[5.61]

$$\vdots \qquad \qquad \vdots$$

$$\lambda_{jk}^{BC} = \frac{1}{I} \sum_{i=1}^{I} \log_e m_{ijk} - \frac{1}{IK} \sum_{i=1}^{I} \sum_{k=1}^{K} \log_e m_{ijk} - \frac{1}{IJ} \sum_{i=1}^{I} \sum_{j=1}^{J} \log_e m_{ijk} + \lambda$$

(see Haberman 1978, 219; Everitt 1977, 88; Reynolds 1977, 139 for further discussion). Although derving the parameter estimates in this way may appear to be rather tedious, in practice it presents no difficulties to the researcher as the computer programs (e.g. BMDP, ECTA, C-TAB) which perform the iterative proportional fitting will automatically compute the parameter estimates.

Furthermore, given the estimated expected frequencies, asymptotic variances of the λ, $\hat{\lambda}_i^A$,... terms can be obtained without further iterative calculations as long as the MLE's of the expected cell frequencies can be expressed in closed-form. For example, in the case of a *saturated* log-linear model, Goodman (1970, 1971a) notes that each of the $\hat{\lambda}^z$ terms (where $\hat{\lambda}^z$ is used here to denote any particular parameter estimator) has an asymptotic variance which can be estimated from the expression

$$\text{Var}(\hat{\lambda}^z) = \sum_{i=1}^{I} \sum_{j=1}^{J} a_{ij}^2/n_{ij} \qquad [5.62]$$

in the two-dimensional case, from the expression

$$\text{Var}(\hat{\lambda}^z) = \sum_{i=1}^{I} \sum_{j=1}^{J} \sum_{k=1}^{K} a_{ijk}^2/n_{ijk} \qquad [5.63]$$

in the three-dimensional case, and from simple extensions of these expressions in the higher-dimensional case. In [5.62] and [5.63] the a_{ijk}, etc. are constants whose values depend upon which particular $\hat{\lambda}^z$ is under consideration and whose values sum to zero[†], i.e.

$$\sum_{i=1}^{I} \sum_{j=1}^{J} \sum_{k=1}^{K} a_{ijk} = 0 \qquad [5.64]$$

in the three dimensional case, and so on. When all variables in the contingency table are dichotomous, equations such as [5.62] and [5.63] reduce to a particularly simple form (see Goodman 1972a; Fienberg 1977, 71). For example, in the three-dimensional $2\times2\times2$ case, [5.63] can be simplified to the form

$$\text{Var}(\hat{\lambda}^z) = \left[\sum_{i=1}^{2} \sum_{j=1}^{2} \sum_{k=1}^{2} (1/n_{ijk}) \right] \Big/ 8^2 \qquad [5.65]$$

and thus all parameter estimates in the $2\times2\times2$ saturated model have the same estimated asymptotic variance.

[†]To appreciate the meaning of these constants, the reader should note that each parameter in expressions such as [5.61] can be expressed as a linear combination of individual $\log_e m_{ijk}$ terms. For example, in the simple $2\times2\times2$ case, the parameter λ_1^A can be written as

$$\lambda_1^A = \frac{1}{4}(\log_e m_{111} + \log_e m_{112} + \log_e m_{121} + \log_e m_{122}) - \frac{1}{8}(\log_e m_{111}$$
$$+ \log_e m_{112} + \log_e m_{121} + \log_e m_{122} + \log_e m_{211} + \log_e m_{212} + \log_e m_{221} + \log_e m_{222})$$

$$= \frac{1}{8}\log_e m_{111} + \frac{1}{8}\log_e m_{112} + \frac{1}{8}\log_e m_{121} + \frac{1}{8}\log_e m_{122} - \frac{1}{8}\log_e m_{211}$$
$$- \frac{1}{8}\log_e m_{212} - \frac{1}{8}\log_e m_{221} - \frac{1}{8}\log_e m_{222}$$

$$= \sum_{i=1}^{2} \sum_{j=1}^{2} \sum_{k=1}^{2} a_{ijk} \, \log_e m_{ijk}$$

For *unsaturated* log-linear models in general, expressions such as [5.62] and [5.63] provide only approximations to the desired asymptotic variances. In fact, such expressions provide overestimates or *upper bounds* for the estimated asymptotic variances in such cases. However, if the log-linear model is unsaturated but is a 'direct' model (i.e. a model for which the MLE's of the expected cell frequencies can be expressed in closed-form) then methods do exist for obtaining the required estimates of the asymptotic variances without further iterative calculations (see Lee 1977; Haberman 1974a, 205), and these methods are implemented by the standard iterative scaling computer programs (e.g. BMDP, C-TAB).

For 'indirect' unsaturated models (i.e. models for which the MLE's of the expected cell frequencies cannot be expressed in closed–form), estimates of the asymptotic variances can be obtained by an iterative procedure (see Haberman 1978, 220). Alternatively, a 'direct' model can be found for which the 'indirect' model is a subset, and then Lee's (1977) method can be applied to the direct model, providing an upper bound for the estimated asymptotic variance. (This procedure is adopted by the standard BMDP log-linear modelling program.)

5.6.2   *The iterative weighted least squares procedure*

This is the procedure employed by the widely used computer program GLIM (Generalized Linear Interactive Modelling) developed by Nelder and his associates at Rothamsted under the sponsorship of the Royal Statistical Society (Baker and Nelder 1978). The procedure was proposed by Nelder and Wedderburn (1972) as a method of obtaining maximum likelihood estimates in a broad class of linear models known as 'generalized linear models' (GLM's). Such models are those which have error distributions belonging to the exponential family (i.e. distributions such as the Normal, the binomial, the Poisson, the χ^2 and the gamma) and which have forms which satisfy three conditions relating to the error structure, the 'linear predictor' and the 'link function' (see Nelder and Wedderburn 1972; Nelder 1974; Baker and Nelder 1978 and in the geographical literature O'Brien, 1983; O'Brien and Wrigley 1984 for further details). Although this is not the appropriate place to consider the characteristics of 'generalized linear models' (we will defer such discussion until Ch. 11), it should be noted that the log-linear models of this chapter can be viewed as merely a special case of the general class of GLM's; a special case in which the error distribution is Poisson (see our discussion in Chapter 1) and the 'link function' is the natural logarithm.

The iterative weighted least squares procedure which Nelder and Wedderburn proposed was developed from an algorithm first used for probit models (Finney 1971). The procedure is very similar to the non-iterative weighted least squares method which we discussed in detail in Chapters 2 and 4. However, in this case, the general matrix expression [4.36] becomes

$$\hat{\boldsymbol{\lambda}} = (\mathbf{X}'\mathbf{V}_z^{-1}\mathbf{X})^{-1}\mathbf{X}'\mathbf{V}_z^{-1}\mathbf{z} \qquad\qquad [5.66]$$

where[†], in the case of a three-dimensional log-linear model for example, $\hat{\boldsymbol{\lambda}}$ is composed of individual parameter estimates $\hat{\lambda}$, $\hat{\lambda}^A$, . . . , $\hat{\lambda}^{BC}$, etc., V_z^{-1} is a matrix of weights which has terms of the form $\hat{m}_{ijk}$ down its principal diagonal and zeros elsewhere, and where z is a vector of modified or 'working' values of the logarithms of the expected cell frequencies which take the form

$$z_{ijk} = \log_e \hat{m}_{ijk} + \frac{n_{ijk} - \hat{m}_{ijk}}{\hat{m}_{ijk}} \qquad [5.67]$$

It can be seen that the solution of [5.66], and a number of elements within it, depends upon the estimates of the expected cell frequencies. Initially, these values are unknown. Thus, a direct solution of [5.66] is not possible, and we must use an iterative procedure. This iterative procedure begins by taking the observed cell frequencies n_{ijk} as a first approximation to the $\hat{m}_{ijk}$ (unless $n_{ijk} = 0$ in which case Nelder recommends setting $\hat{m}_{ijk} = \frac{1}{2}$). From this we get a first approximation to the weights, and we set $z_{ijk} = \log_e \hat{m}_{ijk} = \log_e n_{ijk}$. We then obtain a first approximation of the elements of $\hat{\boldsymbol{\lambda}}$ from [5.66]. These approximate values of the parameter estimates then allow us to calculate a second approximation to the $\hat{m}_{ijk}$, and we can thus obtain a second approximation to the weights, set $z_{ijk} = \log_e \hat{m}_{ijk} + (n_{ijk} - \hat{m}_{ijk})/\hat{m}_{ijk}$, and obtain a second approximation of the elements of $\boldsymbol{\lambda}$ from [5.66]. We continue this procedure until the process converges. The values of the parameters obtained are then the maximum likelihood estimates.

As we noted in [4.37], standard errors of $\hat{\boldsymbol{\lambda}}$ can be obtained directly from [5.66]. We simply take the square roots of the values along the principal diagonal of the matrix

$$\mathbf{V}_{\hat{\lambda}} = (\mathbf{X}'\mathbf{V}_z^{-1}\mathbf{X})^{-1} \qquad [5.68]$$

where $\mathbf{V}_{\hat{\lambda}}$ is an estimator of the variance-covariance matrix of $\hat{\boldsymbol{\lambda}}$. It should be noted that [5.68] provides direct and exact estimates of the asymptotic variances of the parameter estimates for both saturated and non-saturated log-linear models. Consequently, the iterative weighted least squares procedure differs in this respect from the iterative proportional fitting procedure which is often accompanied by procedures which provide only upper bound estimates of the asymptotic variances in the case of unsaturated models (particularly those which are 'indirect' models).

In both equations [5.66] and [5.68] it should be noted that matrix $\mathbf{X}$ is assumed to be a 'cornered effect' design matrix of the type shown in the footnote to page 160. The usual GLIM-type constraints on the parameters of the log-linear models (i.e. constraints of the form [5.17], [5.20] and [5.35]) are assumed to hold.

[†] Nelder and Wedderburn 1972; Nelder 1974; Baker and Nelder 1978, provide further details but, unfortunately for the reader, they write their equations in a general form suitable for all GLM's, rather than in the specific notation of log-linear models. Consequently, the reader must translate their notation into that adopted in this chapter.

5.6.3   *The Newton–Raphson procedure*

This is a very old and widely used procedure for the solution of non-linear equations (see Ostrowski 1966; Goldfeld and Quandt 1972) and it is modifications of this type of classical numerical optimization procedure which are used to find the maximum likelihood estimates of the logistic/logit models which we discussed in Chapter 2. In the context of log-linear models, Haberman (1978, 64, 128, 170, 207) shows that the Newton–Raphson procedure reduces to a series of weighted least-squares problems. As such the procedure is very similar to the Nelder and Wedderburn iterative weighted least squares procedure which is used in GLIM. Consequently, as the procedures differ in minor detail rather than in essence, there is little need to duplicate our previous discussion, and the interested reader is referred to Haberman (1978) who makes extensive use of the Newton–Raphson procedure. Widely available computer programs which employ this procedure include MULTIQUAL developed by Bock and Yates (1973) and FREQ developed by Haberman (1979).

Like Nelder and Wedderburn's iterative weighted least squares procedure, the Newton–Raphson procedure has the advantage over iterative proportional fitting of providing direct and exact estimates of the parameters and the asymptotic variances of the parameter estimates. Haberman (1978) provides a detailed discussion of such issues. The question of whether the Newton–Raphson procedure, Nelder–Wedderburn iterative weighted least squares procedure, or iterative proportional fitting procedure converges most rapidly in any particular case is debatable, and is likely to be unimportant to most geographers and environmental scientists who will simply utilize the available standard computer programs. However, there is some evidence to suggest that the Newton-Raphson procedure handles tables of higher-dimension less efficiently than the iterative proportional fitting procedure, and that it is more difficult to implement in such cases.

5.7   Model selection

In Examples 5.1–5.3 we selected an 'acceptable' model from a hierarchical set of log-linear models by fitting all possible models in the hierarchical set and choosing the most parsimonious model which had a satisfactory fit to be observed data (as measured by the G^2 statistic). For the simple two-dimensional and three-dimensional tables in these examples, this strategy performed satisfactorily. However, in the case of four-dimensional contingency tables where there are 167 possible models in the hierarchical set, or higher-dimensional tables where there can be thousands of possible models, this naive selection strategy is no longer satisfactory. In such cases, systematic, efficient and, if possible, statistically elegant model selection strategies are required. The search for such model selection strategies has generated a considerable literature. Many possibilities have been suggested and these have been adopted by users with varying degrees of generality. For convenience we will consider in this chapter just four of these possible strategies. As such, our discussion is not intended to be exhaustive of the possibilities but is intended merely to illustrate the most widely adopted approaches. We will treat the four selected strategies as distinct entities, although we will see that they share a number of common elements.

Before discussing the four approaches in detail it will be useful, however, if we first give some brief consideration to the general criteria for model selection. Commonly used criteria are parsimony, goodness-of-fit, simplicity of interpretation, convergence with substantive theory and the inclusion of all significant effects.

We have seen in Examples 5.1–5.3 that we normally attempt to select a log-linear model which provides a good fit to the observed contingency table data (i.e. which has a non-significant G^2 statistic) and which is parsimonious in the sense of having no parameters in it which are not necessary to describe the structural relationships which exist between the variables in the contingency table. Sometimes, but not always, this will result in a model which contains all significant effects, i.e. a model for which the change in G^2 due to the addition of any single term is non-significant, while the change in G^2 due to the deletion of any single term is significant. In certain cases, however, it is useful if these criteria are partially relaxed. Previous theory must, to some extent, guide selection and a model which is compatible with theory is usually to be preferred to one that is not. Sometimes this will imply that a model which is consistent with previous substantive results, but which has slightly more terms than the most parsimonious of the possible models with a good fit to the observed data, should be selected instead. This is not to deny, of course, that existing theory may be incorrect, and that the observed data and the models fitted to it may suggest that the theory needs modifying. It merely suggests that the investigator should beware of selecting, on purely statistical grounds, a model whose form is essentially an artifact of the sample variation within the observed contingency table data, when a model which conforms with previous research findings provides a perfectly acceptable alternative. In other words, model selection cannot be a purely mechanical process. Theoretical knowledge and common sense must guide the procedure, and the final selection must reflect a trade-off of the possible model selection criteria. This should be borne in mind by the reader when he utilizes any of the four selection procedures to be discussed below.

5.7.1   Strategy 1.   Stepwise selection

Stepwise procedures for the selection of log-linear models were suggested by Goodman (1971a) and include both the forward selection and backward elimination methods familiar to users of multiple regression models (see Fienberg 1977, 65; Payne 1977, 132 for further details). To illustrate this approach we will consider the example of a hypothetical four-dimensional $3 \times 3 \times 2 \times 2$ table.

We begin the strategy by choosing a significance level, say 5 per cent, and we then examine the goodness-of-fit of the hierarchy of models shown in [5.69] to [5.72] below (i.e. a model containing only main effect terms; a model containing main effect terms and all two-variable interaction terms; a model containing main effect terms and all two-variable and three-variable interaction terms; a saturated model containing main effect terms and all two-variable, three-variable and four-variable terms) where $i = 1, 2, 3; j = 1, 2, 3; k = 1, 2; l = 1, 2$.

$$\log_e m_{ijkl} = \lambda + \lambda_i^A + \lambda_j^B + \lambda_k^C + \lambda_l^D \tag{5.69}$$

$$\log_e m_{ijkl} = \lambda + \lambda_i^A + \lambda_j^B + \lambda_k^C + \lambda_l^D + \lambda_{ij}^{AB} + \lambda_{ik}^{AC} + \lambda_{il}^{AD} + \lambda_{jk}^{BC}$$
$$+ \lambda_{jl}^{BD} + \lambda_{kl}^{CD} \tag{5.70}$$

$$\log_e m_{ijkl} = \lambda + \lambda_i^A + \lambda_j^B + \lambda_k^C + \lambda_l^D + \lambda_{ij}^{AB} + \lambda_{ik}^{AC} + \lambda_{il}^{AD} + \lambda_{jk}^{BC}$$
$$+ \lambda_{jl}^{BD} + \lambda_{kl}^{CD} + \lambda_{ijk}^{ABC} + \lambda_{ijl}^{ABD} + \lambda_{ikl}^{ACD} + \lambda_{jkl}^{BCD} \tag{5.71}$$

$$\log_e m_{ijkl} = \lambda + \lambda_i^A + \lambda_j^B + \lambda_k^C + \lambda_l^D + \lambda_{ij}^{AB} + \lambda_{ik}^{AC} + \lambda_{il}^{AD} + \lambda_{jk}^{BC}$$
$$+ \lambda_{jl}^{BD} + \lambda_{kl}^{CD} + \lambda_{ijk}^{ABC} + \lambda_{ijl}^{ABD} + \lambda_{ikl}^{ACD} + \lambda_{jkl}^{BCD} + \lambda_{ijkl}^{ABCD} \tag{5.72}$$

If only the saturated model [5.72] adequately fits the data then the model selection procedure stops at that point. If, however, both the saturated model and the no four-variable interactions model [5.71] adequately fit the data but the no three-variable interactions model [5.70] does not, then we assume that the 'acceptable' model lies some where between models [5.70] and [5.71]. Similarly, if the saturated model, the no four-variable interactions model [5.71] *and* the no three-variable interactions model [5.70] adequately fit the data but the no two-variable interactions model [5.69] does not, then we assume that the 'acceptable' model lies between models [5.69] and [5.70]. Finally, if all four models [5.69]−[5.72] adequately fit the data, we assume that the 'acceptable' model lies somewhere between model [5.69] and the equiprobability model, $\log_e m_{ijkl} = \lambda$.

Having decided between which pair of models [5.69] to [5.72] the 'acceptable' model lies, we then adopt either a forward selection or backward elimination method to refine the solution. To illustrate this we will assume that we have determined that our 'acceptable' model lies between models [5.70] and [5.71] and that the G^2 values for these two models are 38.23 with sixteen degrees of freedom and 2.43 with four degrees of freedom respectively.

Using the *forward selection* method we would then commence with model [5.70] and add to it each of the sets of three-variable interaction parameters in turn. This might produce the following results:

Model specification	G^2	d.f.	Conditional G^2	d.f.
Model [5.70] + $A \cdot B \cdot C$	15.13	12	$(38.23 - 15.13) = 23.10$	4
Model [5.70] + $A \cdot B \cdot D$	29.99	12	$(38.23 - 29.99) = 8.24$	4
Model [5.70] + $A \cdot C \cdot D$	31.20	14	$(38.23 - 31.20) = 7.03$	2
Model [5.70] + $B \cdot C \cdot D$	28.20	14	$(38.23 - 28.20) = 10.03$	2

We then add to Model (5.70) the three-variable interaction parameters with the largest, statistically significant, conditional G^2 statistic (where the conditional G^2 is defined in equation [5.44]). In our case these are the $A \cdot B \cdot C$ parameters. We then add to Model [5.70] + $A \cdot B \cdot C$ the three-variable interaction parameters with the next largest, statistically significant, conditional G^2 statistic. In our case these are the $B \cdot C \cdot D$

parameters and adding these produces the result

Model specification	G^2	d.f.	Conditional G^2	d.f.
Model [5.70] + $A \cdot B \cdot C + B \cdot C \cdot D$	8.93	10	$(15.13 - 8.93) = 6.20$	2

The conditional G^2 statistic is found to be significant and it appears that we are justified in adding the $B \cdot C \cdot D$ parameter to our model.

At this point, it should be stressed that our tests of the significance of parameters $A \cdot B \cdot C$ and $B \cdot C \cdot D$ are conditional on the other parameters present in the model at that stage (i.e. in the case of parameters $A \cdot B \cdot C$ the test is conditional on the presence of all main effects and all two-variable interaction effects). However, at a later stage, when other parameters (i.e. $B \cdot C \cdot D$) have entered the model, parameters $A \cdot B \cdot C$ may be found to be no longer significant when considered conditional upon all main effects, all two-variable interaction effects *and* the $B \cdot C \cdot D$ effects. At this stage in the process, therefore, we should attempt to delete from the model any three-variable interaction parameters which no longer make a significant contribution. In our case, we can attempt to do this by testing whether it is appropriate to delete the $A \cdot B \cdot C$ parameters. In this case we get the result

Model specification	G^2	d.f.	Conditional G^2	d.f.
Model [5.70] + $B \cdot C \cdot D$	28.20	14	$(28.20 - 8.93) = 19.27$	4

The conditional G^2 statistic is found to be significant so we find, therefore, that we cannot delete the $A \cdot B \cdot C$ parameters.

Next, we add to our current log-linear model specification (Model [5.70] + $A \cdot B \cdot C + B \cdot C \cdot D$) the three-variable interaction parameters with the next largest, statistically significant, conditional G^2 statistic. In our case, these are the $A \cdot C \cdot D$ parameters and adding these produces the result

Model specification	G^2	d.f.	Conditional G^2	d.f.
Model [5.70] + $A \cdot B \cdot C + B \cdot C \cdot D + A \cdot C \cdot D$	2.93	8	$(8.93 - 2.92) = 6.02$	2

This indicates that we are justified in adding the $A.C.D$ parameters to our previous model. We should now test whether we are justified in deleting the $A.B.C$ and/or the $B.C.D$ parameters from the model. In this case, we find (on the basis of conditional G^2 statistics) that neither set of parameters can be deleted, nor can the remaining set of parameters $A.B.D$ be included into the model. We therefore conclude the model selection process and choose

Model [5.70] + $A.B.C + B.C.D + A.C.D$

as our 'acceptable' model.

Using the *backwards elimination* method we basically reverse the previous steps. In other words, we commence with model [5.71] and delete from it each of the sets of three-variable interaction parameters ($B.C.D$, $A.C.D$, $A.B.D$, $A.B.C$) in turn. This might produce the following results

Model specification	G^2	d.f.	Conditional G^2	d.f.
Model [5.71] $- B \cdot C \cdot D$	13.99	6	$(13.99 - 2.43) = 11.56$	2
Model [5.71] $- A \cdot C \cdot D$	12.46	6	$(12.46 - 2.43) = 10.03$	2
Model [5.71] $- A \cdot B \cdot D$	2.93	8	$(\ 2.92 - 2.43) = \ 0.49$	4
Model [5.71] $- A \cdot B \cdot C$	20.50	8	$(20.50 - 2.43) = 18.07$	4

We then delete from Model [5.71] the three-variable interaction parameters associated with the smallest, statistically non-significant, conditional G^2 statistic in the above results. In our case these are the $A \cdot B \cdot D$ parameters and this implies the selection of the log-linear specification, Model [5.70]$+A \cdot B \cdot C+A \cdot C \cdot D+B \cdot C \cdot D$. In this case, we are unable to delete any further sets of parameters as all other conditional G^2 statistics are statistically significant. Neither are we able to add back the $A.B.D$ parameters, and thus we conclude the model selection process and choose

$$\text{Model } [5.70] + A \cdot B \cdot C + B \cdot C \cdot D + A \cdot C \cdot D$$

as our 'acceptable' model.

In this case, both forwards selection and backwards elimination methods have produced the same 'acceptable' model. However, this will not always be the case. In addition, it should be noted that as in stepwise regression procedures, the significance values of these stepwise tests must be interpreted with considerable caution because of the highly dependent nature of the test statistics.

5.7.2   Strategy 2.   Abbreviated stepwise selection

This is the name proposed by Goodman (1971a) for a version of the stepwise selection procedure which uses *standardized values* of the parameter estimates in the saturated log-linear model to determine the point at which the forward selection or backward elimination process will start.

This strategy begins with the fitting of the saturated log-linear model and the computation of the standardized values of the parameter estimates. If we denote any particular parameter estimator $\hat{\lambda}^z$ and its asymptotic variance estimator as $\text{Var}(\hat{\lambda}^z)$, then the standardized value of $\hat{\lambda}^z$ can be calculated as

$$\hat{\lambda}^z / \sqrt{\text{Var}(\hat{\lambda}^z)} \qquad\qquad [5.73]$$

Under the null hypothesis that λ^z is zero, [5.73] is distributed asymptotically as a standard normal distribution with zero mean and unit variance. We can, therefore, compare the computed standardized value from [5.73] with the critical value obtained from the standard normal distribution tables at a given level of probability. In simple terms, we can say that if the computed standardized value is greater than $+ 1.96$ or less than $- 1.96$ we can reject the null hypothesis that λ^z is zero. However, it should be borne in mind that in the saturated model we are concerned with a *set* of parameters rather than just a single λ^z in isolation, and we thus face a simultaneous-inference problem. Goodman (1964, 1969) has discussed this problem and provides appropriate multiple-test procedures.

As an illustration of this procedure we will consider the example of a hypothetical $2\times2\times2\times2$ contingency table. Table 5.19 shows the hypothetical parameter estimates and standardized values for the saturated log-linear model fitted to such a table. These standardized values can then be used to guide our selection of a 'starting' or 'base' model; a model which we subsequently attempt to refine using the forward selection or backwards elimination process. In the case of Table 5.19, the standardized values indicate that all main effect terms, all two-variable interaction terms, and the three-variable $A \cdot B \cdot C$ interaction terms should be included in the initial 'starting' or 'base' model. The $A \cdot B \cdot D$, $B \cdot C \cdot D$ and $A \cdot B \cdot C \cdot D$ terms should not be included in such a base model, and the position of the $A \cdot C \cdot D$ term is unclear.

Table 5.19
Hypothetical parameter estimates and standardized values for saturated model fitted to a $2\times2\times2\times2$ table

Parameter	Parameter estimate	Standardized value	Parameter	Parameter estimate	Standardized value
λ_1^A	−0.508	−13.84	λ_{11}^{BD}	−0.163	−4.64
λ_1^B	−0.237	−6.46	λ_{11}^{CD}	0.149	4.06
λ_1^C	0.464	12.64	λ_{111}^{ABC}	0.198	5.40
λ_1^D	0.234	6.38	λ_{111}^{ABD}	0.036	0.98
λ_{11}^{AB}	0.245	6.68	λ_{111}^{ACD}	−0.071	−1.93
λ_{11}^{AC}	−0.189	−5.15	λ_{111}^{BCD}	−0.008	−0.22
λ_{11}^{AD}	−0.199	−5.42	λ_{1111}^{ABCD}	0.023	0.63
λ_{11}^{BC}	−0.218	−5.94			

Note: Assuming constraints of the form [5.15], [5.19], [5.34] and [5.42] the estimate of λ_2^A is 0.508 and its standardized value 13.84, the estimate of λ_{12}^{AB} is −0.245 and its standardized value −6.68, and so on. In addition, as all variables are dichotomous, all parameter estimates have the same estimated asymptotic variance (see equation [5.65]).

Using the *forward selection* method we would then choose the model

$$\log_e m_{ijkl} = \lambda + \lambda_i^A + \lambda_j^B + \lambda_k^C + \lambda_l^D + \lambda_{ij}^{AB} + \lambda_{ik}^{AC} + \lambda_{il}^{AD} + \lambda_{jk}^{BC}$$
$$+ \lambda_{jl}^{BD} + \lambda_{kl}^{CD} + \lambda_{ijk}^{ABC} \qquad \qquad \textbf{[5.74]}$$

as our 'starting' or 'base' model. (In this case we will assume that this model has a G^2 statistic of 7.10 with four degrees of freedom). We would then fit all other possible hierarchical models which *include* [5.74] and which differ from it by just one additional set of effects; i.e. we would add to [5.74] each of the other three-variable interaction parameters in turn. This might produce the following results

Model specification	G^2	d.f.	Conditional G^2	d.f.
Model [5.74]$+A \cdot B \cdot D$	6.05	3	(7.10−6.05) = 1.05	1
Model [5.74]$+A \cdot C \cdot D$	3.15	3	(7.10−3.15) = 3.95	1
Model [5.74]$+B \cdot C \cdot D$	7.01	3	(7.10−7.01) = 0.09	1

We then add to Model [5.74] the three-variable interaction term with the largest, statistically significant, conditional G^2 statistic. In our case this is $A \cdot C \cdot D$. We would then attempt to add to Model [5.74] $+ A \cdot C \cdot D$, the three-variable interaction term with the next largest, statistically significant, conditional G^2 statistic, but, in this case, there is no such term.

At this point we would then test whether it is appropriate to delete any terms from the current specification. This would involve testing whether it is appropriate to delete $A \cdot B \cdot C$ now that $A \cdot C \cdot D$ has entered the model. We should also consider testing whether it is appropriate to delete $B \cdot D$ as this is the only two-variable interaction term not automatically required by the hierarchical constraints imposed by including $A \cdot B \cdot C$ and $A \cdot C \cdot D$. In this case, we will assume that no parameters can be deleted in this way. We therefore conclude the model selection process and choose Model [5.74] $+ A \cdot C \cdot D$ as our 'acceptable' model.

Using the *backward elimination* method it would be appropriate for us to choose

$$\log_e m_{ijkl} = \lambda + \lambda_i^A + \lambda_j^B + \lambda_k^C + \lambda_l^D + \lambda_{ij}^{AB} + \lambda_{ik}^{AC} + \lambda_{il}^{AD} + \lambda_{jk}^{BC}$$
$$+ \lambda_{jl}^{BD} + \lambda_{kl}^{CD} + \lambda_{ijk}^{ABC} + \lambda_{ikl}^{ACD} \qquad [5.75]$$

as our 'starting' or 'base' model. We would then fit all possible hierarchical models which are *included in* [5.75] and differ from it by just one set of effects. This might produce the following results

Model specification	G^2	d.f.	Conditional G^2	d.f.
Model [5.75] $- A \cdot C \cdot D$	7.10	4	$(7.10 - 3.15) = 3.95$	1
Model [5.75] $- A \cdot B \cdot C$	9.50	4	$(9.50 - 3.15) = 6.35$	1
Model [5.75] $- B \cdot D$	9.40	4	$(9.40 - 3.15) = 6.25$	1

On the basis of these results we find that there are no terms which can be eliminated from [5.75] at this stage. Neither are we able to add any additional terms. Thus the model selection process concludes and we choose [5.75] as our 'acceptable' model.

Although this abbreviated stepwise procedure is a widely used model selection strategy, the reader should be aware that it is not without its problems and critics. Two possible problems with the method are usually recognized. First, if many terms in the saturated model are unnecessary, the standard errors of the parameter estimates may be considerably inflated, leading to small standardized values (and thus possible exclusion) for terms which are in fact necessary in any 'acceptable' model. Second, and particularly important when there are multiple-category variables in the contingency table, the parameter estimates and thus the standardized values will depend upon the coding system ('centred effect' or 'cornered effect') and associated constraints which are employed.

5.7.3   *Strategy 3.   Screening*

This is the name given to a selection procedure which uses Brown's (1976) partial and marginal association tests as an alternative to fitting the saturated model and calculating the standardized values. Screening the parameters of the log-linear model using Brown's tests guides the selection of a 'starting' or 'base' model which is then refined, in the usual way, using forward selection or backward elimination methods.

To understand the screening procedure, it is important to recall that in the preceding discussion of stepwise selection (and also in Example 5.3) we have noted how any test of the significance of a particular term in a log-linear model is conditional upon the other parameters included in the model. As a result, no *single* test is likely to be sufficient to assess the importance of a parameter or set of parameters but, equally, it is impractical to consider computing all possible tests. Brown (1976) has suggested a way of partially circumventing this problem. He suggests that each term in a log-linear model be tested using two test statistics; a test of partial association and a test of marginal association. These tests assess each term in two extreme situations: the first conditonal on all other terms of the same order, and the second conditional on only the lower-order relatives of the term in question. The two tests are representative of the range (they can be thought of as *approximate* lower and upper bounds to the range) of the conditional G^2 values that would be obtained by adding that particular term to a previous specification of the log-linear model. As such, the two tests enable us to screen terms and to decide whether they are significant and necessary in the model, insignificant and unnecessary in the model, or of question-able significance and in need of further investigation.

Brown's test of the hypothesis that the *partial association* of Q variables in a contingency table (where Q here stands for any particular number) is zero, is a test of whether a significant difference exists between the fit of two models. One of these models is the full model of order Q, and the other is the full model minus the Q – variable interaction term of interest. For example, in a four-dimensional table, to test the partial association of variables A and B, the full second-order model $A+B+C+D+A\cdot B+A\cdot C+A\cdot D+B\cdot C+B\cdot D+C\cdot D$ would first be fit-ted, and then the same model with the $A\cdot B\cdot$ interaction term excluded (i.e. $A+B+C+D+A\cdot C+A\cdot D+B\cdot C+B\cdot D+C\cdot D$). The difference in the G^2 values of these two models is then the partial association test statistic. It can be seen that this is a conditional test of the $A\cdot B$ interaction term adjusted for all other terms of the same order.

Brown's test of the hypothesis that the *marginal association* of Q variables is zero is a test that the Q-variable interaction is zero in the marginal sub-table formed by the Q variables. For example, in a four-dimensional table, marginal association between variables A and B is tested by forming the two-dimensional table indexed by A and B and testing the $A\cdot B$ interaction term. The reader should note, however, that this is equivalent to fitting both the model $A+B+A\cdot B$ and also the model $A+B$ to the full four-dimensional contingency table and computing the difference in their G^2 statistics. Clearly, the marginal association test provides a test of the $A\cdot B$ interaction term conditional on only its lower-order relatives. It ignores the effects of the other variables in the contingency table (in our case C and D), and thus risks attributing to the term being tested ($A\cdot B$ in our case) effects which are due to the interactions of the excluded variables or to the interactions of the excluded variables with the included variables.

Although the formal definitions of Brown's tests may be slightly difficult for the reader to fully appreciate at a first reading, the tests are surprisingly simple to illustrate. For example, in the case of a four-dimensional contingency table, the partial association test statistic of the $B \cdot C \cdot D$ interaction term can be computed by fitting the model

$$A + B + C + D + A \cdot B + A \cdot C + A \cdot D + B \cdot C + B \cdot D + C \cdot D + A \cdot B \cdot C \\ + A \cdot B \cdot D + A \cdot C \cdot D + B \cdot C \cdot D \qquad \text{[5.76]}$$

and also the model

$$A + B + C + D + A \cdot B + A \cdot C + A \cdot D + B \cdot C + B \cdot D + C \cdot D + A \cdot B \cdot C \\ + A \cdot B \cdot D + A \cdot C \cdot D \qquad \text{[5.77]}$$

and by subtracting the G^2 statistic of [5.76] from the G^2 statistic of [5.77]. The marginal association test statistic can be computed by fitting the model

$$B + C + D + B.C + B.D + C.D + B.C.D \qquad \text{[5.78]}$$

and also the model

$$B + C + D + B.C + B.D + C.D \qquad \text{[5.79]}$$

and subtracting the G^2 statistic of [5.78] from the G^2 statistic of [5.79]. The significance of the partial and marginal association test statistics is judged with reference to the critical values obtained from the tabulated χ^2 distribution for the appropriate degrees of freedom (difference in number of parameters fitted in the two models compared) and at a specified level of significance.

Using the partial and marginal association tests, model selection proceeds in the following way. First, partial and marginal association test statistics are computed for each term in the saturated log-linear model (though it should be noted that the saturated model is *not* fitted). Although the two test statistics do not bound all possible values of the conditional test statistics for any particular term, they are, nevertheless, representative of the range of such statistics and can be thought of as very approximate lower and upper bounds to the range.† When both the partial and marginal association test statistics for a particular term are *small*, it is unlikely that any other conditional test statistic for that term will be large. Hence, that term would appear not to be needed in the model. When both partial and marginal test statistics for a particular term are *large*, it is unlikely that any other conditional test statistic for that term will be small. Hence, that term would appear to be required in the model. When the partial and marginal test statistics give conflicting results (one large, one small) then it is uncertain whether that term is required in the model, and further study is required to decide the issue. Using Brown's tests in this way, each term can be screened and included in one of three groups: (a) likely to be important and required in the

---

† We would expect that the marginal association test statistic would provide the upper-bound and the partial association the lower-bound, but this is often not the case for multi-variable interactions and their positions may be the reverse of those expected (see Example 5.5).

model: (b) likely to be unimportant and not required in the model: (c) in need of further study.

On the basis of this screening of parameters, a 'starting' or 'base' model is specified and is then refined using the forward selection or backward elimination methods discussed in the previous sections. The base model could be composed either of all terms with significant partial association test statistics, or all terms with significant marginal association test statistics. However, an alternative and commonly adopted procedure is to include in the base model all terms which are significant according to *both* tests. Forward selection is then used to investigate the possibility of adding to this base model the terms which are significant in only one test of association.

As an illustration of this screening strategy, we will consider the example of a hypothetical $3 \times 3 \times 2 \times 2$ contingency table. Table 5.20 shows hypothetical partial association and marginal association statistics for each of the terms in the saturated model. Screening the terms on the basis of these statistics, we can divide them into three groups:

1. Those likely to be important and required in the model (A, B, C, D, $A \cdot B$, $A \cdot C$, $A \cdot D$, $B \cdot D$, $C \cdot D$, $A \cdot B \cdot D$).

2. Those likely to be unimportant and not required in the model ($A \cdot B \cdot C$, $B \cdot C \cdot D$, $A \cdot B \cdot C \cdot D$).

3. Those in need of further study ($B \cdot C$, $A \cdot C \cdot D$).

On the basis of this screening, we then choose as our 'starting' or 'base' model

$$\log_e m_{ijkl} = \lambda + \lambda_i^A + \lambda_j^B + \lambda_k^C + \lambda_l^D + \lambda_{ij}^{AB} + \lambda_{ik}^{AC} + \lambda_{il}^{AD}$$
$$+ \lambda_{jl}^{BD} + \lambda_{kl}^{CD} + \lambda_{ijl}^{ABD} \qquad [5.80]$$

and we then investigate, using the forward selection method, the possibility of adding either or both the terms $B \cdot C$ and $A \cdot C \cdot D$ to [5.80].

Overall, screening provides considerably more information than the saturated model and its standardized values. It gives an idea of the significance of a term by considering the approximate bounds of the range of conditional test statistics for that term, whereas the standardized value provides only a point estimate. However, this extra

Table 5.20
Partial and marginal association test statistics

λ-term	d.f.	Partial association statistic	Marginal association statistic	λ-term	d.f.	Partial association statistic	Marginal association statistic
A	2	28.1	28.1	$B \cdot D$	2	9.4	9.9
B	2	17.6	17.6	$C \cdot D$	1	6.3	7.1
C	1	9.3	9.3	$A \cdot B \cdot C$	4	7.6	7.2
D	1	25.8	25.8	$A \cdot B \cdot D$	4	10.9	11.4
$A \cdot B$	4	14.9	16.8	$A \cdot C \cdot D$	2	4.6	6.1
$A \cdot C$	2	9.1	8.9	$B \cdot C \cdot D$	2	2.5	2.4
$A \cdot D$	2	10.2	12.4	$A \cdot B \cdot C \cdot D$	4	0.2	0.2
$B \cdot C$	2	5.6	7.1				

Note: The critical χ^2 values at the 5 per cent level are: 3.84 (1 d.f.), 5.99 (2 d.f.) and 9.48 (4 d.f.). The test of marginal association is equal to that of partial association for the main effects and the highest order interactions information is gained, as we have seen above, by fitting four models per term (two models being required to compute the partial association statistic, and two to compute the marginal association statistic). Potentially, therefore, screening is a time-consuming and expensive process. However, the reader should note that screening can readily be automated in a computer program and, in fact, the log-linear modelling routines in the widely available BMDP package automatically compute Brown's partial and marginal association tests, and thus facilitate the use of screening as a model selection procedure.

5.7.4   Strategy 4.   Aitkin's simultaneous test procedure

A common characteristic of all selection strategies for log-linear models is that they involve multiple tests of the data. Theoretically we should make some allowance for this, and should adjust the significance levels of our tests to compensate for the multiple testing. If we do not compensate in this way, but treat the significance tests as independent entities, this will greatly increase the probability of attributing significance to what is merely random variation. Unfortunately, the model selection strategies we have discussed above make few attempts at such formal adjustment. Instead, they rely to a large extent on the common sense of the user and on explicit or implicit statements to the effect that: 'the significance levels [of multiple tests of the data] are merely guidelines and should not be interpreted as formal tests of significance' (Benedetti and Brown 1978). Recently, however, major advances have been made towards overcoming these difficulties. Aitkin (1978, 1979, 1980) has shown how it is possible to adjust significance levels to compensate for multiple testing, and he has developed a formal simultaneous test procedure which provides a systematic, internally consistent and efficient approach to model selection. To illustrate Aitkin's procedure we will consider the example of a hypothetical four-dimensional $2 \times 2 \times 2 \times 2$ contingency table.

The first step in the procedure is to fit various members of the hierarchical set of log-linear models; beginning in this case, with a model which contains all main effects. To this, we add, successively, all two-variable interactions, all three-variable interactions and all four-variable interactions. This produces a set of models like those shown in expressions [5.69] to [5.72]. In our hypothetical case, this might produce the results in Table 5.21. The reader should notice that, on the right-hand side of the table, the G^2 goodness-of-fit statistics are partitioned to show the change in G^2 due to a particular source (set of terms).

The second step in the simultaneous test is to determine a Type 1 error rate α for each of the q effects to be tested in the model (where q here refers to any number).[†] Also an overall Type 1 error rate $\gamma = 1 - (1 - \alpha)^q$ for the null hypothesis that all q effects are simultaneously

---

[†] A Type 1 error is that committed when we *incorrectly reject* the null hypothesis. The concept of Type 1 errors is discussed in most introductory statistics texts for geographers (see for example Gregory 1978, 116; Norcliffe 1977 33; Silk 1979, 52; Hammond and McCullagh 1978, 166; King 1969, 73) and the reader should revise this concept if necessary. Kmenta (1971, 122–36) provides a further useful discussion.

Table 5.21
Goodness-of-fit statistics for models fitted to a hypothetical $2\times2\times2\times2$ table

Model	G^2	d.f.	Source	Change in G^2	d.f.
(a) $A+B+C+D$	75.11	11			
(b) Model (a)$+A\cdot B+A\cdot C+A\cdot D$ $+B\cdot C+B.D+C.D$	16.99	5	2-variable interactions	58.12	6
(c) Model (b)$+A\cdot B\cdot C+A\cdot B\cdot D$ $+A\cdot C\cdot D+B\cdot C\cdot D$	2.21	1	3-variable interactions	14.78	4
(d) Model (c)$+A.B.C.D$	0.00	0	4-variable interactions	2.21	1

equal to zero should be chosen. Aitkin recommends that the value of α is chosen such that the overall Type 1 error rate will normally lie between 0.25 and 0.5. In our case, there are $q = 11$ effects of higher-order than the main effects (in the case of higher-dimensional tables, the main effects will rarely be tested as a group, as it is unusual to reduce the mutual independence model to an equiprobability model). As a result, if we use the conventional 5 per cent significance level for each effect i.e. choose α to be 0.05, then the overall rate γ will equal 0.431, which is within Aitkin's recommended range.

Having determined the overall Type 1 error rate, we then determine family Type 1 error rates for the pooled two-variable interactions (γ_2), three-variable interactions (γ_3) and the four-variable interactions (γ_4). These are:

$$\gamma_2 = 1-(0.95)^6 = 0.265, \quad \gamma_3 = 1-(0.95)^4 = 0.185, \quad \gamma_4 = 0.05 \quad \textbf{[5.81]}$$

The reader should note that these family type 1 error rates are linked to the overall type 1 error rate in the following way:

$$\gamma = 1-(0.95)^{11} = 1-(1-\gamma_2)(1-\gamma_3)(1-\gamma_4) = 0.431 \quad \textbf{[5.82]}$$

The next step in the simultaneous test procedure is to begin the reduction of the saturated model to a more parsimonious form. We do this by first testing the four-variable interaction at the γ_4 level of significance. In our case $\gamma_4 = 0.05$, the change in G^2 statistic is 2.21, and the critical value ($\chi^2_{1,0.05}$) from the chi-square distribution tables for one degree of freedom at the 0.05 level of significance is 3.84. The change-in-G^2 statistic for the four-variable interaction is thus not of significance and, consequently, the three-variable interactions are pooled with the four-variable interaction. (If the change-in-G^2 statistic had been significant, the model selection process would have stopped at this point.)

Next, we test the pooled three-variable and four-variable interactions at the $\gamma_{3,4}$ level of significance,[†] where

$$\gamma_{3,4} = 1-(1-\gamma_3)(1-\gamma_4) = 1-(0.95)^5 = 0.226 \quad \textbf{[5.83]}$$

[†] The reader should note that Aitkin made a small adjustment to the original form of the simultaneous test procedure with respect to this point in his 1980 paper. Here we adopt the change he suggested.

In our case, the sum of the three-variable and four-variable change-in-G^2 statistics is 16.99 and the critical value $(\chi^2_{5,0.226})$ from the chi-square distribution tables for five degrees of freedom at the 0.226 level of significance is 6.93. Thus the pooled three-variable and four-variable interactions are significant. Had they not been significant, the three variable interactions would have been eliminated from our model, the two-variable interactions would have been pooled with the three-variable and four-variable interactions and tested at the $\gamma_{2,3,4} = \gamma$ level of significance where

$$\gamma = \gamma_{2,3,4} = 1 - (1 - \gamma_2)(1 - \gamma_{3,4}) = 1 - (0.95)^{11} = 0.431 \qquad [5.84]$$

Having found a significant pooled set of effects at an appropriate significance level, permutations of terms within the set are examined to see which (if any) terms can be eliminated. The condition for elimination is that the change-in-G^2 statistic for the eliminated terms should not exceed the critical value for the pooled set. If this condition is satisfied and no further terms can be eliminated then the procedure is said to have determined a *minimal adequate subset*.

In our case, there are in fact twenty four ways in which we can permutate the three-variable interactions within the pooled set of effects which we have found to be significant. Two possibilities are shown in Table 5.22. In all permutations we eliminate terms from the bottom up until the critical chi-square value (in our case 6.93) is exceeded. In the first permutation we can eliminate $A \cdot B \cdot C \cdot D$ and $B \cdot C \cdot D$ in this way, leaving a minimal adequate subset of $A \cdot B \cdot C$, $A \cdot B \cdot D$ and $A \cdot C \cdot D$. In the second permutation we can eliminate $A \cdot B \cdot C \cdot D$, $A \cdot B \cdot D$ and $B \cdot C \cdot D$, leaving a minimal adequate subset of $A \cdot C \cdot D$ and $A \cdot B \cdot C$. In both cases, notice that we have eliminated a term $B \cdot C \cdot D$ which has a substantial

Table 5.22
Searching for a minimal adequate subset. Two possible permutations of the three-variable interaction terms

Model	G^2	d.f.	Source	Change in G^2	d.f.
(b) = $A + B + C + D + A \cdot B + A \cdot C$					
$\qquad + A \cdot D + B \cdot C + B \cdot D + C \cdot D$	16.99	5			
Permutation 1					
Model (b) + $A \cdot B \cdot C$	15.11	4	$A \cdot B \cdot C$	1.88	1
Model (b) + $A \cdot B \cdot C + A \cdot B \cdot D$	14.21	3	$A \cdot B \cdot D$	0.90	1
Model (b) + $A \cdot B \cdot C + A \cdot B \cdot D + A \cdot C \cdot D$	6.60	2	$A \cdot C \cdot D$	7.61	1
Model (b) + $A \cdot B \cdot C + A \cdot B \cdot D + A \cdot C \cdot D$ $\qquad + B.C.D$	2.21	1	$B \cdot C \cdot D$	4.39	1
Model (b) + $A \cdot B \cdot C + A \cdot B \cdot D + A \cdot C \cdot D$ $\qquad + B \cdot C \cdot D + A \cdot B \cdot C \cdot D$	0.00	0	$A \cdot B \cdot C \cdot D$	2.21	1
Permutation 2					
Model (b) + $A \cdot C \cdot D$	8.59	4	$A \cdot C \cdot D$	8.40	1
Model (b) + $A \cdot C \cdot D + A \cdot B \cdot C$	6.89	3	$A \cdot B \cdot C$	1.70	1
Model (b) + $A \cdot C \cdot D + A \cdot B \cdot C + B \cdot C \cdot D$	3.30	2	$B \cdot C \cdot D$	3.59	1
Model (b) + $A \cdot C \cdot D + A \cdot B \cdot C + B \cdot C \cdot D$ $\qquad + A \cdot B \cdot D$	2.21	1	$A \cdot B \cdot D$	1.09	1
Model (b) + $A \cdot C \cdot D + A \cdot B \cdot C + B \cdot C \cdot D$ $\qquad + A \cdot B \cdot D + A \cdot B \cdot C \cdot D$	0.00	0	$A \cdot B \cdot C \cdot D$	2.21	1

change-in-G^2 statistic, whilst retaining terms with smaller change-in-G^2 statistics. This is clearly not very satisfactory. In addition, it can be seen that different permutations will yield different minimal adequate subsets, and the process can rapidly become very tedious in higher-dimensional tables.

To simplify this process, Aitkin (1980) suggests determining which interactions can be eliminated by concentrating upon just one permutation of the terms within the pooled set. The permutation he suggests we concentrate upon is that determined by fitting the 'all three-variable interactions' model; computing the standardized values of the parameter estimates in this model; entering the three-variable interaction terms in the order determined by the size of their standardized values. In our case, let us assume that having fitted the 'all three-variable interactions' model we find that $A \cdot C \cdot D$ has the largest standardized value, followed by $B \cdot C \cdot D$, $A \cdot B \cdot C$ and $A \cdot B \cdot D$. The permutation we test is, therefore, that shown in Table 5.23. In this permutation, we can eliminate $A \cdot B \cdot C \cdot D$, $A \cdot B \cdot D$ and $A \cdot B \cdot C$ (their combined change in G^2 statistics are less than the critical value of 6.93), and this leaves a minimal adequate subset of $A \cdot C \cdot D$ and $B \cdot C \cdot D$; both elements of which now have higher change-in-G^2 statistics than any of the excluded terms.

Once a satisfactory minimal adequate subset of terms in the pooled set has been obtained, further and final reductions in the model may be attempted, by retaining the terms in the minimal adequate subset plus their lower-order relatives but considering the other lower-order terms (other than the main effects) for exclusion. In our case, the terms in our minimal adequate subset are $A \cdot C \cdot D$ and $B \cdot C \cdot D$ and these have lower-order relatives A, B, C, D, $A \cdot C$, $A \cdot D$, $B \cdot C$, $B \cdot D$, $C \cdot D$. Thus the lower-order term we can consider excluding is $A \cdot B$. Table 5.24 shows that the change-in-G^2 associated with this term is 1.50. Since the order of terms in the model is now data-based (rather than an *a priori* hierarchical order), Aitkin (1980) recommends that the significance of this term should be assessed by a pooled test using the critical value $\chi^2_{11,0.431} = 11.15$. Once again, terms are eliminated from the bottom up in Table 5.24. In this way, it is clear that $A \cdot B$ can be excluded from the model. This leaves a final 'acceptable' model of the form

Table 5.23 Permutation of three-variable interaction terms in order determined by standardized values

$$\log_e m_{ijkl} = \lambda + \lambda_i^A + \lambda_j^B + \lambda_k^C + \lambda_l^D + \lambda_{ik}^{AC} + \lambda_{il}^{AD} + \lambda_{jk}^{BC}$$
$$+ \lambda_{jl}^{BD} + \lambda_{kl}^{CD} + \lambda_{ikl}^{ACD} + \lambda_{jkl}^{BCD} \qquad [5.85]$$

with a G^2 value of 6.30 with four degrees of freedom.

Model	G^2	d.f.	Source	Change in G^2	d.f.
(b) $= A+B+C+D+A\cdot B+A\cdot C$ $+A\cdot D+B\cdot C+B\cdot D+C\cdot D$	16.99	5			
Model (b)$+A\cdot C\cdot D$	8.59	4	$A\cdot C\cdot D$	8.40	1
Model (b)$+A\cdot C\cdot D+B\cdot C\cdot D$	4.80	3	$B\cdot C\cdot D$	3.79	1
Model (b)$+A\cdot C\cdot D+B\cdot C\cdot D+A\cdot B\cdot C$	3.30	2	$A\cdot B\cdot C$	1.50	1
Model (b)$+A\cdot C\cdot D+B\cdot C\cdot D+A\cdot B\cdot C$ $+A\cdot B\cdot D$	2.21	1	$A\cdot B\cdot D$	1.09	1
Model (b)$+A\cdot C\cdot D+B\cdot C\cdot D+A\cdot B\cdot C$ $+A\cdot B\cdot D+A\cdot B\cdot C\cdot D$	0.00	0	$A\cdot B\cdot C\cdot D$	2.21	1

Table 5.24
Further refinement of the
log-linear model

Model	G^2	d.f.	Source	Change in G^2	d.f.
$A+B+C+D+A\cdot C+A\cdot D+B\cdot C$ $+B\cdot D+C\cdot D+A\cdot C\cdot D+B\cdot C\cdot D$	6.30	4			
$A+B+C+D+A\cdot B+A\cdot C+A\cdot D+B\cdot C$ $+B\cdot D+C\cdot D+A\cdot C\cdot D+B\cdot C\cdot D$	4.80	3	$A\cdot B$	1.50	1
$A*B*C*D$	0.00	0	$A\cdot B\cdot C, A\cdot B\cdot D$ $A\cdot B\cdot C\cdot D$	4.80	3

On first encountering the simultaneous test procedure, the reader may consider it to be a rather complex process. However, it presents little difficulty in practice, and is particularly well suited to use with the GLIM computer package. In some cases, particularly if the contingency table is not too large but the sample size is large, the simultaneous test procedure will produce the same final model as the screening or stepwise procedures. However, in these circumstances, it should be remembered that the simultaneous test procedure has a firmer foundation is statistical theory as it gives explicit consideration to the multiple-testing simultaneous-inference problem. In larger tables, the effects of the adjustment of significance levels in the simultaneous test procedure are often seen, and the stepwise or screening procedures often leave terms in the finally selected model which the simultaneous test procedure finds unnecessary.

Given the choice of α, the reduction of the saturated model to a parsimonious form using the simultaneous test procedure becomes a systematic, internally consistent procedure. However, the user should not abandon critical thought and commonsense, and he should not allow the procedure to become merely a rigid, mechanical process.

For further discussion of the relative strengths and weakness of the model selection procedures outlined in this section, and discussion of the difficulties involved in making direct comparison, see Fienberg (1977), Benedetti and Brown (1978), Whittaker and Aitkin (1978), Aitkin (1980), and in the geographical literature Fingleton (1981), Upton (1981).

EXAMPLE 5.4   *Shopping behaviour in Manchester*

An example of the use of Aitkin's simultaneous test procedure to select an appropriate log-linear model is to be found in the work of Fingleton (1981). Fingleton uses the procedure to select the most 'acceptable' model to represent the structural relationships between the variables in the four-dimensional $2\times2\times2\times2$ contingency table shown in Table 5.25. The four variables are: A – age of consumer; B – income of consumer; C – car ownership; D – shopping centre patronage. Although we will see in Chapter 6 that it is more appropriate to regard Variable D (shopping centre patronage) as a response variable and variables A, B, and C as explanatory variables, at this stage no such division is recognized and we will treat all variables as response variables. In addition, although Table 5.25 contains a number a small cell frequencies, including a zero cell frequency, we will say nothing about such issues at this stage and leave our discussion of such matters until Chapter 9.

The first step in the simultaneous test procedure is to fit the various members of the hierarchy of log-linear models shown in Table 5.26. The second step is then to determine a Type 1 error rate α for each of the effects to be tested in the model, an overall Type 1 error rate, and various family Type 1 error rates for pooled two-variable interactions, three-variable interactions and four-variable interactions. Setting $\alpha = 0.05$, these Type 1 error rates and the associated critical values for the chi-square distribution are as follows:

Overall	γ	$= 1 - (0.95)^{11} = 0.431$	$\chi^2_{11,0.431} = 11.15$
Four-variable	γ_4	$= 1 - (0.95)^1 = 0.05$	$\chi^2_{1,0.05} = 3.84$
Three- and four-variable	$\gamma_{3,4}$	$= 1 - (0.95)^5 = 0.226$	$\chi^2_{5,0.226} = 6.93$
Two-, three-, and four-variable	$\gamma_{2,3,4} = \gamma = 0.431$		$\chi^2_{11,0.431} = 11.15$

Table 5.25
A classification of a sample of consumers in a shopping survey in Manchester (cell frequencies from Fingleton 1981)

		Variable D								
		Use nearest shopping centre				Use other shopping centre				
		Variable C				Variable C				
		Car owner	Not car owner			Car owner		Not car owner		
		Variable B		Variable B		Variable B		Variable B		
		Low income	High income	Low income	High income	Low income	High income	Low income	High income	Total
Variable A	Young	17	2	12	3	48	3	57	24	166
	Old	51	1	18	2	105	0	53	11	241
	Total	68	3	30	5	153	3	110	35	407

Table 5.26
Goodness-of-fit statistics for a range of models fitted to Table 5.25

Model	Fitted marginals	G^2	d.f.	Source	Change in G^2	d.f.
(a) $A + B + C + D$	$[A][B][C][D]$	87.76	11			
(b) Model (a) $+ A \cdot B + A \cdot C$ $+ A \cdot D + B \cdot C + B \cdot D + C \cdot D$	$[AB][AC][AD]$ $[BC][BD][CD]$	6.69	5	2-variable interactions	81.07	6
(c) Model (b) $+ A \cdot B \cdot C + A \cdot B \cdot D$ $+ A \cdot C \cdot D + B \cdot C \cdot D$	$[ABC][ABD]$ $[ACD][BCD]$	1.18	1	3-variable interactions	5.51	4
(d) Model (c) $+ A \cdot B \cdot C \cdot D$	$[ABCD]$	0.00	0	4-variable interactions	1.18	1

Comparing the change in G^2 statistics in Table 5.26 with the critical values from the chi-square distribution, we can see that both the four-variable interaction and the pooled three-variable and four-variable interactions have change-in-G^2 statistics which are less than the critical values (1.18 compared to 3.84, and 6.69 compared to 6.93). Consequently, the three-and four-variable interactions are not significant and must be pooled with the two-variable interactions. In contrast, the change-in-G^2 statistic for the pooled two-variable, three-variable, and four-variable interactions ($81.07 + 5.51 + 1.18$) is greater than the critical chi-square value (11.15). Hence, the pooled two-variable, three-variable, and four-variable interactions are significant, and we can examine this pooled set of terms to see which can be eliminated and which must be retained to form a minimal adequate subset.

Table 5.27 shows the particular permutation of terms within the pooled two-variable, three-variable and four-variable interactions which Fingleton uses to determine which terms must be eliminated and which must be retained. From this permutation, terms are eliminated from the bottom up until the critical chi-square value of 11.15 is exceeded. In this way we can eliminate the four-variable interaction ($A.B.C.D$), all the three-variable interactions ($A.B.C$, $A.B.D$, $A.C.D$, $B.C.D$) and the two-variable interactions $B.D$ and $A.D$, and this leaves a minimal adequate subset consisting of $B.C$, $A.B$, $A.C$ and $C.D$. No other terms can be eliminated from the model, and this leaves a final 'acceptable' model of the form

$$\log_e m_{ijkl} = \lambda + \lambda_i^A + \lambda_j^B + \lambda_k^C + \lambda_l^D + \lambda_{ij}^{AB} + \lambda_{ik}^{AC} + \lambda_{jk}^{BC} + \lambda_{kl}^{CD} \quad [5.86]$$

We will return to this example in Chapter 6. For the moment, the reader should simply notice that there is association between the age, income and car ownership variables (A, B and C) but the shopping centre patronage variable (D) is associated with only the car ownership variable (C).

Table 5.27 Permutation of terms used by Fingleton to determine which terms must be eliminated and which retained

Model	G^2	d.f.	Source	Change in G^2	d.f.
(a) = $A+B+C+D$	87.76	11			
Model (a)+$B\cdot C$	46.73	10	$B\cdot C$	41.03	1
Model (a)+$B\cdot C+A\cdot B$	29.16	9	$A\cdot B$	17.57	1
Model (a)+$B\cdot C+A\cdot B+A\cdot C$	16.84	8	$A\cdot C$	12.32	1
Model (a)+$B\cdot C+A\cdot B+A\cdot C+C\cdot D$	9.40	7	$C\cdot D$	7.44	1
Model (a)+$B\cdot C+A\cdot B+A\cdot C+C\cdot D$ $+A\cdot D$	6.90	6	$A\cdot D$	2.50	1
Model (a)+$B\cdot C+A\cdot B+A\cdot C+C\cdot D$ $+A\cdot D+B\cdot D$	6.69	5	$B\cdot D$	0.21	1
Model (a)+$B\cdot C+A\cdot B+A\cdot C+C\cdot D$ $+A\cdot D+B\cdot D+$ all 3-variable and 4-variable interactions	0.00	0	3-variable and 4-variable interactions	6.69	5

EXAMPLE 5.5   _Non-fatal deliberate self-harm in Bristol_

As part of a much wider study of the geography of mental health in Bristol, Griffin (1984) has used log-linear models to study the four-dimensional $3 \times 2 \times 2 \times 2$ contingency table shown in Table 5.28. This table classifies 1,443 individual cases of non-fatal deliberate self-harm admitted to Bristol's three teaching hospitals in 1972–73. The cases were all residents of the City of Bristol at the time of their admission to hospital, and the four variables in the table are: A – age of the individual (young, middle-aged, old); B – sex of the individual (male, female); C – time of admission to hospital ('night-time hours': 20.00–7.59, 'daytime hours': 8.00–19.59); D – location of the home of the individual (inner city, outer city). In this example, we will now utilize Griffin's results to compare Brown's screening procedure and Aitkin's simultaneous test procedure as means of selecting the most 'acceptable' model to represent the structural relationships between the variables in Table 5.28.

Table 5.28
A classification of cases of non-fatal deliberate self-harm admitted to Bristol hospitals in 1972–73

		Variable D								
		Inner city				Outer city				
		Variable C				Variable C				
		Night-time		Daytime		Night-time		Daytime		
		Variable B		Variable B		Variable B		Variable B		
		M	F	M	F	M	F	M	F	Total
Variable A	Young	67	173	49	164	53	108	60	115	789
	Middle-aged	46	65	46	92	42	58	46	104	499
	Old	2	4	30	51	7	9	16	36	155
	Total	115	242	125	307	102	175	122	255	1,443

The first step in the use of the screening procedure is to compute the partial association and marginal association statistics for each of the terms in the saturated model (note that the BMDP program will compute these statistics automatically). Table 5.29 shows these test statistics, and screening the terms on the basis of these statistics enables us to divide the terms into three groups.

Table 5.29
Partial and marginal association test statistics for terms in the saturated model appropriate for Table 5.28

λ-term	d.f.	Partial assoc'n statistic	Marginal assoc'n statistic	λ-term	d.f.	Partial assoc'n statistic	Marginal assoc'n statistic
A	2	466.56	466.56	$B \cdot D$	1	2.07	2.40
B	1	187.91	187.91	$C \cdot D$	1	1.15	1.22
C	1	21.28	21.28	$A \cdot B \cdot C$	2	1.65	1.85
D	1	12.65	12.65	$A \cdot B \cdot D$	2	4.61	4.39
$A \cdot B$	2	8.94	7.83	$A \cdot C \cdot D$	2	11.08	10.81
$A \cdot C$	2	80.51	79.12	$B \cdot C \cdot D$	1	0.13	0.03
$A \cdot D$	2	6.41	7.10	$A \cdot B \cdot C \cdot D$	2	1.51	1.51
$B \cdot C$	1	3.96	2.22				

1. Likely to be important and required in the model ($A,B,C,D,$ $A \cdot B,$ $A \cdot C,$ $A \cdot D,$ $A \cdot C \cdot D$; plus $C \cdot D$ because it is a lower-order relative of $A \cdot C \cdot D$ even though it is not significant on the basis of either partial or marginal test).

2. Likely to be unimportant and not required in the model ($B \cdot C \cdot D,$ $A \cdot B \cdot C,$ $A \cdot B \cdot C \cdot D$).

3. In need of further study ($B \cdot C,$ $A \cdot B \cdot D$; plus $B \cdot D$ because it is a lower-order relative of $A \cdot B \cdot D$ and is not included amongst the 'likely to be important' terms).

The second step in the procedure is to choose the 'starting' or 'base' model on the basis of the previous screening. This implies the choice of the model

$$\log_e m_{ijkl} = \lambda + \lambda_i^A + \lambda_j^B + \lambda_k^C + \lambda_l^D + \lambda_{ij}^{AB} + \lambda_{ik}^{AC}$$
$$+ \lambda_{il}^{AD} + \lambda_{kl}^{CD} + \lambda_{ikl}^{ACD} \qquad \qquad [5.87]$$

and fitting this model to Table 5.28 gives a G^2 goodness-of-fit value of 13.83 with 9 d.f.

Using the forward selection method we then investigate the possibility of adding to [5.87] the 'doubtful' terms $B \cdot C$ and $A \cdot B \cdot D$ plus $B \cdot D$. We begin by adding the $B \cdot C$ term. Fitting such a model gives a G^2 statistic of 10.02 with 8 d.f. The associated conditional G^2 statistic ($13.83 - 10.02 = 3.81$ with 1 d.f.) is not significant at the conventional 5 per cent level (although it is significant at the 10 per cent level), indicating that, at this particular level of significance, it is not statistically justifiable to add the term $B \cdot C$ to our base model [5.87]. Likewise, we can add the terms $A \cdot B \cdot D$ and $B \cdot D$ (where $B \cdot D$ is required because of the hierarchical principle). Fitting such a model gives a G^2 of 7.53 with 6 d.f. The associated conditional G^2 statistic ($13.83 - 7.53 = 6.30$ with 3 d.f.) is not significant at the 5 per cent level (though it is at the 10 per cent level), indicating that, at this particular level of significance, it is not justifiable to add the terms $A \cdot B \cdot D$ and $B \cdot D$ to our base model. Furthermore, we find that it is not justifiable to delete any terms from our base model or to add any additional terms to it. Thus the model selection process concludes and we choose [5.87] as our final 'acceptable' model. However, the reader should note that if we follow the advice of Goodman (1970, 229; 1971a, 39) and add 0.5 to each cell frequency in Table 5.28, adding the $B \cdot C$ term to our base model becomes statistically justifiable at the conventional 5 per cent level (the conditional G^2 now becomes $13.81 - 9.93 = 3.88$ with 1 d.f.). As a result, the final 'acceptable' model becomes

$$\log_e m_{ijk} = \lambda + \lambda_i^A + \lambda_j^B + \lambda_k^C + \lambda_l^D + \lambda_{ij}^{AB} + \lambda_{ik}^{AC} + \lambda_{il}^{AD} + \lambda_{jk}^{BC} + \lambda_{kl}^{CD}$$
$$+ \lambda_{ikl}^{ACD} \qquad \qquad [5.88]$$

Many, perhaps a majority of, users of log-linear models adopt Goodman's advice on this matter and would thus choose [5.88] as the 'acceptable' model.

The first step in the use of Aitkin's simultaneous test procedure is to fit the various members of the hierarchy of log-linear models shown in Table 5.30. The second step is then to determine a Type 1 error rate α for each of the effects to be tested in the model, an overall Type 1 error

Table 5.30
Goodness-of fit statistics
for a range of models fitted
to Table 5.28

Model	G^2	d.f.	Source	Change in G_2	d.f.
(a) $A+B+C+D$	119.70	18			
(b) Model (a)$+A \cdot B+A \cdot C+A \cdot D$ $+B \cdot C+B \cdot D+C \cdot D$	18.76	9	2-variable interactions	100.94	9
(c) Model (b)$+A \cdot B \cdot C+A \cdot B \cdot D$ $+A \cdot C \cdot D+B \cdot C \cdot D$	1.51	2	3-variable interactions	17.25	7
(d) Model (c)$+A \cdot B \cdot C \cdot D$	0.00	0	4-variable interactions	1.51	2

rate, and various family Type 1 error rates for pooled two-variable interactions, three-variable interactions and four-variable interactions. Setting α at the conventional 0.05 level gives an overall Type 1 error rate of 0.431. The family Type 1 error rates for pooled two-variable interactions, three-variable interactions and four-variable interactions, and the associated critical values from the chi-square distribution are as follows.[†]

Overall	γ	$= 1-(0.95)^{11} = 0.431$	$\chi^2_{18,0.431}$	$\simeq 18.30$
Four-variable	γ_4	$\doteq 1-(0.95)^1 = 0.05$	$\chi^2_{2,0.05}$	$= 5.99$
Three- and four-variable	$\gamma_{3,4}$	$= 1-(0.95)^5 = 0.226$	$\chi^2_{9,0.226}$	$\simeq 11.80$
Two-, three- and four-variable	$\gamma_{2,3,4}$	$= \gamma = 0.366$	$\chi^2_{18,0.431}$	$\simeq 18.30$

Comparing the change-in-G^2 statistics in Table 5.30 with the critical values from the chi-square distribution, we can see that the four-variable interaction has a change-in-G^2 statistic which is less than the critical value (1.51 compared to 5.99) but the pooled three-variable and four-variable interactions have a change-in-G^2 statistic (17.25 + 1.51) which is greater than the critical chi-square value (approximately 11.80). Hence, the pooled three-variable and four-variable interactions are significant and we examine this pooled set of terms to see which can be eliminated and which must be retained to form a minimal adequate subset.

Table 5.31 shows the particular permutation of terms within the pooled three-variable and four-variable interactions which Griffin uses to determine which terms must be eliminated and which terms must be retained. From this permutation, terms are eliminated from the bottom up until the critical chi-square value of approximately 11.80 is exceeded. In this way we can eliminate, $A \cdot B \cdot C \cdot D$, $B \cdot C \cdot D$, $A \cdot B \cdot C$ and $A \cdot B \cdot D$ and this leaves a minimal adequate subset consisting of only $A \cdot C \cdot D$.

[†] The reader should note that in this $3 \times 2 \times 2 \times 2$ table there are 18 parameters of higher order than the main effect parameters. There are 2 four-variable parameters and 7 three-variable parameters. However, there are only 11 *types* of interactions (A.B, A.C, A.D, B.C, B.D, C.D, A.B.C, A.B.D, A.C.D, B.C.D, A.B.C.D); 6 two-variable interactions, 4 three-variable interactions, and 1 four-variable interaction. Aitkin uses the number of effects (i.e. interaction types) in determining the Type 1 error rates (see the discussion of this point in Aitkin, 1978).

Table 5.31
Permutation of terms used by Griffin to determine which terms must be eliminated and which retained

Model	G^2	d.f.	Source	Change in G^2	d.f.
(b) $= A+B+C+D+A \cdot B+A \cdot C$ $+A \cdot D+B \cdot C+B \cdot D+C \cdot D$	18.76	9			
Model (b) $+A \cdot C \cdot D$	7.96	7	$A \cdot C \cdot D$	10.80	2
Model (b) $+A \cdot C \cdot D+A \cdot B \cdot D$	3.26	5	$A \cdot B \cdot D$	4.70	2
Model (b) $+A \cdot C \cdot D+A \cdot B \cdot D+A \cdot B \cdot C$	1.64	3	$A \cdot B \cdot C$	1.62	2
Model (b) $+A \cdot C \cdot D+A \cdot B \cdot D+A \cdot B \cdot C$ $+B \cdot C \cdot D$	1.51	2	$B \cdot C \cdot D$	0.13	1
Model (b) $+A \cdot C \cdot D+A \cdot B \cdot D+A \cdot B \cdot C$ $+B \cdot C \cdot D+A \cdot B \cdot C \cdot D$	0.00	0	$A \cdot B \cdot C \cdot D$	1.51	2

Once a satisfactory minimal adequate subset of terms in the pooled set has been obtained, we can attempt further reduction of the model. We do this by retaining in the model the terms in the minimal adequate subset (only $A \cdot C \cdot D$ in our case) plus their lower-order relatives ($A, C, D, A \cdot C, A \cdot D, C \cdot D$ in our case) plus any necessary main effects (B in our case), and considering the other lower-order terms for exclusion ($A \cdot B, B \cdot C, B \cdot D$ in our case). Table 5.32 shows the results of such a consideration. Since the order of the terms in the model is now data based (rather than an *a priori* hierarchical order), Aitkin (1980) recommends that the significance of these terms should be assessed by a pooled test using the critical value $\chi^2_{18,0.431} \simeq 18.30$. Once again, terms are eliminated from the bottom up in this table. In this way, it is clear that we can eliminate the two-variable interaction terms $B \cdot D$ and $B \cdot C$. At this point the model selection process concludes and we choose

$$\log_e m_{ijkl} = \lambda + \lambda_i^A + \lambda_j^B + \lambda_k^C + \lambda_l^D + \lambda_{ij}^{AB} + \lambda_{ik}^{AC}$$
$$+ \lambda_{il}^{AD} + \lambda_{kl}^{CD} + \lambda_{ikl}^{ACD} \qquad \qquad [5.89]$$

as the final 'acceptable' model. This is the same model as that [5.87] selected by the screening process, and this confirms a point made in Section 5.7.4. It was noted there that if the contingency table is not too large but sample size is large, the simultaneous test procedure will

Table 5.32
Further refinement of the log-linear model

Model	G^2	d.f.	Source	Change in G^2	d.f.
$A+B+C+D+A \cdot C+A \cdot D+C \cdot D$ $+A \cdot C \cdot D$	21.66	11			
$A+B+C+D+A \cdot B+A \cdot C+A \cdot D$ $+C \cdot D+A \cdot C \cdot D$	13.88	9	$A \cdot B$	7.83	2
$A+B+C+D+A \cdot B+A \cdot C+A \cdot D$ $+B \cdot C+C \cdot D+A \cdot C \cdot D$	10.02	8	$B \cdot C$	3.81	1
$A+B+C+D+A \cdot B+A \cdot C+A \cdot D$ $+B \cdot C+B \cdot D+C \cdot D+A \cdot C \cdot D$	7.96	7	$B \cdot D$	2.06	1
$A*B*C*D$	0.00	0	$A \cdot B \cdot C, A \cdot B \cdot D,$ $B \cdot C \cdot D, A \cdot B \cdot C \cdot D$	7.96	7

produce the same final model as the screening procedure. However, in larger tables (and also in this example if Goodman's advice is followed and 0.5 is added to each cell of the table) the simultaneous test procedure will often yield a more parsimonious final 'acceptable' model than the screening procedure.

Model [5.89] can be interpreted as showing that the age structure of cases varies significantly between the sexes (AB); that times of admission vary significantly between the age groups (AC); that the age structure of cases varies significantly between the inner city and the outer city (AD); that the inner city and outer city cases vary with respect to their times of admission (CD); that the association (CD) is modified by the age group to which an individual belongs (ACD).[†] For further comments, including a more detailed investigation of the locational component see Griffin (1984).

5.8   The analysis of residuals

In the preceding discussion of model selection strategies it was stressed that model selection should not be allowed to become a mechanical process, and the user should not abandon critical thought and commonsense. A particularly useful aid in this context is the analysis of residuals, for residuals provide us with detailed information on the fit of a model in each cell of the contingency table. In other words, residuals provide us with 'internal goodness-of-fit' information rather than simply a measure of the overall goodness-of-fit such as that which we derive from the G^2 statistic. Formal model selection strategies should, therefore, be supplemented by residual analysis and, at the very least, selection of an 'acceptable' model using stepwise selection, screening or the simultaneous test procedure should be followed by an examination of the residuals of this 'acceptable' model.

The simplest definition of a residual is the difference between an observed cell frequency and the estimate of the expected cell frequency, e.g. in the three-dimensional case

$$n_{ijk} - \hat{m}_{ijk} \qquad \qquad [5.90]$$

However, simple raw residuals of this type provide no information on the proportional or relative size of the discrepancy between the observed and estimated expected frequency. As a result, it is useful to standardize the raw residuals in some way. Perhaps the simplest way to do this is to divide the raw residuals by the square root of the estimated expected cell frequency. This produces a residual known (see Haberman 1973a; 1978, 230) as the *standardized residual*

$$e_{ijk} = (n_{ijk} - \hat{m}_{ijk}) / \sqrt{\hat{m}_{ijk}} \qquad \qquad [5.91]$$

and e_{ijk} has an approximate normal distribution with an asymptotic mean of zero and an asymptotic variance of less than one.

[†] If Goodman's advice is followed and 0.5 is added to each cell, the screening procedure produces an 'acceptable' model of the form [5.88] in which the term $B \cdot C$ is included. This additional term can be interpreted as indicating that the time of admission varies between the sexes.

A theoretically more satisfying way to standardize the raw residuals is to divide them by their own estimated standard errors (i.e. by the square roots of their estimated variances $\hat{c}_{ijk}$). This produces a residual known (see Haberman 1973a; 1978, 231, 275) as the *adjusted residual*

$$r_{ijk} = (n_{ijk} - \hat{m}_{ijk})/\sqrt{\hat{c}_{ijk}} \qquad [5.92]$$

and r_{ijk} has an approximate standard normal distribution with asymptotic mean of zero and asymptotic variance of one. However, computation of adjusted residuals is much more difficult than standardized residuals (see Haberman 1978, 231, 275 for explicit formulas).

If the number of categories in each variable in the contingency table is large and all m_{ijk} are comparable in size, then the standardized residual e_{ijk} is a fairly good approximation to the adjusted residual r_{ijk}. However, it is possible to improve this approximation by using improved standardized residuals such as (see Haberman 1973a; 1978, 230)

$$e^*_{ijk} = \frac{(n_{ijk} - \hat{m}_{ijk})}{\sqrt{\left[\hat{m}_{ijk}\left(1 - \frac{n_{ij+}}{n_{i++}}\right)\left(1 - \frac{n_{i+k}}{n_{i++}}\right)\right]}} \qquad [5.93]$$

The asymptotic variance of e^*_{ijk} is still less than one but it is larger than that of e_{ijk}.

Finally, two alternative definitions of a residual should be noted. The first is the Freeman–Tukey residual (see Bishop, Fienberg and Holland 1975, 137)

$$e^{FT}_{ijk} = \sqrt{n_{ijk}} + \sqrt{n_{ijk} + 1} - \sqrt{4\hat{m}_{ijk} + 1} \qquad [5.94]$$

In this case, we make use of the result that if n_{ijk} follows a Poisson distribution with mean m_{ijk}, then $\sqrt{n_{ijk}} + \sqrt{n_{ijk} + 1}$ is approximately normally distributed with approximate mean of $\sqrt{4\hat{m}_{ijk} + 1}$ and variance of one. Residuals of this type are an option in the BMDP log-linear modelling program, and thus can readily be used by the researcher. The second of the alternative residuals is

$$e^N_{ijk} = \frac{3\left[n_{ijk}^{2/3} - (\hat{m}_{ijk} - 1/6)^{2/3}\right]}{2\hat{m}_{ijk}^{1/6}} \qquad [5.95]$$

This has been proposed by Nelder (1974) and is based upon a suggestion by Anscombe (1953). Nelder claims that it is more nearly normally distributed, but he also notes that it cannot be independent of m_{ijk}, must have a lower bound, and must be skew to the right for small m_{ijk}.

Residuals of the type [5.91] to [5.95] can help us to detect deviations from our fitted log-linear models. Such deviations might take the form of isolated large anomalies ('outliers' or 'rogue cells'), or systematically arranged smaller deviations indicating patterns in the data which we might otherwise have overlooked. As a rough guide to the significance of an individual deviation, we can compare the value of the residual to the critical value obtained from the standard normal distribution. Alternatively, we can make use of the fact that residuals such as [5.91] and [5.94], when summed over all i, j and k (i.e. $\sum_i \sum_j \sum_k e_{ijk}$), are distributed aysmptotically as χ^2 (see Bishop, Fienberg and Holland 1975, 137). However, we must always remember that such methods only provide a rough guide. The approximations to the theoretical distributions have varying degrees of exactness; the residuals in different cells of the table will inevitably be somewhat correlated with one another; and the appeal to asymptotic theory helps us to only a limited extent for those cells which have small observed frequencies. Nevertheless, the systematic examination of residuals is an indispensible aid in the fitting of log-linear models, particularly in the case of higher-dimensional contingency tables.

EXAMPLE 5.6   *Opinions about a television series in urban and rural areas*

Lee (1978) provides a small pedagogic example which illustrates the potential value of residual analysis in log-linear modelling. He presents the hypothetical three-dimensional contingency table shown in Table 5.33 in which sixty-six randomly selected adults are cross-classified according to their sex (variable A), their opinion about the content of a television series (variable B – divided into two categories 'approved' or 'unapproved') and their location (variable C – divided into two categories 'rural' and 'urban'). In Chapter 9 we will argue that it is perhaps more appropriate to regard variable B (opinion about the television series) as a response variable and variables A and C as explanatory variables. However, at this stage we will simply accept Lee's treatment of all variables as response variables.

Lee fits a hierarchical set of log-linear models to the data of Table 5.33, three members of which are shown in Table 5.34. On the basis of the goodness-of-fit statistics, the only model which provides an adequate fit to the data appears to be the saturated model.

Table 5.33 Lee's hypothetical contingency table

Variable A		Variable C			
		Rural		Urban	
		Variable B		Variable B	
		Approved	Unapproved	Approved	Unapproved
	Female	3	7	6	12
	Male	5	15	17	1

Table 5.34 Three members of the hierarchical set of models fitted to Table 5.33

Model specification	G^2	d.f.
$\log_e m_{ijk} = \lambda + \lambda_i^A + \lambda_j^B + \lambda_k^C + \lambda_{ij}^{AB} + \lambda_{ik}^{AC} + \lambda_{jk}^{BC} + \lambda_{ijk}^{ABC}$	0.00	0
$\log_e m_{ijk} = \lambda + \lambda_i^A + \lambda_j^B + \lambda_k^C + \lambda_{ij}^{AB} + \lambda_{ik}^{AC} + \lambda_{jk}^{BC}$	8.03*	1
$\log_e m_{ijk} = \lambda + \lambda_i^A + \lambda_j^B + \lambda_k^C$	27.78*	4

*indicates G^2 in upper 5 per cent tail of χ^2 distribution with d.f. as indicated.

However, Lee notes that in fitting the models in Table 5.34 only the overall goodness-of-fit has been considered; there has been no evaluation of internal goodness-of-fit information. To remedy this, he thus calculates, using expression [5.91], the standardized residuals for each cell of the contingency table under the model of mutual independence (i.e. $A+B+C$). This gives the results shown in Table 5.35.

On average, the rural area standardized residuals are much smaller than those of the urban area. As a result, Lee suggests that it is appropriate to try the model

$$\log_e m_{ijk} = \lambda + \lambda_i^A + \lambda_j^B + \lambda_k^C + \lambda_{ik}^{AC} + \lambda_{jk}^{BC} \qquad [5.96]$$

This is model type 3 of Table 5.11; a conditional independence model which implies that sex and opinion about the television series are independent given location. The G^2 value for this model is 16.54 with two degrees of freedom. Hence, its overall fit is not adequate. However, the standardized residuals from this model (shown in Table 5.36) are suggestive. In the rural area they are practically zero, indicating that in this area the interaction between sex and opinion is insignificant. However, in the urban area the standardized residuals are large, indicating that in this area the interaction between sex and opinion is highly significant. It is clear, therefore, that a three-variable interaction $A \cdot B \cdot C$ exists because the interactions between sex and opinion are significantly different in the two locations. As a result, the term λ_{ijk}^{ABC} cannot be excluded from our model, and the saturated model is the only model in the hierarchical set to provide an adequate fit. However, rather than simply accept the saturated model as the only adequate representation of the data in Table 5.33, Lee suggests that the residuals in Table 5.36 indicate that it may be more appropriate to divide Table 5.33 into two parts (rural and urban) and to fit different models to the two sub-tables. In the rural sub-table, the model of independence between sex and opinion will prove adequate, whereas in the urban sub-table a model which allows interaction between sex and opinion will be necessary.

Table 5.35 Standardized residuals from model of mutual independence

	Rural		Urban	
	Approved	Unapproved	Approved	Unapproved
Female	−1.22	0.10	−0.44	1.37
Male	−1.09	1.93	2.33	−3.01

Table 5.36 Standardized residuals from model [5.96]

	Rural		Urban	
	Approved	Unapproved	Approved	Unapproved
Female	0.20	−0.12	−1.62	2.16
Male	−0.14	0.09	1.62	−2.16

CHAPTER 6

Categorical response variable, categorical explanatory variables: the log-linear model approach

Now that some of the major features of log-linear models have been introduced we can return from the problems of cell (g) of Table 1.1 to the problems of cell (f). In Chapter 4 we noted that two types of models are now widely used to handle the problems of cell (f); linear logit models and log-linear models. Log-linear models for cell (f) problems are a special case of the general class of log-linear models; a special case in which we utilize any response/explanatory distinction which might exist among the variables. In so doing, we concentrate upon assessing the effects of the explanatory variables on the response variable(s) rather than (as in Ch. 5) concentrating upon describing the structural relationships among the response variables. In this short chapter we will see how the cell (g) log-linear models of Chapter 5 must be modified to take account of the division of variables into explanatory and response, and we will see how log-linear models for cell (f) problems are closely linked to linear logit models despite a superficially different appearance.

6.1 Fixed marginal totals

In Section 5.5 we noted that when estimating the expected cell frequencies for a particular log-linear model, certain marginal totals in the estimated table are constrained to be equal to the corresponding marginal totals in the observed table, and these constrained marginal totals are summarized in the 'fitted marginals' notation for that model. For example, in the three-dimensional model:

$$\log_e m_{ijk} = \lambda + \lambda_i^A + \lambda_j^B + \lambda_k^C + \lambda_{ij}^{AB} \qquad [6.1]$$

the 'fitted marginals' are $[AB]\,[C]$ and this implies that the estimated expected cell frequencies are constrained as follows:

$$\sum_{k=1}^{K} \hat{m}_{ijk} = \sum_{k=1}^{K} n_{ijk} \qquad [6.2]$$

$$\sum_{i=1}^{I} \sum_{j=1}^{J} \hat{m}_{ijk} = \sum_{i=j}^{I} \sum_{j=1}^{J} n_{ijk} \qquad [6.3]$$

where expression [6.2] summarizes the [AB] constraints and [6.3] the [C] constraints. In terms of the parameters of the log-linear model, what this implies is that the log-linear model must include all λ-terms corresponding to the constrained marginal totals and all lower-order relatives of these terms (because of the hierarchical principle).

Now, for some sets of data, certain observed marginal totals are fixed by the sampling design. For example, in the three-dimensional table shown in Table 6.1, instead of having just a fixed total sample size, N (as in all our examples in Ch. 5), we might have column totals fixed by the sampling design.[†] In other words, in Table 6.1 the [BC] marginal totals n_{+11}, n_{+21}, n_{+12} and n_{+22} might each be fixed *a priori* by the sampling design to a value of 125. Consequently in any analysis of Table 6.1 these [BC] marginal totals must be maintained at their *a priori* fixed values. Thus any log-linear model fitted to Table 6.1 *must* include the λ_{jk}^{BC} term and its lower-order relatives, and these cannot be excluded from any finally selected 'acceptable' model.

Table 6.1
A three-dimensional $2 \times 2 \times 2$ contingency table in which the [BC] marginal totals are fixed

		Variable C			
		$k = 1$		$k = 2$	
Variable A		Variable B		Variable B	
		$j = 1$	$j = 2$	$j = 1$	$j = 2$
	$i = 1$	n_{111}	n_{121}	n_{112}	n_{122}
	$i = 2$	n_{211}	n_{221}	n_{212}	n_{222}
Fixed marginal Total [BC]		$n_{+11}=125$	$n_{+21}=125$	$n_{+12}=125$	$n_{+22}=125$ $N=500$

6.2   Log-linear models for mixed explanatory/response variable situations

Log-linear models appropriate for the mixed explanatory/response variable problems of cell (f) of Table 1.1 have essentially the same general form as those we have utilized in Chapter 5 for the problems of cell (g). The same principles of hierarchical structuring, parameter estimation, model testing, reduction to a parsimonious form, and so on, apply. The difference is that our attention is now focussed on the way in which the response variable is affected by the explanatory variables.

[†] Fixed column or row totals will occur when the underlying sample model is product multinomial, rather than multinomial as in the case of our examples in Chapter 5 (see Section 1.4 for further discussion).

Consequently, we treat the structural relationships between the explanatory variables as fixed or given 'facts of life' and, as a result, treat as *fixed* the marginal totals of the contingency tables corresponding to the explanatory variables. On the basis of what we discovered in Section 6.1 this implies that we consider only those models in which all interactions between the explanatory variables (and the lower-order relatives of these interactions) are automatically included.

A simple example will help to clarify the previous statements. Imagine a three-dimensional table in which variable C is regarded as a response variable, and variables A and B as explanatory variables. In all the log-linear models we consider, we must automatically include the interaction between A and B and the lower-order relatives of this interaction. In other words, we must include the parameters λ_{ij}^{AB}, λ_{j}^{B} and λ_{i}^{A} in all the log-linear models we fit, for these take account of the 'facts of life' (the relationships between the explanatory variables) and they are included in the models simply to ensure that the marginal totals in the estimated table are constrained to equal the fixed 'facts of life'. These parameters cannot be excluded from any of the log-linear models which we fit, irrespective of their statistical 'significance' or 'non-significance'. Table 6.2 shows all possible members of the hierarchical set of log-linear models appropriate for such a table when variable C is regarded as the response variable and A and B as explanatory variables.

Table 6.2 The hierarchical set of log-linear models for a table in which C is the response variable and A and B are explanatory variables

Model specification	Fitted marginals
$\log_e m_{ijk} = \lambda + \lambda_i^A + \lambda_j^B + \lambda_{ij}^{AB} + \lambda_k^C + \lambda_{ik}^{AC} + \lambda_{jk}^{BC} + \lambda_{ijk}^{ABC}$	$[ABC]$
$\log_e m_{ijk} = \lambda + \lambda_i^A + \lambda_j^B + \lambda_{ij}^{AB} + \lambda_k^C + \lambda_{ik}^{AC} + \lambda_{jk}^{BC}$	$[AB]\,[AC]\,[BC]$
$\log_e m_{ijk} = \lambda + \lambda_i^A + \lambda_j^B + \lambda_{ij}^{AB} + \lambda_k^C + \lambda_{ik}^{AC}$	$[AB]\,[AC]$
$\log_e m_{ijk} = \lambda + \lambda_i^A + \lambda_j^B + \lambda_{ij}^{AB} + \lambda_k^C + \lambda_{jk}^{BC}$	$[AB]\,[BC]$
$\log_e m_{ijk} = \lambda + \lambda_i^A + \lambda_j^B + \lambda_{ij}^{AB} + \lambda_k^C$	$[AB]\,[C]$
$\log_e m_{ijk} = \lambda + \lambda_i^A + \lambda_j^B + \lambda_{ij}^{AB}$	$[AB]$

The structure of log-linear models for a mixed explanatory/response variable problem can be clearly seen. Conventionally we write the parameters which *must* be included as the first terms in the models. All our attention, and the thrust of the model selection strategies we discussed in Section 5.7, is then directed at the latter terms in the model, i.e. at the parameters which concern the relationship between the response and explanatory variables.

EXAMPLE 6.1   *A hypothetical three-dimensional table*

Williams (1976) illustrates a number of basic features of log-linear modelling using the hypothetical data shown in Table 6.3. He labels the three-variables in a non-geographical fashion as: A – work status; B – education; C – sex. However, as this is a purely illustrative labelling and not important to the purpose of this example, we can drop it and consider the three variables as three arbitrary, geographically relevant variables.

Table 6.3
Wiliams's hypothetical
three-dimensional contingency table

		Variable C			
		k = 1		k = 2	
		Variable B		Variable B	
		j = 1	j = 1	j = 2	j = 2
Variable A	i = 1	10	6	15	5
	i = 2	5	15	8	6

Assuming that all variables are response variables and that Table 6.3 represents a cell (g) problem, Williams fits thc sct of log-linear models shown in Table 6.4. Simply on the basis of the G^2 goodness-of-fit values it can be seen that the most parsimonious model with an acceptable fit to the data is the fifth model in Table 6.4 which takes the form:

$$\log_e m_{ijk} = \lambda + \lambda_i^A + \lambda_j^B + \lambda_k^C + \lambda_{ij}^{AB} \qquad [6.4]$$

This is the type 4 model of Table 5.11; a multiple independence model which implies that the joint variable AB is independent of the third variable C. However, if we compare the fourth and fifth models in Table 6.4 using the conditional G^2 statistic defined in equation [5.44] (i.e. 5.27–0.91 = 4.81 with 1 d.f.) we find that it is not statistically justifiable to choose the more parsimonious fifth model [6.4] in preference to the fourth model. Hence, our 'acceptable' model takes the form

$$\log_e m_{ijk} = \lambda + \lambda_i^A + \lambda_j^B + \lambda_k^C + \lambda_{ij}^{AB} + \lambda_{jk}^{BC} \qquad [6.5]$$

This is the type 3 model of Table 5.11, a conditional independence model which implies that variable A is independent of variable C given variable B.

Table 6.4
Some members of the
hierarchical set of log-
linear models fitted to
Table 6.3, assuming all
variables are response variables

Model	Fitted marginals	G^2	d.f.
$A+B+C+A\cdot B+A\cdot C+B\cdot C+A\cdot B\cdot C$	[ABC]	0.0	0
$A+B+C+A\cdot B+A\cdot C+B\cdot C$	[AB] [AC] [BC]	0.59	1
$A+B+C+A\cdot B+A\cdot C$	[AB] [AC]	4.72	2
$A+B+C+A\cdot B+B\cdot C$	[AB] [BC]	0.91	2
$A+B+C+A\cdot B$	[AB] [C]	5.72	3
$A+B+C+B\cdot C$	[BC] [A]	7.88*	3
$A+B+C$	[A] [B] [C]	12.70*	4

* indicates G^2 in upper 5 per cent tail of χ^2 distribution with d.f. as indicated.

For our purposes, in this example, the most interesting aspect of Williams's hypothetical illustration is that it can, and has been, reformulated as a mixed explanatory/response variable problem. Upton (1978b) re-examined Williams's data and argued that variable A should be treated as a response variable and variables B and C as explanatory variables. Table 6.5 shows the hierarchical set of log-linear models fitted to the data of Table 6.3 under this assumption. The reader should notice how $B \cdot C$, the interaction between the explanatory variables, and the lower-order relatives (B and C) of this interaction, are included in all the models fitted.

Table 6.5
The hierarchical set of log-linear models fitted to Table 6.3 assuming variable A is the response variable and variables B and C are explanatory variables

Model	Fitted marginals	G^2	d.f.
$B+C+B \cdot C+A+A \cdot B+A \cdot C+A \cdot B \cdot C$	$[ABC]$	0.0	0
$B+C+B \cdot C+A+A \cdot B+A \cdot C$	$[BC]\,[AB]\,[AC]$	0.59	1
$B+C+B \cdot C+A+A \cdot B$	$[BC]\,[AB]$	0.91	2
$B+C+B \cdot C+A+A \cdot C$	$[BC]\,[AC]$	6.43*	2
$B+C+B \cdot C+A$	$[BC]\,[A]$	7.88*	3
$B+C+B \cdot C$	$[BC]$	7.94	4

* indicates G^2 in upper 5 per cent tail of χ^2 distribution with d.f. as indicated.

In this particular case, Upton's conclusions agree with those of Williams. On the basis of the G^2 values in Table 6.5, the most 'acceptable' model is found to be the third, i.e. the most 'acceptable' model takes the form [6.5]. However, this exact equivalence between the 'acceptable' model derived from Williams's cell (g), all-response variable treatment, and Upton's cell (f), mixed explanatory/response variable treatment is somewhat unusual. The reader should remember that the focus of the two analyses differs; the first is concerned with parsimonious description of the structural relationships among the response variables, whilst the second is concerned with assessing the effects of the explanatory variables on the response variable. Usually, this will give rise to, at least, slightly different 'acceptable' models being selected under the two different treatments.

EXAMPLE 6.2   *Shopping behaviour in Manchester (continued)*

In Example 5.4 we considered Fingleton's (1981) analysis of a four-dimensional $2 \times 2 \times 2 \times 2$ table shown in Table 5.25. The four variables in this table are: A – age of consumer; B – income of consumer; C – car ownership; D – shopping centre patronage. In Example 5.4 these were all treated as response variables. However, it is more logical in geographical terms to treat variable D, shopping centre patronage, as a response variable, and to treat variables A, B and C as explanatory variables. Upton and Fingleton (1979) and Fingleton (1981) have fitted log-linear models to the data of Table 5.25 under such an assumption, and Fingleton (1981) has used Aitkin's simultaneous test procedure to select the most 'acceptable' model.

Table 6.6 shows the first step in this simultaneous test procedure; the fitting of the various members of the hierarchy of log-linear models (the reader should compare this table to Table 5.26, noting how the interactions between the explanatory variables A, B and C are included in all models in Table 6.6 and, he should remember that $A^*B^*C = A + B + C + A \cdot B + A \cdot C + B \cdot C + A \cdot B \cdot C$). The second step is then to determine a Type 1 error rate α for each of the effects to be tested in the model, an overall Type 1 error rate, and various family Type 1 error rates for pooled two-variable interactions, three-variable interactions and four-variable interactions. Setting $\alpha = 0.05$, these Type 1 error rates and the associated critical values from the chi-square distribution are as follows:

Overall	$\gamma = 1 - (0.95)^7$	$= 0.302$	$\chi^2_{7, 0.302} = 8.36$
Four-variable	$\gamma_4 = 1 - (0.95)^1$	$= 0.05$	$\chi^2_{1, 0.05} = 3.84$
Three- and four-variable	$\gamma_{3,4} = 1 - (0.95)^4$	$= 0.185$	$\chi^2_{4, 0.185} = 6.20$
Two-, three- and four-variable	$\gamma_{2,3,4} = \gamma = 0.302$		$\chi^2_{7, 0.302} = 8.36$

Comparing the change-in-G^2 statistics in Table 6.6 with the critical values from the chi-square distribution we can see that both the four-variable and pooled three- and four-variable interactions have change-in-G^2 statistics which are less than the critical values (1.18 compared to 3.84, and $2.76 + 1.18 = 3.94$ compared to 6.20). Consequently, the three- and four-variable interactions are not significant and must be pooled with the two-variable interactions. In contrast, the change-in-G^2 statistic for the pooled two-variable, three-variable and four-variable interactions ($10.13 + 2.76 + 1.18$) is greater than the critical chi-square value (8.36). Hence, the pooled two-variable, three-variable and four-variable interactions are significant, and we can examine this pooled set of terms to see which can be eliminated, and which must be retained to form a minimal adequate subset.

Table 6.7 shows the particular permutation of terms within the pooled two-variable, three-variable and four-variable interactions which Fingleton uses to determine which terms must be eliminated and which terms retained. From this permutation, terms are eliminated from the bottom up until the critical chi-square value of 8.36 is exceeded. In this way we can eliminate $A \cdot B \cdot C \cdot D$, $A \cdot B \cdot D$, $A \cdot C \cdot D$, $B \cdot C \cdot D$ and also $B \cdot D$ and $A \cdot D$. This leaves a minimal adequate subset consisting of just $C \cdot D$.

Table 6.6
Goodness-of-fit statistics for a range of models fitted to Table 5.25 when variable D is regarded as a response variable and variables A, B and C as explanatory variables

Model	Fitted marginals	G^2	d.f.	Source	Change in G^2	d.f.
(a) $A^*B^*C + D$	[ABC] [D]	14.07	7			
(b) $A^*B^*C + D + A \cdot D$ $+ B \cdot D + C \cdot D$	[ABC] [AD] [BD] [CD]	3.94	4	2-variable interactions	10.13	3
(c) $A^*B^*C + D + A \cdot D$ $+ B \cdot D + C \cdot D + A \cdot B \cdot D$ $+ A \cdot C \cdot D + B \cdot C \cdot D$	[ABC] [ABD] [ACD] [BCD]	1.18	1	3-variable interactions	2.76	3
(d) $A^*B^*C + D + A \cdot D + B \cdot D$ $+ C \cdot D + A \cdot B \cdot D + A \cdot C \cdot D$ $+ B \cdot C \cdot D + A \cdot B \cdot C \cdot D$	[ABCD]	0.00	0	4-variable interactions	1.18	1

Table 6.7
Permutation of terms used by Fingleton to determine which terms must be eliminated and which retained

Model	G^2	d.f.	Source	Change in G^2	d.f.
$A*B*C+D$	14.07	7			
$A*B*C+D+C\cdot D$	6.64	6	CD	7.43	1
$A*B*C+D+C\cdot D+A\cdot D$	4.14	5	AD	2.50	1
$A*B*C+D+C\cdot D+A\cdot D+B\cdot D$	3.94	4	BD	0.20	1
$A*B*C+D+C\cdot D+A\cdot D+B\cdot D$	0.00	0	$A\cdot B\cdot D$, $A\cdot C\cdot D$,	3.94	4
+ all relevant 3-variable			$B\cdot C\cdot D$, $A\cdot B\cdot C\cdot D$		
and 4-variable interactions					

No other terms can be eliminated from the model and this leaves a final 'acceptable' model of the form:

$$\log_e m_{ijk} = \lambda + \lambda_i^A + \lambda_j^B + \lambda_k^C + \lambda_{ij}^{AB} + \lambda_{ik}^{AC} + \lambda_{jk}^{BC} + \lambda_{ijk}^{ABC} + \lambda_l^D + \lambda_{kl}^{CD} \quad \textbf{[6.6]}$$

In contrast to the final 'acceptable' model in Example 5.4, the three-explanatory-variable interaction term λ_{ijk}^{ABC} has been retained in model [6.6] as it is one of the parameters which take account of the 'facts of life' and which cannot be eliminated. In terms of the parameters which concern the relationship between the response and explanatory variables (the parameters on which our attention focusses), these indicate that there is no significant relationship between shopping centre patronage (variable D) and age (variable A) or income (variable B), but there is a relationship (λ_{kl}^{CD}) between car ownership and shopping centre patronage.

EXAMPLE 6.3   *Relationship between ethnic origin, birthplace, age and occupation in Canada in 1871*

Darroch and Ornstein (1980) present an important national level study of the relationship between ethnicity and occupational structure in nineteenth-century Canada. The study is based upon a national sample of 10,000 individual household records drawn at random from the microfilms of the manuscripts of the Canadian census of 1871. As part of this study, Darroch and Ornstein use cell (f) type log-linear models to analyse the extent to which differences in the native/foreign born proportions and in the age composition of the various ethnic-origin groups to be found in Canada in 1871 might be responsible for important observed differences among their occupational structures.

A total of 6,035 men in the sample had occupations listed on the census manuscripts. These men were then classified according to four variables: (E) – ethnic origin (divided into six categories: French, Irish Catholic, Irish Protestant, English, Scottish and German); (B)– birthplace (divided into two categories: European or North American born); (A) – age (divided into four categories: under 25, 25–34, 35–54, over 55); (O) – occupation (divided into eight categories: farmer, manufacturer-merchant, professionals-government, white collar-clerical, artisans, semi-skilled, labourers, servants). The cross-classification of these four variables gives rise to a four-dimensional $6 \times 2 \times 4 \times 8$ contingency table which has 384 cells.

Darroch and Ornstein analyse this large contingency table by assuming that occupation is the response variable and that ethnic origin, birthplace and age are explanatory variables. They then fit a hierarchical set of cell (f) type log-linear models to this contingency table, all members of which include the interaction between the explanatory variables E, B and A and the lower-order relatives of this interaction (i.e. all models include the terms: $E^*B^*A = E + B + A + E \cdot B + E \cdot A + B \cdot A + E \cdot B \cdot A$). Some members of the hierarchical set of models are shown in Table 6.8. On the basis of the G^2 goodness-of-fit statistics plus an investigation of certain conditional G^2 statistics, Darroch and Ornstein select the fifth model in Table 6.8 as the most 'acceptable' model, i.e. a model of the form:

$$E + B + A + E \cdot B + E \cdot A + B \cdot A + E \cdot B \cdot A + O + O \cdot E + O \cdot B + O \cdot A \quad [6.7]$$

It can be seen that the model has a relatively simple form. All the explanatory variables (ethnic-status, birthplace and age) have significant effects on occupational structure. However, these effects are largely independent of one another, i.e. they are represented by simple pairwise relationships ($O \cdot E$, $O \cdot B$ and $O \cdot A$) and are not a function of third or fourth variables. This represents an important substantive result. It implies that at the national level the important relationship between ethnic origin and occupation observed in nineteenth-century Canada[†] does *not* vary across the four age groups or between the native-born North Americans and the European immigrants. That is to say, ethnic differences in occupations are not an artifact of age or

Table 6.8
Some members of the hierarchical set of models fitted to the four-dimensional contingency table of ethnic-origin (E) by birthplace (B) by age (A) by occupation (O) for men in Canada in 1871

Model	Fitted marginals	G^2	d.f.
$E^*B^*A + O + O \cdot E + O \cdot B + O \cdot A + O \cdot E \cdot B$ $+ O \cdot E \cdot A + O \cdot B \cdot A + O \cdot E \cdot B \cdot A$	[OEBA]	0.0	0
$E^*B^*A + O + O \cdot E + O \cdot B + O \cdot A + O \cdot E \cdot A$	[EBA] [OEA] [OB]	117.7	161
$E^*B^*A + O + O \cdot E + O \cdot B + O \cdot A + O \cdot E \cdot B$	[EBA] [OEB] [OA]	178.9	231
$E^*B^*A + O + O \cdot E + O \cdot B + O \cdot A + O \cdot B \cdot A$	[EBA] [OBA] [OE]	191.0	245
$E^*B^*A + O + O \cdot E + O \cdot B + O \cdot A$	[EBA] [OE] [OB] [OA]	215.6	266
$E^*B^*A + O + O \cdot E + O \cdot A$	[EBA] [OE] [OA]	288.5	273
$E^*B^*A + O + O \cdot E + O \cdot B$	[EBA] [OE] [OB]	471.2*	287
$E^*B^*A + O + O \cdot B + O \cdot A$	[EBA] [OB] [OA]	407.6*	301
$E^*B^*A + O + O \cdot E$	[EBA] [OE]	521.6*	294
$E^*B^*A + O + O \cdot A$	[EBA] [OA]	487.9*	308
$E^*B^*A + O + O \cdot B$	[EBA] [OB]	662.6*	322
$E^*B^*A + O$	[EBA] [O]	719.0*	329

*indicates G^2 in upper 5 per cent tail of χ^2 distribution with d.f. as indicated.

[†] Certain ethnic origin groups were more highly represented in certain occupational categories than others: e.g. 28 per cent of Irish Catholics were labourers or semi-skilled workers compared to the national figure of 18 per cent; the Scottish-origin groups were most distinctly farmers; and the English-origin Church of England group had the smallest proportion in farming and the largest proportion in the professions, white-collar occupations and merchant occupations.

immigrant/native-proportion differences among the ethnic groups. Darroch and Ornstein suggest that this implies that demographic processes (i.e. a convergence of the age structures of the various ethnic groups and a decline in the proportion of foreign born over time) would be unlikely to alter a pattern of ethnic-group occupational differences that in 1871 appeared to be uniform across the categories of age and birthplace.

Finally, although we will not report the detailed results here, it will be of interest to geographers to note that Darroch and Ornstein extend their analysis to consider the extent of urban – rural and regional variations in the relationship between ethnicity and occupational structure. They find that the relationship differs between urban and rural areas and amongst regions. Hence, they imply that a geographical dimension to studies of the ethnic division of labour, and to the wider question of the 'vertical mosaic' (i.e. the assimilation of successive waves of European immigrants into the social and economic structure of nineteenth-century Canada) is essential. However, they do suggest that the urban – rural difference in ethnic occupational distributions follows the same pattern in each region.

6.3   Log-linear models as logit models

Although the log-linear models for cell (f) problems have an appearance which differs considerably from the linear logit models for cell (f) problems which we considered in Chapter 4, it is a simple matter to demonstrate that they are mathematically equivalent. To illustrate what to the reader may seem a somewhat remarkable statement let us first consider the case of a simple three-dimensional $2 \times J \times K$ contingency table in which variable A (a dichotomous variable) is regarded as a response variable and variables B and C are regarded as explanatory variables. We will then move on to consider the case in which A is a multiple-category response variable.

6.3.1   The dichotomous response variable case

As we have seen above, all log-linear models for a three-dimensional $2 \times J \times K$ contingency table in which variable A is the response variable and variables B and C are explanatory variables must include the parameters λ_{jk}^{BC}, λ_j^B and λ_k^C for these take account of the 'facts of life', i.e. the relationships between the explanatory variables. The saturated log-linear model, therefore, takes the form:

$$\log_e m_{ijk} = \lambda + \lambda_j^B + \lambda_k^C + \lambda_{jk}^{BC} + \lambda_i^A + \lambda_{ij}^{AB} + \lambda_{ik}^{AC} + \lambda_{ijk}^{ABC} \qquad [6.8]$$

Now it will be recalled that in our previous discussions of dichotomous linear logit models we were concerned, primarily, with what we termed the *odds* (more specifically the log odds) of selecting one particular response category over the other response category (e.g. $P_{1|i}/P_{2|i}$) given the values of the explanatory variables. In this $2 \times J \times K$ table case, therefore, where A is the response variable, it is reasonable for us to be interested in what the odds are of being in each category of variable A given the particular combinations of categories of the explanatory variables, B and C. In other words, it is reasonable for us to be interested in the odds m_{1jk}/m_{2jk} for each combination j and k or, more specifically, in the log odds $\log_e(m_{1jk}/m_{2jk}) = \log_e m_{1jk} - \log_e m_{2jk}$. In the case of the saturated model [6.8] it can be seen that:

$$\log_e m_{1jk} = \lambda + \lambda_j^B + \lambda_k^C + \lambda_{jk}^{BC} + \lambda_1^A + \lambda_{1j}^{AB} + \lambda_{1k}^{AC} + \lambda_{1jk}^{ABC} \qquad [6.9]$$

and

$$\log_e m_{2jk} = \lambda + \lambda_j^B + \lambda_k^C + \lambda_{jk}^{BC} + \lambda_2^A + \lambda_{2j}^{AB} + \lambda_{2k}^{AC} + \lambda_{2jk}^{ABC} \qquad [6.10]$$

By subtracting [6.10] from [6.9] we can therefore derive an expression for the required log odds:

$$\log_e \frac{m_{1jk}}{m_{2jk}} = \log_e m_{1jk} - \log_e m_{2jk}$$

$$= (\lambda_1^A - \lambda_2^A) + (\lambda_{1j}^{AB} - \lambda_{2j}^{AB}) + (\lambda_{1k}^{AC} - \lambda_{2k}^{AC}) + (\lambda_{1jk}^{ABC} - \lambda_{2jk}^{ABC})$$

$$[6.11]$$

The reader should notice that in this subtraction the terms in [6.9] and [6.10] which involve only the explanatory variables cancel out (e.g. $\lambda_{jk}^{BC} - \lambda_{jk}^{BC}$).

Now, if we assume a 'centred effect' coding system and associated constraints [5.15] and [5.19] it follows that $\lambda_1^A = -\lambda_2^A$, $\lambda_{1j}^{AB} = -\lambda_{2j}^{AB}$, and so on. Consequently, the terms in brackets in [6.11] can be seen to be, $(\lambda_1^A - \lambda_2^A) = (\lambda_1^A + \lambda_1^A) = 2\lambda_1^A$, and so on (see Reynolds 1977, 132; Payne 1977, 137; Fienberg 1977, p. 78). This implies that [6.11] can be rewritten as:

$$\log_e \frac{m_{1jk}}{m_{2jk}} = 2\lambda_1^A + 2\lambda_{1j}^{AB} + 2\lambda_{1k}^{AC} + 2\lambda_{1jk}^{ABC} \qquad [6.12]$$

or, by defining

$$\omega = 2\lambda_1^A \qquad\qquad \omega_j^B = 2\lambda_{1j}^{AB}$$

$$\omega_k^C = 2\lambda_{1k}^{AC} \qquad\qquad \omega_{jk}^{BC} = 2\lambda_{1jk}^{ABC} \qquad\qquad [6.13]$$

it can be rewritten as

$$\log_e \frac{m_{1jk}}{m_{2jk}} = \omega + \omega_j^B + \omega_k^C + \omega_{jk}^{BC} \qquad [6.14]$$

If we assume a 'cornered effect' coding system and associated GLIM-type constraints [5.17] and [5.20], it follows that λ_1^A, λ_{1j}^{AB}, λ_{1k}^{AC} and λ_{1jk}^{ABC} in [6.11] all equal zero. Consequently, model [6.11] can be rewritten as:

$$\log_e \frac{m_{1jk}}{m_{2jk}} = -\lambda_2^A - \lambda_{2j}^{AB} - \lambda_{2k}^{AC} - \lambda_{2jk}^{ABC} \qquad [6.15]$$

or by defining

$$\omega = -\lambda_2^A \qquad \omega_j^B = -\lambda_{2j}^{AB}$$
$$\omega_k^C = -\lambda_{2k}^{AC} \qquad \omega_{jk}^{BC} = -\lambda_{2jk}^{ABC}$$

[6.16]

we can see that [6.15] can also be rewritten in the form of model [6.14].

Model [6.14] is clearly a linear logit model of the type we considered in Chapter 4. It represents an *equivalent* way of expressing the saturated log-linear model [6.8], and it thus illustrates the mathematical relationship between log-linear and linear logit models for cell (f) problems. To help the reader gain further insight into this equivalence, it may be useful to rewrite [6.14] in the linear logit model notation of Chapter 4 (see equation[4.1]). In this notation [6.14] becomes:

$$\log_e \frac{P_{1|g}}{P_{2|g}} = \beta_1 + \beta_2 X_{g2} + \beta_3 X_{g3} + \beta_4 X_{g2} X_{g3}$$

[6.17]

where the explanatory variables, X, are all dummy variables (coded according to either the 'centred effect' or 'cornered effect' system) and where g denotes the group or sub-population defined on the basis of the cross-classification of the explanatory variables. In order to relate the terms in [6.17] and [6.14] we define the subscript g in [6.17] as equivalent to jk in [6.14], the subscript 2 in [6.17] as equivalent to B in [6.14], and subscript 3 in [6.17] as C in [6.14]. Consequently we can define

$$\beta_1 = \omega \qquad \beta_2 X_{g2} = \omega_j^B$$
$$\beta_3 X_{g3} = \omega_k^C \qquad \beta_4 X_{g2} X_{g3} = \omega_{jk}^{BC}$$

[6.18]

It must be acknowledged, however, that the relationship between the notational systems of linear logit models and log-linear models can easily become a rather confusing issue for the reader. For readers who are interested in the topic, further information is given in the footnote below.[†] Other readers should simply accept the notational equivalence of [6.14] and [6.17], and concentrate upon the relationship between cell (f) log-linear and linear logit models established above.

We have observed, then, that model [6.14] is the saturated linear logit model equivalent to the saturated log-linear model [6.8]. In a similar fashion (and working through all the stages outlined above), the alternative log-linear models for this particular problem:

$$\log_e m_{ijk} = \lambda + \lambda_j^B + \lambda_k^C + \lambda_{jk}^{BC} + \lambda_i^A + \lambda_{ij}^{AB} + \lambda_{ik}^{AC}$$

[6.19]

$$\log_e m_{ijk} = \lambda + \lambda_j^B + \lambda_k^C + \lambda_{jk}^{BC} + \lambda_i^A + \lambda_{ij}^{AB}$$

[6.20]

$$\log_e m_{ijk} = \lambda + \lambda_j^B + \lambda_k^C + \lambda_{jk}^{BC} + \lambda_i^A + \lambda_{ik}^{AC}$$

[6.21]

$$\log_e m_{ijk} = \lambda + \lambda_j^B + \lambda_k^C + \lambda_{jk}^{BC} + \lambda_i^A$$

[6.22]

[†] For certain readers, it will help to facilitate comparison between [6.14] and [6.17] if we write both models out in matrix form. Assuming, for convenience, that we are dealing with a $2\times2\times2$ problem,

become in linear logit from:

$$\log_e \frac{m_{1jk}}{m_{2jk}} = \omega + \omega_j^B + \omega_k^C \qquad\qquad [6.23]$$

$$\log_e \frac{m_{1jk}}{m_{2jk}} = \omega + \omega_j^B \qquad\qquad [6.24]$$

$$\log_e \frac{m_{1jk}}{m_{2jk}} = \omega + \omega_k^C \qquad\qquad [6.25]$$

$$\log_e \frac{m_{1jk}}{m_{2jk}} = \omega \qquad\qquad [6.26]$$

---

i.e. in our $2 \times J \times K$ discussion above $J = 2$ and $K = 2$, and assuming a 'centred effect' coding system then [6.14] and [6.17] take the respective matrix forms:

$$
\begin{bmatrix}
\log_e(m_{111}/m_{211}) \\
\log_e(m_{112}/m_{212}) \\
\log_e(m_{121}/m_{221}) \\
\log_e(m_{122}/m_{222})
\end{bmatrix}
=
\begin{bmatrix}
1 & 1 & 1 & 1 \\
1 & 1 & -1 & -1 \\
1 & -1 & 1 & -1 \\
1 & -1 & -1 & 1
\end{bmatrix}
\begin{bmatrix}
\omega \\
\omega_1^B \\
\omega_1^C \\
\omega_{11}^{BC}
\end{bmatrix}
$$

$$
\begin{bmatrix}
\log_e(P_{1|1}/P_{2|1}) \\
\log_e(P_{1|2}/P_{2|2}) \\
\log_e(P_{1|3}/P_{2|3}) \\
\log_e(P_{1|4}/P_{2|4})
\end{bmatrix}
=
\begin{bmatrix}
X_{11} & X_{12} & X_{13} & X_{12}X_{13} \\
X_{21} & X_{22} & X_{23} & X_{22}X_{23} \\
X_{31} & X_{32} & X_{33} & X_{32}X_{33} \\
X_{41} & X_{42} & X_{43} & X_{42}X_{43}
\end{bmatrix}
\begin{bmatrix}
\beta_1 \\
\beta_2 \\
\beta_3 \\
\beta_4
\end{bmatrix}
$$

$$
=
\begin{bmatrix}
1 & 1 & 1 & 1 \\
1 & 1 & -1 & -1 \\
1 & -1 & 1 & -1 \\
1 & -1 & -1 & 1
\end{bmatrix}
\begin{bmatrix}
\beta_1 \\
\beta_2 \\
\beta_3 \\
\beta_4
\end{bmatrix}
$$

Assuming a 'cornered effect' coding system, [6.14] and [6.17] take the respective matrix forms

$$
\begin{bmatrix}
\log_e(m_{111}/m_{211}) \\
\log_e(m_{112}/m_{212}) \\
\log_e(m_{121}/m_{221}) \\
\log_e(m_{122}/m_{222})
\end{bmatrix}
=
\begin{bmatrix}
1 & 1 & 1 & 1 \\
1 & 1 & 0 & 0 \\
1 & 0 & 1 & 0 \\
1 & 0 & 0 & 0
\end{bmatrix}
\begin{bmatrix}
\omega \\
\omega_1^B \\
\omega_1^C \\
\omega_{11}^{BC}
\end{bmatrix}
$$

$$
\begin{bmatrix}
\log_e(P_{1|1}/P_{2|1}) \\
\log_e(P_{1|2}/P_{2|2}) \\
\log_e(P_{1|3}/P_{2|3}) \\
\log_e(P_{1|4}/P_{2|4})
\end{bmatrix}
=
\begin{bmatrix}
X_{11} & X_{12} & X_{13} & X_{12}X_{13} \\
X_{21} & X_{22} & X_{23} & X_{22}X_{23} \\
X_{31} & X_{32} & X_{33} & X_{32}X_{33} \\
X_{41} & X_{42} & X_{43} & X_{42}X_{43}
\end{bmatrix}
\begin{bmatrix}
\beta_1 \\
\beta_2 \\
\beta_3 \\
\beta_4
\end{bmatrix}
$$

$$
=
\begin{bmatrix}
1 & 1 & 1 & 1 \\
1 & 1 & 0 & 0 \\
1 & 0 & 1 & 0 \\
1 & 0 & 0 & 0
\end{bmatrix}
\begin{bmatrix}
\beta_1 \\
\beta_2 \\
\beta_3 \\
\beta_4
\end{bmatrix}
$$

These matrix expressions serve to illustrate to the reader a rather difficult notational issue. It can be seen that in [6.14], as in the log-linear model [6.8] from which it was derived, the elements of the associated design matrix **X** are conventionally not written out explicitly. However, in linear logit models such as [6.17] the elements of the design matrix are, conventionally, made explicit.

respectively. As in model [6.14], the ω-parameters in the linear logit models [6.23] to [6.26] must be interpreted in the manner discussed in Section 4.4 rather than in log-linear model fashion discussed in Section 5.2. Estimation of the ω-parameters in [6.14], [6.23] to [6.26] can be undertaken directly using weighted least squares on the dichotomous linear logit models themselves. Alternatively, the equivalent log-linear model [6.8] can be fitted using the iterative proportional fitting, iterative weighted least squares, or Newton–Raphson procedures, and the maximum likelihood estimates of the λ-parameter derived in this way can then be used to compute the estimates of the ω-parameters by substituting into the expressions [6.13] or [6.16].

The reader should notice that in deriving the linear logit models [6.14] and [6.23] to [6.26], we have treated as fixed the marginal totals of the contingency table corresponding to the explanatory variables. In other words, we have conditioned on the marginal totals involving the explanatory variables. The importance of this assumption of fixed marginal totals must be stressed. The reason is that (as we have seen above) the parameters which involve *only* the explanatory variables (λ_{jk}^{BC}, λ_j^B, λ_k^C in our case) cancel out in the derivation of the linear logit model. Consequently, a model such as

$$\log_e m_{ijk} = \lambda + \lambda_j^B + \lambda_k^C + \lambda_{jk}^{BC} + \lambda_i^A + \lambda_{ij}^{AB} \qquad [6.27]$$

in which the marginal totals corresponding to the explanatory variables have been fixed will produce the same logit model [6.24] as the model

$$\log_e m_{ijk} = \lambda + \lambda_i^A + \lambda_j^B + \lambda_k^C + \lambda_{ij}^{AB} \qquad [6.28]$$

in which the marginal totals corresponding to the explanatory variables have *not* been fixed. (To see this subtract $\log_e m_{2jk} = \lambda + \lambda_2^A + \lambda_j^B + \lambda_k^C + \lambda_{2j}^{AB}$ from $\log_e m_{1jk} = \lambda + \lambda_1^A + \lambda_j^B + \lambda_k^C + \lambda_{1j}^{AB}$ and utilize [6.13] or [6.16] to produce the logit model [6.24].) It follows, therefore, that if the ω-parameters of the logit model [6.24] are estimated using the technique of fitting the equivalent log-linear model, estimating its λ-parameters, and substituting these into expression [6.13] or [6.16], then the two log-linear models [6.27] and [6.28] will produce different estimates of the ω-parameters. To avoid such problems, the reader must remember the lessons of this chapter. The appropriate log-linear model from which to derive the estimates of the ω-parameters of the linear logit model is the one which includes *all* terms corresponding to the explanatory variables, i.e. a model in which we treat as fixed the marginal totals corresponding to the explanatory variables.

6.3.2   *The multiple-category response variable case*

Let us now turn to the case in which we have a three-dimensional contingency table in which A is regarded as the response variable and in which A now has three rather than two categories. In this case we must recall the discussion of Section 2.9, and concern ourselves with two sets of odds m_{1jk}/m_{3jk} and m_{2jk}/m_{3jk}, or more specifically with two sets of log odds $\log_e(m_{1jk}/m_{3jk})$ and $\log_e(m_{2jk}/m_{3jk})$. In the case of the saturated model, it can be seen that

$$\log_e m_{1jk} = \lambda + \lambda_j^B + \lambda_k^C + \lambda_{jk}^{BC} + \lambda_1^A + \lambda_{1j}^{AB} + \lambda_{1k}^{AC} + \lambda_{1jk}^{ABC} \qquad [6.29]$$

$$\log_e m_{2jk} = \lambda + \lambda_j^B + \lambda_k^C + \lambda_{jk}^{BC} + \lambda_2^A + \lambda_{2j}^{AB} + \lambda_{2k}^{AC} + \lambda_{2jk}^{ABC} \qquad [6.30]$$

$$\log_e m_{3jk} = \lambda + \lambda_j^B + \lambda_k^C + \lambda_{jk}^{BC} + \lambda_3^A + \lambda_{3j}^{AB} + \lambda_{3k}^{AC} + \lambda_{3jk}^{ABC} \qquad [6.31]$$

By subtracting [6.31] from [6.29] and from [6.30] it is a simple matter to derive expressions for the required log-odds. These take the form of a system of two related linear logit models:

$$
\begin{aligned}
\log_e \frac{m_{1jk}}{m_{3jk}} &= (\lambda_1^A - \lambda_3^A) + (\lambda_{1j}^{AB} - \lambda_{3j}^{AB}) + (\lambda_{1k}^{AC} - \lambda_{3k}^{AC}) \\
&\quad + (\lambda_{1jk}^{ABC} - \lambda_{3jk}^{ABC}) \\
\log_e \frac{m_{2jk}}{m_{3jk}} &= (\lambda_2^A - \lambda_3^A) + (\lambda_{2j}^{AB} - \lambda_{3j}^{AB}) + (\lambda_{2k}^{AC} - \lambda_{3k}^{AC}) \\
&\quad + (\lambda_{2jk}^{ABC} - \lambda_{3jk}^{ABC})
\end{aligned}
\qquad [6.32]
$$

and for both 'centred effect' and 'cornered effect' coding systems the linear logit models [6.32] can be rewritten as:[†]

$$
\begin{aligned}
\log_e \frac{m_{1jk}}{m_{3jk}} &= \omega_1 + \omega_{j1}^B + \omega_{k1}^C + \omega_{jk1}^{BC} \\
\log_e \frac{m_{2jk}}{m_{3jk}} &= \omega_2 + \omega_{j2}^B + \omega_{k2}^C + \omega_{jk2}^{BC}
\end{aligned}
\qquad [6.33]
$$

Notice that in defining these models we are following the convention of previous chapters and using the final category (3) of variable A as the arbitrary base or anchor category.

As in the dichotomous case above, it is a simple matter to take the other possible log-linear models for this problem (i.e. those shown in [6.19] to [6.22]) and define linear logit models similar to those in [6.33].

[†] If we assume a 'centred effect' coding system and associated constraints [5.15] and [5.19] then $\sum_i \lambda_i^A = 0$, etc. Consequently we must define

$$\omega_1 = (\lambda_1^A - \lambda_3^A), \qquad \omega_{j1}^B = (\lambda_{1j}^{AB} - \lambda_{3j}^{AB}), \text{ etc.}$$
$$\omega_2 = (\lambda_2^A - \lambda_3^A), \qquad \omega_{j2}^B = (\lambda_{2j}^{AB} - \lambda_{3j}^{AB}), \text{ etc.}$$

If we assume a 'cornered effect,' coding system and associated GLIM-type constraints [5.17] and [5.20], then $\lambda_1^A = 0$, $\lambda_{1j}^{AB} = 0$, etc. Consequently we must define:

$$\omega_1 = -\lambda_3^A \qquad \qquad \omega_{j1}^B = -\lambda_{3j}^{AB}, \text{ etc.}$$
$$\omega_2 = (\lambda_2^A - \lambda_3^A), \qquad \omega_{j2}^B = (\lambda_{2j}^{AB} - \lambda_{3j}^{AB}), \text{ etc.}$$

If we assume a 'cornered effect' coding system and constraints [5.16] and [5.20] (i.e. original GLIM-type, up to Release 2) then $\lambda_f^A = 0$, $\lambda_{fj}^{AB} = 0$, etc. Consequently we must define:

$$\omega_1 = \lambda_1^A \qquad \qquad \omega_{j1}^B = \lambda_{1j}^{AB} \text{ etc.}$$
$$\omega_2 = \lambda_2^A \qquad \qquad \omega_{j2}^B = \lambda_{2j}^{AB}, \text{ etc.}$$

Once again, the resulting ω-parameters can be estimated directly using weighted least squares on the linear logit models themselves or, alternatively, the equivalent log-linear models can be fitted in the standard way discussed in Chapter 5 and the estimates of the λ-parameters derived in this way can then be used to compute the estimates of the ω-parameters. If the latter method is used, the researcher must take particular care to make sure just what coding and associated parameter constraint system his computer program is using. Only in this way will he be able to substitute his λ-parameter estimates into the correct expressions for the ω-parameters given in the footnote to the previous page. In addition, it must, as always, be remembered that the appropriate log-linear models from which to derive estimates of the ω-parameters in the linear logit models are the ones (like [6.19] to [6.22]) which treat as fixed the marginal totals corresponding to the explanatory variables, i.e. which include *all* terms corresponding to the explanatory variables.

The extension of our three response-category linear logit models [6.33] to the multiple response-category case is straightforward. Given that we have a three-dimensional category table in which variables B and C are regarded as explanatory variables and variable A is a response variable with I categories, then the saturated log-linear model can be expressed as the system of linear logit models:

$$\log_e \frac{m_{1jk}}{m_{Ijk}} = \omega_1 + \omega_{j1}^B + \omega_{k1}^C + \omega_{jk1}^{BC}$$

$$\vdots \qquad\qquad \vdots \qquad\qquad\qquad\qquad\qquad\qquad\qquad \text{[6.34]}$$

$$\log_e \frac{m_{(I-1)jk}}{m_{Ijk}} = \omega_{I-1} + \omega_{jI-1}^B + \omega_{kI-1}^C + \omega_{jkI-1}^{BC}$$

This can be summarized, if required, as

$$\log_e \frac{m_{ijk}}{m_{Ijk}} = \omega_i + \omega_{ji}^B + \omega_{ki}^C + \omega_{jki}^{BC} \quad i = 1, 2, \dots I-1 \qquad \text{[6.35]}$$

and it is a relatively simple matter to express [6.35] in the linear logit notation of Chapter 4.[†]

In a similar fashion, the extension of the models discussed in this chapter to four-dimensional or higher-dimensional tables is straightforward. For example, in the case of a four-dimensional table where A is the response variable and B, C and D are the explanatory variables, the saturated log-linear model takes the form:

$$\log_e m_{ijkl} = \lambda + \lambda_i^A + \lambda_j^B + \lambda_k^C + \lambda_l^D + \lambda_{ij}^{AB} + \lambda_{ik}^{AC} + \lambda_{il}^{AD} + \lambda_{jk}^{BC}$$

$$+ \lambda_{jl}^{BD} + \lambda_{ijk}^{ABC} + \lambda_{ijl}^{ABD} + \lambda_{ikl}^{ACD} + \lambda_{jkl}^{BCD} + \lambda_{ijkl}^{ABCD} \qquad \text{[6.36]}$$

[†] In the linear logit model notation of Chapter 4, [6.35] becomes

$$\log_e \frac{P_{r|g}}{P_{R|g}} = \beta_{1r} + \beta_{2r}X_{g2} + \beta_{3r}X_{g3} + \beta_{4r}X_{g2}X_{g3} \quad r = 1,2, \dots R-1$$

The reader who is interested should explore this notational equivalence, beginning with expressions [6.17], [4.11], [4.7], [4.20].

In the dichotomous response variable case (where $i = 1,\ 2$) the saturated log-linear model [6.36] can be expressed in an equivalent linear logit manner as:

$$\log_e \frac{m_{1jkl}}{m_{2jkl}} = \omega + \omega_j^B + \omega_k^C + \omega_l^D + \omega_{jk}^{BC} + \omega_{jl}^{BD} + \omega_{kl}^{CD} + \omega_{jkl}^{BCD} \qquad [6.37]$$

In the multiple-cateogy response variable case ($i = 1,\ 2, \ldots I$) the saturated log-linear model [6.36] can be expressed in an equivalent linear logit form as:

$$\log_e \frac{m_{ijkl}}{m_{Ijkl}} = \omega_i + \omega_{ji}^B + \omega_{li}^D + \omega_{jki}^{BC} + \omega_{jli}^{BD} + \omega_{kli}^{CD}$$

$$+ \omega_{jkli}^{BCD} \qquad i = 1,\ 2 \ldots I-1 \qquad [6.38]$$

All other principles discussed above then apply with equal force to these higher-dimensional contingency table models.

6.3.3 *Discussion*

Some readers will undoubtedly have found certain aspects of Sections 6.3.1 and 6.3.2 to be rather difficult to comprehend on a first reading. Nevertheless, it is worthwhile for such readers to persevere with their efforts, for these sections have an important message. Basically, they state that, despite superficially different appearances, log-linear models and linear logit models for cell (f) problems are mathematically related. The implications of this are important to our discussion because it demonstrates that there is a formal linkage between two major elements of the unified approach to categorical data analysis. In terms of the cells of the bottom row of Table 1.1 we have seen that the influence of logistic, linear logit and probit models can be viewed as pushing towards the right from cells (d) and (e), and the influence of log-linear models can be viewed as pushing towards the left from cell (g). In cell (f) this produces a 'zone of transition' in which both linear logit and log-linear models are widely used. What we have demonstrated, in Sections 6.3.1 and 6.3.2, is that this cell represents a true transition zone in which these two major traditions in categorical data analysis merge into one another; rather than a zone of discontinuity in which log-linear and linear logit models lie adjacent to each other but are not linked. The linkage we have demonstrated in this section justifies our treatment of the categorical data models in Chapters 2–6 as an integrated *family* of statistical models, and in Chapter 7 we will see that this integration can be exploited effectively in the development of computer software.

Unfortunately, this important methodological linkage which occurs in cell (f) of Table 1.1 is not achieved without cost. The main message may have become clouded for some readers by certain aspects of these differing statistical traditions. In particular, the notation system which we have used in Chapters 2, 3 and 4 (and which we employ again in Chapters 8 and 10), and the notation system which we have used in Chapters 5 and 6 (and which we will use again in Chapter 9) produce expressions for *equivalent* linear logit models (e.g. [6.14] and [6.17]) which look confusingly different (though it is a simple matter to demonstrate that such expressions are the same, e.g. see footnote, p. 226). In addition, the estimation procedures and the associated computer programs which we employ to estimate such linear logit models differ considerably.

Little can be done to help the reader with such problems, except: (a) to reassure him that everything possible has been done in this book to achieve a standard and consistent notation system which reduces linkage problems to a minimum, whilst at the same time being maximally consistent with the other literature in the field; (b) to attempt to convince him that such problems will progressively reduce as his knowledge of categorical data analysis develops; (c) to stress that the important message of this section is the linkage of the major elements in the family of statistical models for categorical data. The viewpoint adopted in this book is that it is worth persevering with a unified view of these models rather than indulge (as sometimes occurs in the social science research literature) in the rather arid discussion of which types of models are 'better' than others or which are merely special cases of others.

CHAPTER 7

Computer programs for categorical data analysis

There are two commonly held positions concerning the value of including a discussion of available computer programs in statistical methods texts such as this. The first is that such programs will have become outdated and are likely to have been superseded long before the book becomes widely used. As such, any discussion is likely to be at best redundant and at worst misleading. The second, and the position held in this book, is that many non-statisticians will learn about new statistical methods most effectively by supplementing the theoretical discussion of textbooks by 'hands-on' experience of data analysis using the best available computer programs. As such, some guidance to the characteristics and likely development of such programs is necessary. This is the aim of this short chapter. Although some of the details will inevitably become outdated, a number of the major programs for categorical data analysis (GLIM, BMDP) are now sufficiently well established and well supported to suggest that the reader will be able to turn to them in confidence that they will have kept abreast of any developments which have occurred since the publication of this book.

7.1 A classification of available programs

Using the cells of Table 1.1 as one axis, and the computational method of parameter estimation employed by the programs as the other, a classification of some of the most important and widely available computer programs can be presented in the form of Table 7.1 (see also O'Brien and Wrigley 1980).

7.2 Programs which use function maximization algorithms

There are a large number of programs of this type designed to fit dichotomous or multiple-category logit or probit models. Many were

Table 7.1
A classification of computer programs for categorical data analysis

Computational method of estimation	Cells of basic classification		
	d and e	f	g
Direct maximum likelihood using Function maximization algorithms	BLOGIT QUAIL 4.0 FREQ CHOMP Nerlove & Press		Nerlove & Press FREQ
Weighted least squares iterative and non-iterative	GLIM (and GLIM macros) BMDP routine PLR	GLIM (and GLIM macros) GENCAT NONMET BMDP routine PLR	GLIM (and GLIM macros)
Iterative proportional fitting		BMDP routine P4F ECTA C-TAB	BMDP routine P4F ECTA C-TAB

developed by researchers in transportation or econometrics and a number of these have common origins in the work of Cragg in the late 1960s (Cragg and Lisco 1968) and Manski, McFadden, Wills and Varian in the early 1970s (see Manski 1974, and the early versions of QUAIL). Further details of some of the most widely used are as follows.

(a) BLOGIT (Crittle and Johnson 1980; Hensher and Johnson 1981a) This is a widely known and widely used program due to its extensive description in the textbook by Hensher and Johnson (1981a) and its distribution by the Australian Road Research Board. Basically, it allows the fitting of dichotomous or multiple-category logit models, but it also has several special and useful features which permit Box–Tukey transformations of the explanatory variables (see Section 10.5), which allow linear restrictions to be imposed on the parameters, and so on. Data input and data transformation is couched in terms of the generic variable/alternative specific variable terminology discussed in Section 2.11.1, and several statistics (ρ^2, maximized log-likelihood values, 't'-statistics, prediction-success tables) which can be used to assess the fit of the model (see Sections 2.6 and 2.7) are automatically provided in the output. The program currently has certain limitations relating to its rather inflexible program control and command structure, to deficiencies in the explanation provided in the user's manual, and to its computational efficiency. However, extension and upgrading of the program is ongoing, (e.g. an improved user's manual is now available, see Hensher 1984).

(b)   QUAIL 4.0   (Berkman and Brownstone 1979). This is a suite of programs developed by McFadden, Wills, Berkman and Brownstone throughout the 1970s at Berkeley and which is now distributed by the Department of Economics at Princeton. It has been designed to handle any statistical problem which has a categorical response variable and its routines focus on the fitting of dichotomous and multiple-category logit models. In addition, it has an experimental multiple response category probit routine which uses the Clark–Daganzo method (see the discussion of CHOMP below). It is backed up by detailed user's and technical manuals, though these are written at too advanced a level for many would-be users. One of the main limitations of QUAIL 4.0 is that it is a large package distributed in either IBM or CDC compatible form and for universities without such machines its implementation involves considerable (and generally unwarranted) effort. (British readers should note that it is available on the London CDC system and many British universities can access this directly or through regional computer networks.)

(c) **CHOMP** (Daganzo and Schonfeld 1978). This is the most widely used program for fitting the multiple response category version of the probit model (see Section 10.3.1). It can be obtained from the Institute of Transportation Studies, University of California, Berkeley. Although the so-called multinomial probit (MNP) model has many useful properties, it is extremely intractable in computational terms. CHOMP attempts to overcome such difficulties by using the Clark (1961) approximation to reduce the estimation problem to one of sequential univariate integration, and it is the only computational method which appears to be practical for problems involving more than three or four response categories. Nevertheless, the accuracy of the estimation procedure when the Clark approximation is adopted is still controversial (Sheffi, Hall and Daganzo 1982), and for problems involving just three or four response categories a direct numerical integration procedure is likely to provide a more accurate method (see the program TROMP by Sparman and Daganzo, 1980).

Many other programs of this type exist. The majority are unpublished and are either purpose-built routines which use numerical optimization algorithms taken from the many well-established subroutine libraries (e.g. in Britain the NAG or AERE Harwell subroutine libraries), or are specialized adaptations of earlier programs. Some are published, such as: Chung's (1978) dichotomous logistic/logit model program for the Canadian Geological Survey; the multiple category logistic/logit model program of Jones (1975); the MULTIQUAL program of Bock and Yates (1973); and the program of Nerlove and Press (1973) which was written for the RAND Corporation in the early 1970s and which represents an early attempt to provide a unified basis for the analysis of categorical data using logit and log-linear models. One of the more recently published programs of this type is Haberman's program called FREQ which uses the Newton–Raphson algorithm to compute maximum likelihood estimates for any logit or log-linear model. A modification of FREQ produces another useful program called LAT which can be used for the estimation of latent structure models.

7.3   Programs based upon weighted least squares algorithms

7.3.1   *Iterative weighted least squares*

(a) GLIM (Baker and Nelder 1978). The Generalized Linear Interactive Modelling (GLIM) package developed by the Working Party on Statistical Computing of the Royal Statistical Society and NAG (Numerical Algorithms Group) is rapidly becoming perhaps the best known and most widely used package for fitting categorical data models. GLIM is designed to fit any member of the family of 'generalized linear models' suggested by Nelder and Wedderburn (1972) and (as shown in Ch. 11) this family includes a majority of both the conventional continuous data linear models and categorical data models which the geographer or environmental scientist is likely to require. It also allows users to specify (in the form of GLIM 'macros') special-purpose models if the default configuration of the package is insufficient for their needs. The third version of the package is currently available at over 500 computer centres worldwide and the basic user's manual is backed up by the regular *GLIM Newsletter* which contains many useful user-supplied macros. A more extensive version of the package, GLIM 4, is to be released in 1985 and will form part of a new statistical data analysis system which is provisionally known as PRISM.

In terms of the models discussed in Chapters 1–6, GLIM 3 allows the fitting of dichotomous logit and probit models, and also hierarchical log-linear models. Maximum likelihood estimates are derived using an iterative weighted least squares procedure which was illustrated for the case of log-linear models in Section 5.6.2. The major limitation of GLIM 3 is its inability to fit (at least in its default configuration) multiple response category logit and probit models. However, the new version, GLIM 4, will be extended to include the multinomial 'error' distribution as standard (see Ch. 11) and this will allow users to fit multiple category logistic/logit models in a much more direct fashion than is currently possible.

The wider statistical data analysis system, PRISM (see Fig. 7.1), is likely to include: GLIM 4; AOV 1 – a general analysis of variance program specifically designed to handle balanced and partially-balanced research designs (e.g. the randomized blocks, Latin squares, Graeco-

Fig. 7.1
The PRISM data analysis system. (Name of system is provisional and may be changed)

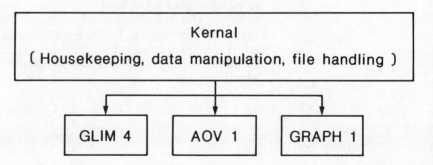

The structure of PRISM

Latin squares, incomplete blocks, and split-plot designs which are often valuable in environmental science research); and GRAPH 1 – a set of graphics subroutines for simple data displays (histographs, pie-charts, scatterplots) and more sophisticated uses (generalized text drawing, picture segments, and graphical input-output).

(b)   BMDP Stepwise Logistic – Routine PLR   (Engleman 1979). This is a readily accessible example of the many stepwise logistic/logit programs which are available. It uses an iterative weighted least squares algorithm to derive the parameter estimates of a dichotomous model, has a number of useful options, and has clear documentation. The BMDP package is one of the most widely distributed in the world and like GLIM it is fully supported. However, the reader should note that there are certain well-known dangers in using any automated stepwise model selection procedure, and for this reason stepwise logistic/logit models were not discussed in Chapter 2. Considerable care must be exercised when using such a stepwise procedure with non-experimental data which have the typical limitations described in Section 8.1.

7.3.2   *Non-iterative weighted least squares*

(a)   GENCAT   (Landis *et al* 1976). This is the most recent version of a program designed to implement the general categorical data methodology of Grizzle, Starmer and Koch (1969) discussed and illustrated in Chapters 4 and 8. It replaces the earlier CATLIN and MODCAT programs (Forthofer *et al* 1971) and both batch mode and interactive versions are available. (Example listings for geographical problems are provided in Wrigley 1980a, b.) The considerable flexibility that the GSK approach provides is offset to some extent by the increased knowledge and expertise required of the user. Analysis proceeds by matrix manipulation and all matrices, both design and contrast, must be specified by the user as input to GENCAT. Nevertheless, GENCAT allows a rich selection of models to be fitted to the problems of cell (f) of Table 1.1, and some of the many extensions of the GSK approach suggested by Koch and his associates over the past ten years are particularly useful for handling the behavioural and attitudinal measurements commonly collected by human geographers.

(b)   NONMET   (Kritzer 1981). This is essentially another version of GENCAT aimed at implementing the GSK approach to categorical data problems. It has the same advantages and many of the same limitations as GENCAT.

7.4   Programs which use iterative proportional fitting algorithms

(a)   ECTA   (Fay and Goodman 1975). Goodman's Everyman's Contingency Table Analysis (ECTA) program was one of the first log-linear modelling programs to become widely available to social scientists. In its original form it is a simple-to-use batch-mode program which uses the

Deming–Stephan iterative proportional fitting algorithm to produce maximum likelihood estimates of expected cell frequencies, log-linear model parameters and standard errors, and which allows for data manipulation and the calculation of quadratic rather than linear effects. It is a program which will often be found to have been modified to meet local requirements (e.g. some users have added the option of screening tests).

(b) C–TAB   (Haberman 1973b). This program represented the standard alternative to ECTA in the mid 1970s. Its basic Deming–Stephan algorithm is also available as *Applied Statistics* algorithm AS 51 (Haberman 1972) and this can be used to construct a purpose-built log-linear modelling program if necessary.

(c) BMDP routine P4F   (Brown 1981). The increasing use of log-linear models in many disciplines resulted in an important new contingency table/log-linear modelling section (routines P1F, P2F, P3F) being included in the 1977 edition of the major BMDP (Biomedical Computer Programs. P-Series) package. These routines remained unchanged in the 1979 edition of BMDP but were extended and revised in the 1981 edition where they now appear as the single routine P4F. The user will find the P4F routine to be simple-to-use and supported by an excellent user's manual. At its heart is Haberman's (1972) iterative proportional fitting algorithm but its strength lies in a wide range of valuable options which it offers including: many contingency table display and manipulation facilities; automatic fitting of all possible log-linear models for two- and three-dimensional tables; automatic tests of marginal and partial association to screen parameters (see Section 5.7.3); stepwise addition or deletion of interactions from user-specified models; automatic derivation of the estimates of the log-linear model parameters and their standard errors (see Section 5.6.1); the printing of standardized and Freeman–Tukey residuals (see Section 5.8); fitting of models to incomplete two-dimensional and multidimensional tables with structural zeros (see Section 9.3 and note that this structural zero element is a much improved feature of P4F, i.e. of the 1981 BMDP edition compared to the 1977 and 1979 editions); identification of 'outlier' or 'rogue' cells and assessment of whether these cells represent statistically significant anomalies (see Section 9.4). The P4F routine has the advantage of being part of a major, fully-supported package available at most university computer centres. It effectively supersedes earlier programs such as ECTA and C–TAB and can be expected to be extended to include any important future developments in log-linear modelling.

PART 3

Extensions of the basic statistical models

Now that we have discussed the basic family of statistical model for categorical data analysis, we can turn in this third part of the book to a consideration of some useful extensions of the models and procedures of Part 2. Although it was stressed in Chapter 1 that maximum benefit is to be gained from a strict sequential reading of the book, Part 3 is something of an exception to that statement. Some readers, particularly those who are using the book to provide the statistical background to courses in discrete choice modelling, may wish to omit Part 3 on a first reading and move directly to Part 4. Other readers, with specialized requirements, may wish to consider just Chapter 8 or just Chapter 9. In contrast to Part 2, where the chapters are firmly interlinked, selective and non-sequential reading of this type is possible in Part 3. The order of its two constituent chapters is arbitrary, and the sections of each chapter are, by necessity, less intergrated into a developmental structure than those in Part 2.

CHAPTER 8

Special topics in logistic/logit modelling

In Chapters 2, 3 and 4 an introduction to the form, estimation and testing of the logistic/logit models appropriate for the problems of cells (d), (e) and (f) was provided. Despite the length of that introduction, there were many points in the text at which more detailed discussion was deferred to a later stage. In this chapter we take up these issues and consider some recently developed methods for assessing the fit of logistic/logit regression models and making the fit more 'resistant' to extreme observations, some logistic/logit model analogues to classical spatial analysis methods, some logistic/logit models for ordered categories and systems of equations, some extensions of the weighted least squares linear-logit models for cell (f) problems, and an alternative treatment of the error structures in such cell (f) logit models.

8.1 Logistic regression diagnostics and resistant fitting

Many of the classical articles on logistic/logit regression models assume that such models are fitted to high-quality data obtained under experimental conditions. For most geographers and environmental scientists this will not be the case. In many situations their data will be observational data obtained from sample survey procedures. Such data can often be rather 'poor' in the sense of being subject to errors in the collection process, having a configuration of data points which does not resemble that of a designed experiment, and having troublesome outlying observations. In conventional regression modelling a considerable amount is now known about the effect on least squares estimation of outlying observations and extreme points in the 'design space' or **X** matrix (the so-called 'high-leverage' points; see for example Belsley *et al.* 1980), and methods have been derived for making model fitting less sensitive (more 'resistant') to outlying observations and extreme design points. Fortunately, such methods have recently been extended to the context of logistic/logit models by Pregibon (1981, 1982) and some of the most important aspects of his work will now be considered.

When using logistic/logit models in practical data analysis the geographer and environmental scientist must remain sensitive to the fact that a few data points can have a potentially large influence on the maximum likelihood fit of his models. It is, therefore, extremely useful to have diagnostics which indicate the effect of individual observations (or any small part of the data) on both the overall fit of the model and on individual parameter estimates. Pregibon (1982) suggests the use of three simple diagnostics (other more refined procedures are discussed in more detail in Pregibon 1981).

1. A diagnostic plot of the approximate standardized change in the maximum likelihood fit ($DFIT_i$) caused by the deletion of an individual observation i. The statistic $DFIT_i$ takes from the form[†]

$$DFIT_i = \hat{\boldsymbol{\beta}}_i^{*\prime}\, \mathbf{V}_{\hat{\boldsymbol{\beta}}}^{-1}\, \hat{\boldsymbol{\beta}}_i^* \qquad [8.1]$$

where $\hat{\boldsymbol{\beta}}_i^* = \hat{\boldsymbol{\beta}}$ (all observations) $- \hat{\boldsymbol{\beta}}$ (all observations except i) and $\mathbf{V}_{\hat{\boldsymbol{\beta}}}$ is the estimator of the variance-covariance matrix of $\hat{\boldsymbol{\beta}}$. In the case where individual observations have a large influence, they will stand out on the diagnostic plot of $DFIT_i$ versus i, as shown for observations 8 and 24 in Fig. 8.1.

Fig. 8.1
Illustrative example of diagnostic plot of the standardized change in the maximum likelihood fit caused by the deletion of individual observations i (adapted from Pregibon 1982).

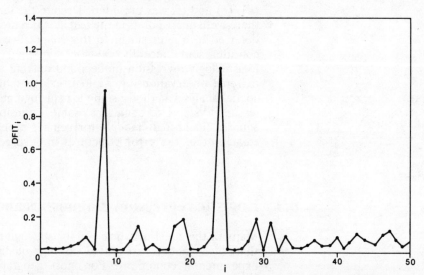

2. A set of diagnostic plots of the standardized change ($DBETA_{ik}$) in individual parameter estimates k caused by the deletion of an individual observation i. In this case the statistic $DBETA_{ik}$ takes the form

$$DBETA_{ik} = \frac{\hat{\beta}_k \text{ (all observations)} - \hat{\beta}_k \text{ (all except } i)}{\text{Standard Error } [\hat{\beta}_k \text{ (all obs.)}]} \qquad [8.2]$$

Once again, as shown in Fig. 8.2, individual observations which have a large influence will stand out on the diagnostic plot. For example,

---

[†] Computationally more efficient formulae for equations [8.1] and [8.2] are given in Pregibon (1981), and used in a geographical example by Wrigley and Dunn (1984). See also Wrigley (1984).

Fig. 8.2
Illustrative examples of standardized change in $\hat{\beta}_2$ and $\hat{\beta}_3$ caused by the deletion of individual observations i (adapted from Pregibon 1982)

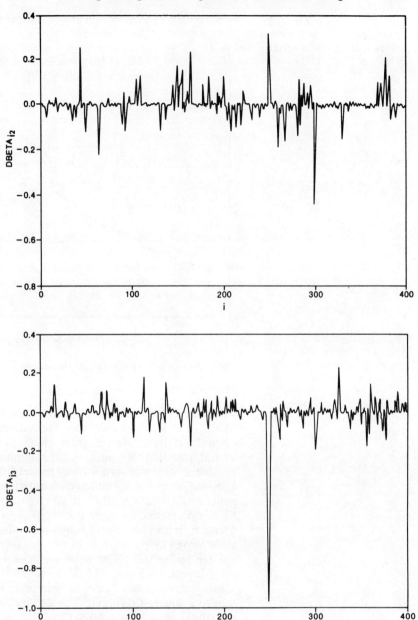

Fig. 8.2(b) shows that just a single observation (number 250) out of 400 in total alters $\hat{\beta}_3$ by nearly one standard error. This would represent a rather disturbing conclusion if found in a research application.

3. A diagnostic plot of the signed square-root components of 'deviance' (SSDs) such as that shown in Fig. 8.3. To understand this plot it is first necessary to know that 'deviance' is a term used by Nelder and Wedderburn (1972) to describe a generalized goodness-

Fig. 8.3
Example of a normal prob-
ability plot of SSDs (after
Pregibon 1982, 491)

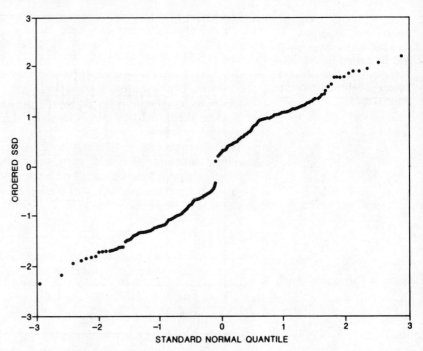

of-fit measure, which takes the form

$$D = -2 \left[\log_e \Lambda(\hat{\boldsymbol{\beta}}) - \log_e \Lambda_{\text{MAX}} \right] \qquad [8.3]$$

where $\log_e \Lambda(\hat{\boldsymbol{\beta}})$ denotes the maximized log likelihood of the fitted model, and $\log_e \Lambda_{\text{MAX}}$ denotes the maximized log likelihood based upon fitting each observation exactly. Furthermore, the reader should note that 'deviance' is the common term used to describe the goodness-of-fit measure for any model fitted in the GLIM program and that, in practice, we have already used one special form of this generalized measure (i.e. it can readily be shown that G^2, the likelihood-ratio chi-square goodness-of-fit statistic used in Chapters 5 and 6, is the 'deviance' statistic for those particular contingency table problems).

Each individual observation i makes a contribution to the over-all 'deviance'. Thus $D = \Sigma_{i=1}^{H} d_i$, and d_i (which is termed a 'deviance component') can be interpreted as approximately equivalent to a squared residual.[†] The asymptotic marginal behaviour of a deviance component is similar to a $\chi^2_{(1)}$ variable and this leads to the suggestion that (in the case where the two residual categories are reasonably equal) it is valuable to consider attaching a sign to each deviance component and then plotting $\pm \sqrt{d_i}$, the signed square-root deviance components (SSDs) as in Fig. 8.3.

In the case of Fig. 8.3, the normal probability plot of the SSDs shows no extremely large deviances, but the hole in the centre of the plot is shown by Pregibon (1982) to be due to a poor fit of the model under investigation.

[†] The deviance for a dichotomous logistic/logit model is given by Nelder (1974, 327). A deviance component can be defined by deleting the summation signs from that expression.

The three diagnostic plots discussed above indicate the large effect which a few data points can have on the maximum likelihood fit of logistic/logit models. As a result, it is extremely useful to search for model fitting procedures which are less sensitive to extreme observations. Pregibon (1982) has suggested two possible 'resistant' fitting procedures. Both of these focus upon a modification of the 'deviance' function, as it can readily be shown that the conventional maximization of the log likelihood function is formally equivalent to minimizing the 'deviance' function.

The first procedure simply involves replacement of the usual deviance function $D = \sum_i d_i$ with one which is less sensitive to observations poorly accounted for by the model. This can be achieved by using a tapering function $\lambda(d)$ for large contributions. The deviance function then becomes

$$D_\lambda = \sum_i \lambda(d) \qquad\qquad [8.4]$$

where λ, $(0 \leq \lambda(d) \leq d)$ is some differentiable monotone nondecreasing (or strictly increasing) function such as

$$\lambda(d) = \begin{cases} d & d \leq H \\ 2(dH)^{1/2} - H & \text{otherwise} \end{cases} \qquad [8.5]$$

where H is an adjustable constant (Pregibon suggests that taking $H = 1.345^2$ leads to estimates with approximately 95 per cent asymptotic relative efficiency). Conventional maximum likelihood estimation occurs when $\lambda(d) = d$ in [8.4]. For other choices of $\lambda(d)$, the effect on the estimation process is to limit the influence of observations poorly accounted for by the model and to expend more effort on improving the fit at other observations.

Although this first procedure provides resistance to poorly fitted observations, it provides little resistance to the so-called 'high-leverage' points, i.e. to observations which exert undue influence on the fit. As a result, Pregibon suggests a second resistant fitting procedure which involves replacing the usual deviance function by

$$D_{\pi\lambda} = \sum_i \pi(x_i)\, \lambda\left[\frac{d_i}{\pi(x_i)} \right] \qquad [8.6]$$

The function $\pi(x_i)$ is included to deal with the 'high leverage' points and it should be chosen in such as way that d/π is large for those observations with undue influence on the fit, and small otherwise. (Pregibon suggests some possible choices of $\pi(x_i)$ which have theoretical and intuitive justification.)

If Pregibon's resistant fitting procedures are to be used in practical data analysis, they will require the researcher to modify the standard logistic/logit programs discussed in Chapter 7. As the resistant procedures are expressed in terms of deviance functions it will prove most convenient to modify the standard GLIM program and Pregibon (1982) offers GLIM 'macros' which facilitate this adaptation.†

---

† More recently (following the completion of the manuscript) the author has provided a detailed empirical application of the logistic model diagnostics and resistant fitting procedures (Wrigley and Dunn, 1984). This paper also provides the GLIM macros and other computational information necessary to use these methods.

8.2   Some logistic/logit model analogues to classical spatial analysis models

In the uses of conventional regression models in geography and environmental science, a number of special formulations have played an important role. Two classes of such models are the traditional trend surface and space-time forecasting models and, together with a majority of classical spatial analysis models, they have conventionally assumed the availability of high level metric data. As an example of how many classical spatial analysis models can be modified to handle low level categorical data, we will now consider the logistic model analogues to these traditional trend surface and space-time models.

8.2.1   Trend surface models

The categorical variable analogue to the traditional polynomial trend surface model can be termed the polynomial *probability surface* model (see Wrigley 1977a, b; Bielawski and Waters 1983). In the dichotomous case it is simply a model of the form:

$$P_{1|i} = \frac{e^{x'_i \beta}}{1 + e^{x'_i \beta}} \qquad [8.7]$$

where

$$x'_i \beta = \beta_1 + \beta_2 U_i + \beta_3 V_i + \beta_4 U_i^2 + \beta_5 U_i V_i + \beta_6 V_i^2 + \ldots \qquad [8.8]$$

and where U_i and V_i are the geographical coordinates of locality i.[†] In the multiple response category case it is simply a model of the form [2.102] where

$$x'_{ri} \beta_r = \beta_{1r} + \beta_{2r} U_i + \beta_{3r} V_i + \beta_{4r} U_i^2 + \beta_{5r} U_i V_i + \beta_{6r} V_i^2 + \ldots \qquad [8.9]$$

As in traditional trend surface analysis, the probability surface model is used to decompose, or separate into two components, the information conveyed by a map of observations on a spatially distributed variable. These components are known as the trend (or regional) and local components and they are mapped to reveal the smoothly varying surfaces underlying the map of original observations and the local variability around these so-called trend surfaces. In the case of categorical variables, the contours of the trend surface are probabilities and thus the trend (regional) surface maps are termed probability surface maps.

Figures 8.4 and 8.5 show a simple illustration of probability surface mapping in which a multiple response category model and maximum likelihood estimation of parameters has been used. Figure 8.4 is a hypothetical map of responses to a question on shopping habits

---

[†] Although this power series polynomial is the most widely used in traditional trend surface analysis, other functional forms (e.g. double Fourier series, orthogonal polynomial) have also been used and can equally well replace [8.8] and [8.9] in the case of probability surface models.

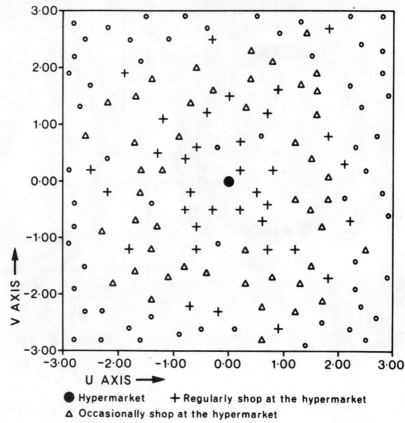

Fig. 8.4
Responses to a survey of shopping habits

● Hypermarket   + Regularly shop at the hypermarket
△ Occasionally shop at the hypermarket
○ Never shop at the hypermarket

conducted during a sample survey of 144 households in the area surrounding a recently opened hypermarket. Although there is some evidence of systematic spatial variation in these responses, the picture is by no means clear. By using the probability surface method, however, it is possible to extract the underlying trend (or regional) surfaces. These are shown in the form of simple line printer maps in Fig. 8.5(a), (b) and (c), and they are respectively the second order probability surfaces of regularly shopping at the hypermarket, occasionally shopping at the hypermarket and never shopping at the hypermarket. Inspection of Fig. 8.5 (a) shows that the underlying regional structure of regularly shopping at the hypermarket is a 'dome' centred on the hypermarket with probability contours falling away rapidly in all directions. In contrast, Fig. 8.5 (c) shows the underlying structure of never shopping at the hypermarket to be a flat-bottomed basin centred on the hypermarket with probability contours rising in all directions. Finally, Fig. 8.5 (b) shows that the underlying regional structure of occasionally shopping at the hypermarket has a 'ring doughnut' or horseshoe type of structure, with the highest predicted probabilities 0.5 to 0.6 occurring in a broken ring or horseshoe some distance from the hypermarket. Immediately surrounding the hypermarket the probability of occasionally shopping at the hypermarket falls to between 0.3 and 0.4, a

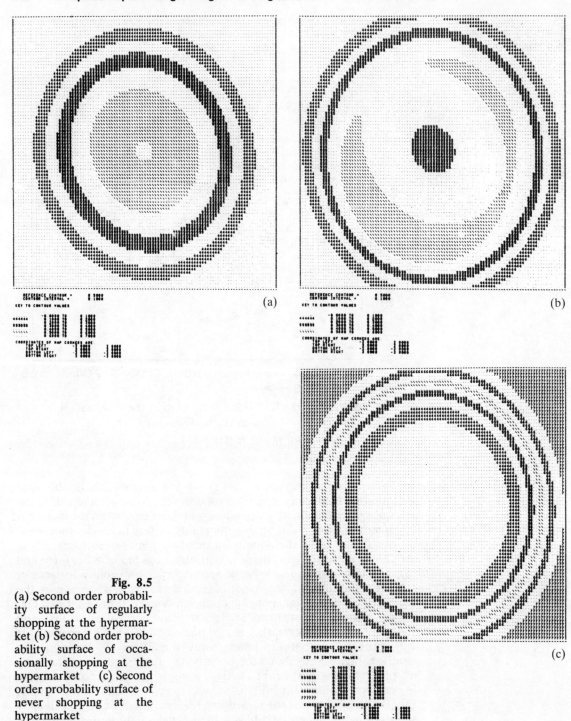

Fig. 8.5
(a) Second order probability surface of regularly shopping at the hypermarket (b) Second order probability surface of occasionally shopping at the hypermarket (c) Second order probability surface of never shopping at the hypermarket

level similar to that found in a ring beyond the horseshoe of higher probabilities.

Traditional trend surface analysis has been found to have a limited, but occasionally valuable, role to play in the environmental sciences (seen Unwin 1975), particularly in certain areas of geological research (see Davis 1973; Harbaugh *et al.* 1977). The use of a logistic model to extract the broad trend of a map as in Fig. 8.5 may prove to have equivalent (albeit limited) utility in those situations where the tradition-al trend surface model is inappropriate due to the categorical nature of the response variable. Moreover, the use of a logistic model to extract the broad trend of a map in this way has useful links to the more sophisticated spatial analysis procedures proposed by Besag (1974, 1975), particularly his so-called *auto-logistic* model (see also Haining 1982) and the reader should be sensitive to the need to complement or replace the probability surface method by these procedures where appropriate.

8.2.2   *Space-time models*

In the case of traditional space-time forecasting models (see Haggett *et al.* 1977, 517–40; Cliff 1977) categorical variable analogues take, in the simplest dichotomous case, the form

$$P_{1|i,t} = \frac{e^{x'_{i(t-1)}\boldsymbol{\beta}}}{1+e^{x'_{i(t-1)}\boldsymbol{\beta}}} \qquad\qquad [8.10]$$

where

$$x'_{i(t-1)}\boldsymbol{\beta} = \beta_1 + \beta_2 X_{i(t-1)} + \beta_3 \sum_j w_{ij} X_{j(t-1)} \qquad [8.11]$$

Models of this form postulate that the probability with which the individual at locality i selects the first response category at time t, depends upon the value of the explanatory variable X at locality i in the previous time period $t-1$, and also upon a weighted sum of values of X in the time period $t-1$ in the j localities surrounding (and thus influencing) events at locality i at time t. The w_{ij} terms are pre-specified non-negative structural weights with w_{ij} proportional to the influence of j on i.

Although it is a simple enough matter to specify the form of categorical analogues to the conventional space-time forecasting models (see Section 10.5 for further discussion) there are few published empirical applications of such models. One study which is available concerns the process of housing deterioration in Indianapolis. In this, Odland and Balzer (1979) use a model of the form [8.10] where

$$x'_{i(t-1)}\boldsymbol{\beta} = \beta_1 + \beta_2 L^0 X_{i(t-1)} + \beta_3 L^1 X_{i(t-1)} + \beta_4 L^2 X_{i(t-1)}$$
$$+ \beta_5 L^3 X_{i(t-1)} \qquad [8.12]$$

and in which $P_{1|i,t}$ is the probability that at least one new housing condemnation has taken place in area i of the city during time period t (the first six months of 1977). The explanatory variables in [8.12] are the numbers of condemned housing structures present at the end of period $t-1$ (the last six months of 1976) in areas of the city which are various spatial lags from area i. Using the notation of Haggett *et al.* (1977, 526) the spatial lag operator L in [8.12] is defined as

$$L^0 X_{i(t-1)} = X_{i(t-1)}$$

[8.13]

$$L^S X_{i(t-1)} = \sum_j w_{ij} X_{j(t-1)} \qquad S > 0$$

where S denotes the spatial lag which is S steps away from area i, and the summation is over the j areas at spatial lag S from i.

Using maximum likelihood estimation of the parameters, Odland and Balzer fitted a series of such models to investigate contagion effects in neighbourhood deterioration. Their results suggest that deterioration is likely to be a contagious process and that condemned houses are likely to exert negative effects on the other housing structures in their vicinity. By estimating models for a series of different area sizes they noted that in Indianapolis the negative effects of condemned housing have a limited range. For a system of hexagonal areas with smallest diameter 400 meters, only the parameter estimates associated with the zero and first order spatial lags were found to be significantly different from zero. More recently, Odland has extended this work (see Example 8.2).

EXAMPLE 8.1   *Aircraft noise disturbance around Manchester Airport*

In October and November 1973, in connection with the proposed introduction of a new runway at Manchester Airport, a social survey of perceived noise disturbance was conducted in Cheadle and Gatley; an area close by and to the northeast of the airport. The sample size of the survey was 1,069 households from which 846 usable questionnaires were returned. To illustrate the empirical application of probability surface mapping, the method was applied to extract regional surfaces of aircraft noise annoyance from responses to the survey question: 'How much does the noise from aircraft annoy or bother you?'. Figures 8.6(a), (b) and (c) show the third-order probability surface maps which were selected as being the best representations of the regional structures of being: (a) highly annoyed by aircraft noise; (b) moderately annoyed by aircraft noise; (c) only a little or not at all annoyed by aircraft noise. Further details of the maximum likelihood estimation and the selection of an appropriate order of surface to represent the noise disturbance data are to be found in Wrigley (1977a).

It is of interest to compare Fig. 8.6(a), (b) and (c), which represent the underlying regional trends in the perception of aircraft noise disturbance, with Fig. 8.7, which shows the aircraft routeings operating at the time of the survey. Of particular importance are routes 506 and 507, particularly the approach route 507, which carried by far the largest proportion of the aircraft traffic passing over the area at the time of the survey. This major traffic route is clearly picked out in Fig. 8.6(a) by a ridge of relatively high probability of being highly annoyed by aircraft noise, whilst Fig. 8.6(c) shows the reversal of this pattern with a valley of low probability of being only a little annoyed by aircraft noise along the traffic route.

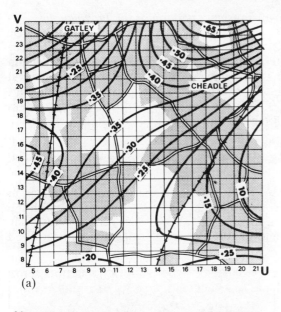

(a)

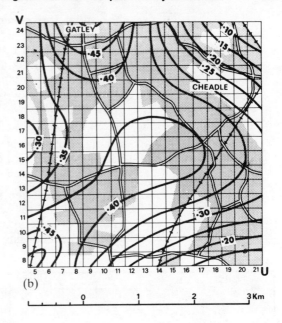

(b)

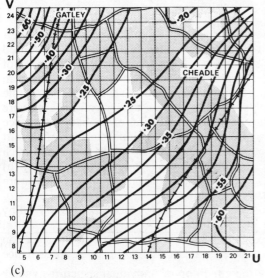

(c)

Fig. 8.6
Third order probability surfaces: (a) highly annoyed by aircraft noise; (b) moderately annoyed by aircraft noise; (c) only a little or not at all annoyed by aircraft noise

Fig. 8.7
Aircraft routeings at the time of the survey

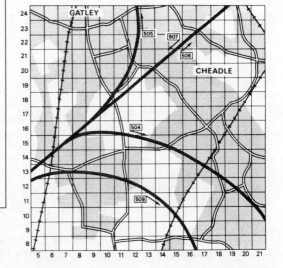

EXAMPLE 8.2   *The space-time pattern of housing deterioration in Indianapolis*

Odland and Barff (1982) have recently extended Odland's preliminary analysis of housing deterioration in Indianapolis discussed in Section 8.2. Their work attempts to combine methods for analysing categorical data with the logic of some existing tests for space-time interactions (Mantel 1967; Glick 1979; Cliff and Ord 1981). For pairs of maps such as those shown in Fig. 8.8 they use logit models of the form (see Ch. 4)

$$\log_e \frac{P_{\text{OBS}|g}}{P_{\text{ALL}|g}} = E_A \left[\log_e \frac{f_{\text{OBS}|g}}{f_{\text{ALL}|g}} \right] = \beta_1 + \beta_2 S_g + \beta_3 T_g + \beta_4 S_g T_g \quad \textbf{[8.14]}$$

to confirm that the space-time pattern of housing condemnations in central Indianapolis is consistent with a diffusion process in which the chances of deterioration are greater at locations near to previously deteriorated housing, and consistent with the operation of localized

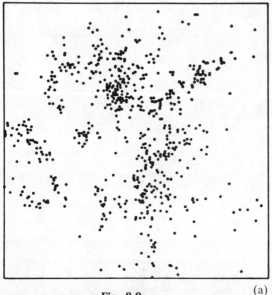
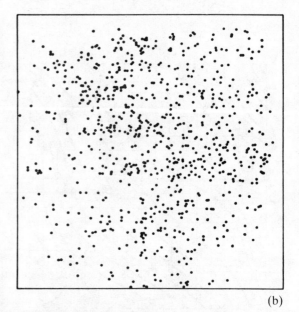

(a)   (b)

Fig. 8.8
An example of the pairs of maps used in Odland and Barff's analysis (after Odland and Barff, 1982) (a) OBS - Locations where housing condemnations occurred in central Indianapolis during 1977 and 1978; (b) ALL - Example of a simulated random pattern of housing condemnations in central Indianapolis in 1977 and 1978 when all housing structures are assumed to be equally likely to be condemned

externalities which make the condition of individual structures vulnerable to the condition of neighbouring buildings.

In [8.14] the two distributions of points (maps) in Fig. 8.8 are treated as different categories of events. The maps represent static pictures of a two-year process (observed or simulated) of housing condemnations, and to represent this dynamic process all *pairs* of points (house condemnations) on each map which are located within 1 kilometer of each other and which are separated by a time interval of 2 years or less are grouped into a table of the form shown in Table 8.1. Each cell of Table 8.1 represents a subpopulation g, and any significant difference between the two maps in Fig. 8.8 can be expected to be reflected by a pattern in which the proportions ($f_{\text{OBS}|g}$ and $f_{\text{ALL}|g}$) of point pairs from each map vary significantly between the cells of Table 8.1. Spatial and temporal clustering of condemnations in the observed map (Fig. 8.8a) would be reflected by a greater proportion of point-pairs from Fig. 8.8a than from Fig. 8.8b falling into the upper left-hand cells of Table 8.1, i.e. $f_{OBS|g}$ would be significantly greater than $f_{ALL|g}$ in these upper left-hand cells.

Table 8.1
Sub-populations used in Odland and Barff's analysis

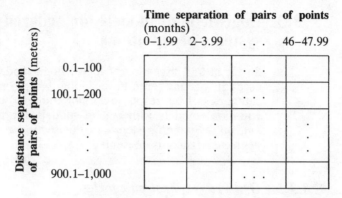

Distance separation of pairs of points (meters)	Time separation of pairs of points (months)			
	0–1.99	2–3.99	. . .	46–47.99
0.1–100			. . .	
100.1–200			. . .	
. . .	.	.		.
. . .	.	.		.
900.1–1,000			. . .	

In model [8.14] S_g is the average of the distances separating point-pairs (observed or simulated house condemnations) in sub-population g of Table 8.1, T_g is the average of the time intervals separating point-pairs in sub-population g, and $S_g T_g$ is a space-time interaction term. If the spacing and timing of observed housing condemnations (Fig. 8.8a) and the simulated random pattern of housing condemnations (Fig. 8.8b) do not differ, then it can be expected that the parameters β_1, β_2, β_3 and β_4 will assume the following values (within the limits of statistical significance)

$$\beta_1 = \log_e \frac{N_{OBS}/(N_{OBS} + N_{ALL})}{1 - [N_{OBS}/(N_{OBS} + N_{ALL})]} \qquad [8.15]$$

$$\beta_2 = \beta_3 = \beta_4 = 0$$

where N_{OBS} and N_{ALL} are the number of space–time point-pairs measured on each map. On the other hand, if housing deterioration follows the pattern hypothesized by Odland and Barff then housing condemnations (Fig. 8.8a) should occur more closely together in space and time than in the simulated random pattern (Fig. 8.8b) and this will be reflected in a statistically significant non-zero value for the space–time interaction term, β_4.

When Odland and Barff compared Fig. 8.8(a) and 8.8(b) using model [8.14], they derived the following parameter estimates (standard errors in parentheses)

$$\hat{\beta}_1 = 1.54340 \quad \hat{\beta}_2 = -1.02340 \quad \hat{\beta}_3 = -0.02426 \quad \hat{\beta}_4 = 0.02689 \qquad [8.16]$$
$$\phantom{\hat{\beta}_1 = }(0.0546) (0.0790) (0.0043) (0.0062)$$

These statistically significant parameter estimates, together with similar results derived from the comparison of Fig. 8.8(a) with other simulated maps, confirm that in the late 1970s the space–time pattern of housing condemnations in central Indianapolis was consistent with the operation of a contagious process in which the chances of deterioration were greater at locations near to previously deteriorated housing, and consistent with the operation of localized externalities which increased the risk of deterioration at locations near to condemned structures.

8.3   Logistic/logit models for ordered categories and systems of equations

So far in our discussion of logistic/logit models we have: (a) taken no account of any natural ordering which might exist amongst the categories of the response variable, and (b) concentrated upon 'single equation' models rather than models for multiple (possibly *simultaneous*) equation systems. In this section a brief assessment of these neglected issues is presented.

8.3.1   Ordered response categories

In situations where the categories of the response variable have a natural ordering, logistic/logit models which preserve the ordinal nature of the response without imposing an arbitrary scoring system are required. Cox (1970, 104) has suggested that one way to achieve this is to appeal to the existence of an underlying continuous and perhaps unobservable random (latent) variable. (An example of such a latent variable is the concept of 'tolerance' used in bioassay, see Cox 1970; Finney 1971. Tolerances are not directly observable but increasing tolerance is revealed through an increase in the probability of survival at particular dosage levels.) In this way, the ordered response categories can be thought of as contiguous intervals on a continuous scale.

In what is currently the most detailed assessment of ordinal response variable models, McCullagh (1980) has utilized this concept to develop logit models of the form

$$\log_e \frac{P^*_{r|i}}{1 - P^*_{r|i}} = \theta_r - \sum_{k=2}^{K} \beta_k X_{ik} \qquad (1 \leqslant r < R) \qquad \text{[8.17]}$$

In this model it is assumed that there are R ordered categories of the response variable (Y_i) with probabilities $P_{1|i}, P_{2|i}, \ldots P_{R|i}$ respectively. Letting $P^*_{r|i}$ represent the cumulative summation $P_{1|i} + P_{2|i} + \ldots + P_{r|i}$ (where $r < R$), the odds of a response category r or less in the ordering (that is to say, $Y_i \leqslant r$) is the ratio $P^*_{r|i}/(1 - P^*_{r|i})$. In this model response categories are envisaged as contiguous intervals on a continuous scale separated by unknown points of division ('cut points') denoted $\theta_1, \ldots, \theta_r, \ldots \theta_R$. If such a continuous underlying variable exists then interpretation of [8.17] with reference to this scale is direct. However, if no sensible latent variable can be assumed to exist, the parameters of such models are still interpretable in terms of the particular categories recorded.

In the extensive and valuable discussion which accompanies McCullagh's paper, a number of commentators (among them Fienberg and Pregibon) point out that an alternative class of logit models can be derived for response categories with a natural ordering. These alternative models are based upon what are referred to as 'continuation ratios' (see Fienberg 1980, 110; Fienberg and Mason 1978, 35). There are a number of minor variations in the possible definitions of 'continuation ratios'. Here we follow Pregibon (1980) and use a definition which involves

$$P_{r|i}^{**} = \text{Prob } (Y_i = r | Y_i > r-1) \qquad\qquad [8.18]$$

Model [8.17] then becomes

$$\log_e \frac{P_{r|i}^{**}}{1 - P_{r|i}^{**}} = \theta_r + \sum_{k=2}^{K} \beta_k X_{ik} \qquad\qquad [8.19]$$

In other words, we consider logits based upon the conditional probabilities of selecting the rth response category given that the response category selected is greater than $r-1$ in the ordering. Pregibon (1982) points out that this amounts to factoring the multinomial response Y_i into conditionally independent binomials, and that this conditional independence can be usefully exploited in the estimation of the models. He claims that such models are computationally more convenient than those proposed by McCullagh and can readily be implemented using a standard computing package such as GLIM.[†]

8.3.2 Systems of logistic/logit models

Most geographers and environmental scientists who have completed the preliminary multivariate statistics course assumed by this book will be aware of the distinction between single-equation regression models and multiple-equation systems of regression models. In single-equation regression models a set of explanatory variables determines a response variable and there is no 'feedback' relationship between the response variable and the explanatory variables. In other words, the single-equation model explains causality in one direction and does not explain the interdependencies that may exist among the explanatory variables themselves or how these explanatory variables are related to other variables. On the other hand, multiple-equation systems of regression models, in their most general form, allow a simultaneous evaluation of all the interrelationships between a set of variables. Causality is no longer assumed to run in one direction. The response variable in one equation in the system becomes an (endogenous) explanatory variable in another, and so on.

A special form of equation system often used by social scientists is known as a *recursive* equation system. In conventional regression form, a simple example of a three-equation recursive system in which there are just two exogenous explanatory variables, Z_1 and Z_2, would look as follows:

[†] The reader should note, however, that McCullagh has distributed a special program called PLUMB which conveniently implements his ordered response category models.

$$Y_1 = \alpha_1 \qquad\qquad\quad + \beta_1 Z_1 + \beta_2 Z_2 + \varepsilon_1$$

$$Y_2 = \alpha_2 + \beta_3 Y_1 \qquad\quad + \beta_4 Z_1 + \beta_5 Z_2 + \varepsilon_2 \qquad\qquad [8.20]$$

$$Y_3 = \alpha_3 + \beta_6 Y_1 + \beta_7 Y_2 \quad + \beta_8 Z_1 + \beta_9 Z_2 + \varepsilon_3$$

Although this is a multiple-equation system, the causal linkages all run in one direction: Y_1 precedes Y_2 which precedes Y_3. Given values of the exogenous variables Z_1 and Z_2, we can solve directly for Y_1. Then, knowing Y_1, Z_1 and Z_2, we can solve for Y_2, and so on. This simple structure allows the models in the system to be estimated using the ordinary least squares (OLS) procedure. Equivalent recursive systems of logit models are discussed by Fienberg (1977, 94–104). He notes that recursive systems of logit models 'mimic' recursive systems of conventional regression models. That is to say, the parameters in each of the logit equations in the recursive system should be estimated by the usual methods, and the MLEs for the parameters in the system are just the MLEs of the parameters for each equation in the system when viewed separately. (This type of system underlies the models discussed in Section 9.6.3.)

Unlike recursive systems, *simultaneous* equation systems are more difficult to handle. In conventional regression model form, a simple three-equation simultaneous system looks as follows:

$$Y_1 = \alpha_1 \qquad\quad + \beta_1 Y_2 + \beta_2 Y_3 + \beta_3 Z_1 + \beta_4 Z_2 + \varepsilon_1$$

$$Y_2 = \alpha_2 + \beta_5 Y_1 \qquad\quad + \beta_6 Y_3 + \beta_7 Z_1 + \beta_8 Z_2 + \varepsilon_2 \qquad [8.21]$$

$$Y_3 = \alpha_3 + \beta_9 Y_1 + \beta_{10} Y_2 \qquad + \beta_{11} Z_1 + \beta_{12} Z_2 + \varepsilon_3$$

In such a system, the response variable of one equation becomes an explanatory variable in another, and vice versa. If none of the parameters in [8.21] is zero, [8.21] is a completely interdependent system since, it can be seen, that it is impossible to solve for any single endogenous variable, say Y_2, without simultaneously solving all three equations. In such a system OLS is no longer a satisfactory estimation procedure.

Systems of logistic/logit models with the same type of simultaneous structure have been investigated by Nerlove and Press (1973), Schmidt and Strauss (1975b), Amemiya (1975), and others. Schmidt and Strauss, in particular, have provided a lucid discussion of a simple simultaneous system of logit models in which there are two dichotomous response variables, Y_i and Q_i, a set of exogenous variables x_i affecting Y_i, and a set of exogenous variables x^*_i affecting Q_i. The simultaneous system of models then takes the form

$$\log_e \left[\frac{\text{Prob}(Y_i = 1 | Q_i)}{\text{Prob}(Y_i = 0 | Q_i)} \right] = x'_i \boldsymbol{\beta} + \alpha Q_i$$

$$[8.22]$$

$$\log_e \left[\frac{\text{Prob}(Q_i = 1 | Y_i)}{\text{Prob}(Q_i = 0 | Y_i)} \right] = x_i^{*\prime} \boldsymbol{\beta}^* + \delta Y_i$$

Furthermore, it can be shown that a 'symmetry' condition holds which implies that $\alpha = \delta$.

From [8.22] it is straightforward (but somewhat tedious) to demonstrate that the joint probabilities must take the form:

$$\text{Prob}(Y_i = 0, \ Q_i = 0) = 1/\Delta_i$$
$$\text{Prob}(Y_i = 0, Q_i = 1) = \exp(x_i^{*\prime}\boldsymbol{\beta}^*)/\Delta_i$$
$$\text{Prob}(Y_i = 1, \ Q_i = 0) = \exp(x_i^{\prime}\boldsymbol{\beta})/\Delta_i$$
$$\text{Prob}(Y_i = 1, \ Q_i = 1) = \exp(x_i^{\prime}\boldsymbol{\beta} + x_i^{*\prime}\boldsymbol{\beta}^* + \alpha)/\Delta_i$$

[8.23]

where

$$\Delta_i = 1 + \exp(x_i^{*\prime}\boldsymbol{\beta}^*) + \exp(x_i^{\prime}\boldsymbol{\beta}) + \exp(x_i^{\prime}\boldsymbol{\beta} + x_i^{*\prime}\boldsymbol{\beta}^* + \alpha)$$

and knowledge of these probabilities permits the statement of the likelihood function necessary to obtain the maximum likelihood estimates of $\boldsymbol{\beta}$, $\boldsymbol{\beta}^*$ and α. This likelihood function is given by Schmidt and Strauss (1975b, 747) and is essentially a simple extension of that discussed in Section 2.4.2.

The extension of [8.22] from the case in which Y_i and Q_i are dichotomous response variables to the case where they are multiple-category response variables with $r = 1, \ldots, R$ and $r = 1, \ldots, R^*$ categories respectively, is straightforward. In this case the simultaneous system of logit models takes the form

$$\log_e\left[\frac{\text{Prob}(Y_i = r|Q_i)}{\text{Prob}(Y_i = R|Q_i)}\right] = x_{ri}^{\prime}\boldsymbol{\beta}_r + \sum_{k=1}^{R-1} \alpha_{kr}Q_{ki}^* \quad (\text{r} = 1, \ldots, R-1)$$

[8.24]

$$\log_e\left[\frac{\text{Prob}(Q_i = r|Y_i)}{\text{Prob}(Q_i = R^*|Y_i)}\right] = x_{ri}^{*\prime}\boldsymbol{\beta}_r^* + \sum_{k=1}^{R^*-1} \alpha_{rk}Y_{ki}^* \quad (r = 1, \ldots, R^*-1)$$

where Q_{ki}^* and Y_{ki}^* are dummy variables such that

$$Q_{ki}^* = \begin{cases} 1 \text{ if } Q_i = k \\ 0 \text{ otherwise} \end{cases} \quad (k = 1, \ldots, R-1)$$

[8.25]

$$Y_{ki}^* = \begin{cases} 1 \text{ if } Y_i = k \\ 0 \text{ otherwise} \end{cases} \quad (k = 1, \ldots, R^*-1)$$

In model [8.24] the analogue of the 'symmetry' condition ($\alpha = \delta$) in [8.22] has already been imposed.

Models similar to [8.22] and [8.24] can be developed for cases in which there are more than two (dichotomous or multiple-category) response variables. The models are essentially the same but algebraically they become much more complex as the number of response variables in the simultaneous system increases.

As a brief illustration of models of the type [8.22] and [8.24] we can consider Schmidt and Strauss's (1975b) application of the system of models [8.24] to the problem of predicting the occupation and industry of employment of workers in the United States, when choice of occupation is assumed to depend upon, among other things, choice of industry, and choice of industry depends upon occupation. Using a sample of 936 individuals from the U.S. Department of Labour, 1967 Survey of Economic Opportunity data file, they fitted a simultaneous system of logit models of the following form, where education, experience, race and sex are the exogenous variables.

$$\log_e \frac{P_{S\&G|i}}{P_{P\&D|i}} = \beta_{11} + \beta_{21} \text{ Education}_i + \beta_{31} \text{ Experience}_i + \beta_{41} \text{ Race}_i$$
$$+ \beta_{51} \text{ Sex}_i + \alpha_{11} \text{ 'Skilled Occupation'}$$
$$+ \alpha_{21} \text{ 'Professional Occupation'}$$

$$\log_e \frac{P_{\text{SKILL}|i}}{P_{\text{MEN}|i}} = \beta^*_{11} + \beta^*_{21} \text{ Education}_i + \beta^*_{31} \text{ Experience}_i + \beta^*_{41} \text{ Race}_i$$
$$+ \beta^*_{51} \text{ Sex}_i + \alpha_{11} \text{ 'Service \& Government'}$$

[8.26]

$$\log_e \frac{P_{\text{PROF}|i}}{P_{\text{MEN}|i}} = \beta^*_{12} + \beta^*_{22} \text{ Education}_i + \beta^*_{32} \text{ Experience}_i + \beta^*_{42} \text{ Race}_i$$
$$+ \beta^*_{52} \text{ Sex}_i + \alpha_{21} \text{ 'Service \& Government'}$$

In this system of models $P_{S\&G|i}/P_{P\&D|i}$ gives the odds that individual i will be in 'Service and Government' industry rather than 'Production and Distribution' industry, and $P_{\text{SKILL}|i}/P_{\text{MEN}|i}$ and $P_{\text{PROF}|i}/P_{\text{MEN}|i}$ gives the odds that individual i will be in a 'Skilled' or 'Professional' occupation respectively rather than a 'Menial' occupation.

The logic of models of the type [8.22] and [8.24] can be extended to the situation in which one of the response variables (say Y_i) is categorical but the other (Q_i) is *continuous*. Simultaneous models of this type have been considered by Schmidt and Strauss (1976) and are referred to as 'mixed' logit models. In the analogue to model [8.22] where Y_i is a dichotomous response variable but Q_i is a continuous response variable, the 'mixed' logit model takes the form:

$$\log_e \left[\frac{\text{Prob}(Y_i = 1|Q_i)}{\text{Prob}(Y_i = 0|Q_i)} \right] = x'_i \boldsymbol{\beta} + \alpha Q_i$$

[8.27]

$$Q_i|Y_i \sim N(x^{*\prime}_i \boldsymbol{\beta}^* + \delta Y_i, \sigma^2)$$

In essence, [8.27] is a logit model mixed with a normal regression model. The first equation in [8.27] is a standard logit expression conditional on Q_i. It takes the same form as that in [8.22] except that Q_i is now continuous. The second equation implies that the distribution of Q_i given Y_i is a normal random variable with mean equal to a linear function of x^*_i and Y_i, and variance equal to σ^2.

As in the case of model [8.22] it is possible to extend [8.27] to the case where Y_i is a multiple-category response variable. Furthermore, it can be shown (Olsen 1978) that, as in the case of models [8.22] and [8.24], there is a necessary 'symmetry' condition for the model [8.27] which implies that $\alpha\sigma^2 = \delta$.

Other 'mixed' models with both categorical and continuous response variables in the equation system have also been proposed. Schmidt (1978), for example, has proposed an alternative to model [8.27] in which Y_i is viewed as being affected not simply by the value of the continuous variable Q_i but by the difference in the value of Q_i conditional upon the choice of response category, i.e. by $(Q_i|Y_i = 1) - (Q_i|Y_i = 0)$ instead of just Q_i. This results in a model of the form:

$$\log_e \frac{\text{Prob}(Y_i = 1)}{\text{Prob}(Y_i = 0)} = x'_i \boldsymbol{\beta} + \alpha[(Q_i|Y_i = 1) - (Q_i|Y_i = 0)]$$

$$Q_i = x_i^{*'} \boldsymbol{\beta}^* + (Y_i x_i^{*'}) \boldsymbol{\delta} + \varepsilon_i \qquad \text{[8.28]}$$

All terms in this model have the same definitions as previously, except that $(Y_i x_i^{*'})$ is a vector of interactions found by multiplying the vector of exogenous explanatory variables $x_i^{*'}$ by the scalar dichotomous variable Y_i, and $\boldsymbol{\delta}$ is now the corresponding vector of parameters relating to these interaction terms. In addition, the error term ε_i is assumed to be independent of Y_i and $x_i^{*'}$. Schmidt believes that models of this form have considerable potential in situations where an individual's decision about a qualitative choice is affected by a differential in a continuous variable, and he cites the interesting case of migration, where the decision to migrate might depend on (amongst other things) the income differential attainable by migration.

Finally, it should be noted that Heckman (1977, 1978) has taken a rather different approach to the 'mixed' model problem. In this he assumes that there is an unobservable (latent) continuous variable S_i which reflects 'sentiment' for or against choice of a particular response category and that actual response category choices ($Y_i = 1$ or 0) are generated by the latent variable crossing particular thresholds. Using the same definition of terms as above, the model then takes the form

$$S_i = x'_i \boldsymbol{\beta} + \alpha Q_i + \gamma_1 Y_i + \varepsilon_{1i}$$

$$Q_i = x_i^{*'} \boldsymbol{\beta}^* + \delta S_i + \gamma_2 Y_i + \varepsilon_{2i} \qquad \text{[8.29]}$$

$$Y_i = \begin{cases} 1 & \text{if } S_i > 0 \\ 0 & \text{if } S_i \leqslant 0 \end{cases}$$

where ε_{1i} and ε_{2i} are error terms. The parameters γ_1, γ_2 and α of [8.29] obey certain constraints and as a result the model can be written in a reduced form. However, details of this and of the estimation of the model are beyond the scope of this chapter. For our purposes, it is Heckman's use of the concept of an underlying continuous latent variable crossing thresholds which is important. A transport planning example of the use of such a model is provided by Westin and Gillen (1978): see also Hensher and Johnson (1981a, 274–81).

8.4 Some extensions of the general matrix formulation of cell (f) linear logit models

Chapter 4 concluded with an outline of a general matrix formulation of weighted least squares linear logit models for cell (f) problems, and this provided an introduction to the so-called GSK approach to categorical data analysis. It was shown how the simple sequence of matrix operations

$$\bar{L} = K[\log_e (Af)] \qquad\qquad [8.30]$$

(where f is a vector of observed proportions and A and K are operator matrices) can be used to generate the necessary observed logit vectors $\bar{L}$ for all linear logit models, where such models can be expressed in the general matrix form

$$E_A (\bar{L}) = X\beta \qquad\qquad [8.31]$$

In addition, it was noted how the addition of further operator matrices, logarithmic operations, or exponentiations to [8.30] can define a wide range of other more complex (non-logit) functions of the observed response proportions. In this way, a wide range of problems in categorical data analysis can be considered within a unified framework.

In this section we will not consider the very general compounded functions

$$F(f) = A_1 * \exp [K\log_e (Af)] \qquad\qquad [8.32]$$

$$F(f) = A_2 * \log_e [A_1 * \exp (K\log_e(Af)] \qquad\qquad [8.33]$$

$$F(f) = A_3 * \exp [A_2 * \log_e (A_1 * \exp (K\log_e (Af)))] \qquad\qquad [8.34]$$

which have been suggested (see Forthofer and Koch 1973; Landis *et al.* 1976, 201) and which can be formed by adding additional operator matrices (A_1, A_2, A_3), exponentiations and logarithmic operations to [8.30]. Instead, we will restrict our attention to logit models and will consider three examples in which the sequence of matrix operations [8.30] plays a much more important role than in the simple dichotomous logit model illustration of Section 4.7. Furthermore, this will serve to introduce two useful extensions of the cell (f) linear logit models discussed in Chapter 4.

8.4.1   A multiple response category model

The simplest extension of the dichotomous logit model illustration in Section 4.7 is a multiple response category logit model such as that discussed in Example 4.2. In the case of Table 4.4 the column vector f in [8.30] takes the form:

$$
\underset{12 \times 1}{f} = \begin{bmatrix} f_{1|1} \\ f_{2|1} \\ f_{3|1} \\ . \\ . \\ . \\ f_{2|4} \\ f_{3|4} \end{bmatrix} = \begin{bmatrix} 118/459 \\ 207/459 \\ 134/459 \\ . \\ . \\ . \\ 1006/2419 \\ 760/2419 \end{bmatrix} \qquad\qquad [8.35]
$$

and the operator matrices $\mathbf{A}$ and $\mathbf{K}$ take the form

$$
\mathbf{A} = \begin{bmatrix} 100 & \dots & 0 \\ 010 & \dots & 0 \\ 001 & \dots & 0 \\ \dots & & \cdot \\ \dots & & \cdot \\ \dots & & \cdot \\ 000 & \dots & 1 \end{bmatrix} \quad \mathbf{K} = \begin{bmatrix} 1 & 0{-}1 & 0 & 0 & 0 & 0 & \dots & 0 & 0 & 0 \\ 0 & 1{-}1 & 0 & 0 & 0 & 0 & \dots & 0 & 0 & 0 \\ 0 & 0 & 0 & 1 & 0{-}1 & \dots & 0 & 0 & 0 \\ 0 & 0 & 0 & 0 & 1{-}1 & \dots & 0 & 0 & 0 \\ & & \cdot & & \cdot & & & \cdot & \\ & & \cdot & & \cdot & & & \cdot & \\ & & \cdot & & \cdot & & & \cdot & \\ 0 & 0 & 0 & 0 & 0 & 0 & \dots & 1 & 0{-}1 \\ 0 & 0 & 0 & 0 & 0 & 0 & \dots & 0 & 1{-}1 \end{bmatrix} \quad \textbf{[8.36]}
$$

12×12 8×12

Using the rules of matrix multiplication, the reader should check that f, $\mathbf{A}$ and $\mathbf{K}$ of [8.35] and [8.36] produce the observed logit vector required for model [4.24]

$$
\begin{bmatrix} \bar{L}_{13|1} \\ \bar{L}_{23|1} \\ \cdot \\ \cdot \\ \cdot \\ \bar{L}_{13|4} \\ \bar{L}_{23|4} \end{bmatrix} = \begin{bmatrix} \log_e f_{1|1}/f_{3|1} \\ \log_e f_{2|1}/f_{3|1} \\ \cdot \\ \cdot \\ \cdot \\ \log_e f_{1|4}/f_{3|4} \\ \log_e f_{2|4}/f_{3|4} \end{bmatrix} = \begin{bmatrix} \log_e f_{1|1} - \log_e f_{3|1} \\ \log_e f_{2|1} - \log_e f_{3|1} \\ \cdot \\ \cdot \\ \cdot \\ \log_e f_{1|4} - \log_e f_{3|4} \\ \log_e f_{2|4} - \log_e f_{3|4} \end{bmatrix} \quad \textbf{[8.37]}
$$

when they are substituted into [8.30].

8.4.2   A repeated-measurement research design example

Table 8.2 shows an example of a cell (f) type contingency table which is more complex than those encountered in Chapter 4. The artificial data in Table 8.2 are assumed to relate to 511 South Wales manufacturing firms which have relocated during the past five years. The data can be assumed to have been derived from a questionnaire in which the managers of the firms assessed the importance of various factors in influencing the choice of a South Wales location. The questions posed can be assumed to have taken the general form:
   'At the time of your recent move was (location factor q) an important/unimportant influence in your decision to choose a South Wales location?'.
Table 8.2 is based on the dichotomous responses (important/unimportant) to three location factors:
1. access to regional development grants;

2. good motorway connections to the major U.K. domestic markets of the South East, West Midlands and North West;

3. access to the regional international airport and seaports.

It can be seen that, in Table 8.2, instead of a set of simple response categories relating to a single variable, there are now a set of what are best described as 'response profiles'. These represent the eight possible combinations of responses to the three location factors.

Table 8.2
Hypothetical data from an industrial survey of mobile firms in South Wales

	Sub-populations			Response profiles								
	Foreign owned	*Size*	*Market orient- ation*	*Factor* 1: I *Factor* 2: I *Factor* 3: I	I I U	I U I	I U U	U I I	U I U	U U I	U U U	*Total*
$g = 1$:	Yes	Larger	Export	27	2	15	9	2	0	5	1	61
$g = 2$:	Yes	Larger	Home	9	12	1	7	3	5	1	2	40
$g = 3$:	Yes	Smaller	Export	9	8	14	7	2	1	3	1	45
$g = 4$:	Yes	Smaller	Home	4	10	2	5	2	6	1	4	34
$g = 5$:	No	Larger	Export	11	9	31	8	2	1	5	2	69
$g = 6$:	No	Larger	Home	3	20	1	9	3	9	1	3	49
$g = 7$:	No	Smaller	Export	12	10	25	18	8	4	3	10	90
$g = 8$:	No	Smaller	Home	14	49	2	24	3	11	2	18	123

Factor 1 I ⎫
Factor 2 I ⎬ = Availability of regional development grants and good motorway connections to major domestic markets important in determining location, access to regional international
Factor 3 U ⎭ airport and seaports unimportant.

A set of location factor questions within an industrial survey such as this can be regarded as an example of a 'repeated-measurement research design'. Such a design involves subjects (firms) selected from various sub-populations (types of firm) being exposed to a series of different measurement conditions or stimuli (location factor questions) and being classified in terms of a response variable with a given and constant number of levels (important/unimportant). In recent years, Koch and his associates (Koch *et al.* 1977; Lehnen and Koch 1974b) have presented a general methodology for the analysis of repeated measurement research designs of this type and have shown how this methodology links into the general GSK approach to the analysis of categorical data. In this section, however, we will take a much more limited view and will consider just the form of the matrices which are necessary to derive the logit vector for a subsequent weighted least squares logit model analysis of Table 8.2.

For each of the three location factors ($q = 1,2,3$) interest centres upon the proportion of firms of type (sub-population) g who consider location factor q to have been important in influencing their choice of a South Wales location (this proportion can be denoted $f_{1|g(q)}$ or more simply $f_{g(q)}$, and the proportion who consider it to have been unimportant (which can be denoted $1 - f_{1|g(q)}$ or more simply $1 - f_{g(q)}$). It is then logical to attempt to account for differences in the ratio of these proportions $(f_{g(q)}/1 - f_{g(q)})$ across different location factors and different types of firm, as a function of the varying characteristics of the firms (i.e. as a function of the explanatory variables, foreign ownership, size, and market orientation).[†] A linear logit model can be formulated which achieves this objective. This takes the form,

$$E_A(\bar{L}_{g(q)}) = E_A\left[\log_e \frac{f_{g(q)}}{1-f_{g(q)}} \right] = \beta_1 + \sum_{k=2}^{K} \beta_k X_{g(q)k} \qquad [8.38]$$

where $X_{g(q)k}$ denotes the particular level of explanatory variable k ($k=2$ – foreign ownership; $k = 3$ – size; $k = 4$ –market orientation, . . .) for firm type g and location factor q, and $\bar{L}_{g(q)}$ is a shorthand expression for $\bar{L}_{12|g(q)}$.

The required observed logit vector, $\bar{L}$, in the general matrix expression [8.31] of the linear logit model [8.38], takes the form:

$$\underset{1 \times 24}{\bar{L}'} = [\bar{L}_{1(1)}, \bar{L}_{1(2)}, \bar{L}_{1(3)}, \bar{L}_{2(1)}, \dots, \bar{L}_{8(2)}, \bar{L}_{8(3)}] \qquad [8.39]$$

where the column vector $\bar{L}$ has been written as a row vector $\bar{L}'$ for purposes of compactness. This logit vector can be generated using the general sequence of matrix operations [8.30] where f, $\mathbf{A}$ and $\mathbf{K}$ now take much more complex forms than in the two previous illustrations. The vector of observed proportions takes the form:

$$\underset{1 \times 64}{f'} = [f'_1, f'_2, \dots f'_7, f'_8] \qquad [8.40]$$

a typical element of which is

$$\underset{1 \times 8}{f'_7} = (1/90) \, [12, \, 10, \, 25, \, 18, \, 8, \, 4, \, 3, \, 10] \qquad [8.41]$$

The operator matrix $\mathbf{K}$ takes the form:

$$\underset{24 \times 48}{\mathbf{K}} = \begin{bmatrix} 1-1 & 0 & 0 & \dots & 0 & 0 \\ 0 & 0 & 1-1 & \dots & 0 & 0 \\ & \cdot & \cdot & & \cdot \\ & \cdot & \cdot & & \cdot \\ & \cdot & \cdot & & \cdot \\ 0 & 0 & 0 & 0 & \dots & 1-1 \end{bmatrix} = \mathbf{I}_8 \otimes \begin{bmatrix} 1-1 & 0 & 0 & 0 & 0 \\ 0 & 0 & 1-1 & 0 & 0 \\ 0 & 0 & 0 & 0 & 1-1 \end{bmatrix} \qquad [8.42]$$

Where $\mathbf{I}_8$ denotes an identity matrix of order 8 and $\otimes$ denotes the direct matrix product. Finally, the operator matrix $\mathbf{A}$ takes the form:

[†] There are a number of other possible ways of analysing this table. For example, if the response profiles are labelled $r=1, \dots 8$, a conventional multiple-category logit model $E_A[\log_e(f_{r|g}/f_{8|g})] = x'_g \boldsymbol{\beta}_r$, $r=1, \dots, 7$ (see Section 4.1 and Example 4.2) can be used. Alternatively, a log-linear model (see Upton 1981) could be used. Each of these models is likely to provide a different perspective on the data but these perspectives will share common features.

$$\mathbf{A}_{48 \times 64} = \mathbf{I}_8 \otimes \mathbf{A}^*, \qquad \mathbf{A}^*_{6 \times 8} = \begin{bmatrix} 1 & 1 & 1 & 1 & 0 & 0 & 0 & 0 \\ 0 & 0 & 0 & 0 & 1 & 1 & 1 & 1 \\ 1 & 1 & 0 & 0 & 1 & 1 & 0 & 0 \\ 0 & 0 & 1 & 1 & 0 & 0 & 1 & 1 \\ 1 & 0 & 1 & 0 & 1 & 0 & 1 & 0 \\ 0 & 1 & 0 & 1 & 0 & 1 & 0 & 1 \end{bmatrix} \qquad [8.43]$$

8.4.3 *Paired comparison experiment examples*

Another example of the general class of repeated measurement research designs considered by Koch *et al.* (1977) as being part of the wider GSK approach to the analysis of categorical data, is a paired comparison experiment. In such an experiment, R choice alternatives $A_1, \ldots, A_r, \ldots, A_R$ are presented in pairs to a set of respondents who are asked to indicate a preference for one member of each pair. One of the most widely used models (Bradley and Terry 1952) for paired comparisons then postulates the existence of an underlying continuum on which the relative 'worths' of the alternatives can be located. If the preference parameter π_r denotes the 'worth' of alternative r as viewed by the respondents (where $\pi_r \geqslant 0$, $r=1, \ldots R$ and $\Sigma_r \pi_r = 1$) then the Bradley-Terry model postulates that in response to the pair (r, s) the probability that alternative r is preferred to alternative s is:

$$\text{Prob} (A_r \to A_s) = \frac{\pi_r}{\pi_r + \pi_s} \qquad [8.44]$$

Alternatively the criterion on which the choice is based can be made explicit and the model written as

$$\text{Prob} (A_{r;\alpha} \to A_{s;\alpha}) = \frac{\pi_{r;\alpha}}{\pi_{r;\alpha} + \pi_{s;\alpha}} \qquad [8.45]$$

Table 8.3
Shop preferences on the basis of price levels

to indicate the probability that alternative r is preferred to s on the basis of criterion α

(r, s)	Preference frequency $n_{r\|rs}$	$n_{s\|rs}$	Total (n_{rs})	(r, s)	Preference frequency $n_{r\|rs}$	$n_{s\|rs}$	Total (n_{rs})
(A,B)	10	20	30	(B,F)	17	13	30
(A,C)	8	21	29	(C,D)	10	19	29
(A,D)	6	24	30	(C,E)	17	12	29
(A,E)	10	20	30	(C,F)	19	10	29
(A,F)	14	15	29	(D,E)	18	11	29
(B,C)	13	16	29	(D,F)	21	9	30
(B,D)	9	21	30	(E,F)	18	11	29
(B,E)	14	15	29				

Key: n_{rs}—number of consumers queried about pair (r, s); $n_{r\|rs}$—number of consumers preferring r, amongst those queried about (r, s).

An example of the results of a simple paired comparison experiment is shown in Table 8.3. Here the hypothetical responses of 441 consumers to paired comparisons of six shops $(A, \ldots, F)$ on the basis of their perceived price levels are reported (for further details see Wrigley 1980a). By fitting a Bradley–Terry model to these data, estimates of the preference parameters $\pi_{r;\alpha}$ can be derived, and these will provide an indication of how the consumers subjectively evaluate the relative merits of the shops on the basis of the price criterion.

To use the weighted least squares linear logit model approach to fit a Bradley–Terry model to these data, an estimate of the probability Prob $(A_{r;\alpha} \rightarrow A_{s;\alpha})$ must first be derived from the data. A good estimate is available from

$$f_{r|rs;\alpha} = \frac{n_{r|rs;\alpha}}{n_{rs}} \qquad [8.46]$$

the observed proportion of consumers who choose r when queried about the pair (r,s) on the basis of criterion α. However, as only one criterion of choice is included in Table 8.3, the subscript α can be dropped for convenience.

A linear logit model can then be formulated which can be used to derive estimates of the Bradley–Terry preference parameters. This takes the form:

$$E_A(\bar{L}_{rs}) = E_A \left[\log_e \frac{f_{r|rs}}{f_{s|rs}} \right] = \log_e \pi_r - \log_e \pi_s = \beta_{rR} - \beta_{sR} \qquad [8.47]$$

and it is based upon the fact that the logit in the unobserved probabilities is the 'asymptotic expectation' of the logit in the observed proportions, and that under the Bradley–Terry model [8.44]

$$E_A(\bar{L}_{rs}) = \log_e \frac{\text{Prob } (A_r \rightarrow A_s)}{\text{Prob } (A_s \rightarrow A_r)} = \log_e \frac{\pi_r}{\pi_s}$$

$$= \log_e \pi_r - \log \pi_s = \beta_{rs} \qquad [8.48]$$

As we saw in Section 2.9.1, there are certain constraints which apply to parameters of such logit models and this reduces the number of parameters which need to be estimated. In this case it can be shown that $\beta_{rs} = \beta_{rR} - \beta_{sR}$ and, therefore, only $R-1$ parameters β_{rR}, $r = 1, \ldots R-1$ need to be estimated. Estimates of the β parameters can be derived by application of weighted least squares methods, and estimates $\hat{\pi}_r$, $r = 1, \ldots R$, of the required Bradley–Terry preference parameters can then be recovered from the $\hat{\beta}$ values by use of the equations

$$\hat{\pi}_r = \frac{e^{\hat{\beta}_{rR}}}{1 + \sum\limits_{s=1}^{R-1} e^{\hat{\beta}_{sR}}} \qquad r = 1, \ldots R-1$$

$$\hat{\pi}_R = \frac{1}{1 + \sum\limits_{s=1}^{R-1} e^{\hat{\beta}_{sR}}} \qquad [8.49]$$

In matrix terms the linear logit model [8.47] takes the general form

$$E_A(\bar{L}) = \mathbf{X}\boldsymbol{\beta} \tag{8.50}$$

where

$$\underset{1 \times 15}{\bar{L}'} = [\bar{L}_{AB}, \bar{L}_{AC}, \ldots, \bar{L}_{EF}], \qquad \underset{1 \times 5}{\boldsymbol{\beta}'} = [\beta_{AF}, \beta_{BF}, \ldots, \beta_{EF}] \tag{8.51}$$

and

$$\underset{15 \times 5}{\mathbf{X}} = \begin{bmatrix}
1 & -1 & 0 & 0 & 0 \\
1 & 0 & -1 & 0 & 0 \\
1 & 0 & 0 & -1 & 0 \\
1 & 0 & 0 & 0 & -1 \\
1 & 0 & 0 & 0 & 0 \\
0 & 1 & -1 & 0 & 0 \\
0 & 1 & 0 & -1 & 0 \\
0 & 1 & 0 & 0 & -1 \\
0 & 1 & 0 & 0 & 0 \\
0 & 0 & 1 & -1 & 0 \\
0 & 0 & 1 & 0 & -1 \\
0 & 0 & 1 & 0 & 0 \\
0 & 0 & 0 & 1 & -1 \\
0 & 0 & 0 & 1 & 0 \\
0 & 0 & 0 & 0 & 1
\end{bmatrix} \tag{8.52}$$

Table 8.4
Shop preferences on basis of price level, quality of goods and stock range

	Pattern of preferences								
(r, s)	PQR rrr	PQR srr	PQR rsr	PQR ssr	PQR rrs	PQR srs	PQR rss	PQR sss	Total (n_{rs})
(A,B)	6	10	3	6	7	15	5	9	61
(A,C)	6	9	6	4	7	15	4	9	60
(A,D)	3	7	4	6	8	16	5	11	60
(B,C)	8	6	10	7	6	10	6	8	61
(B,D)	5	8	5	8	4	15	5	10	60
(C,D)	7	7	4	6	8	13	6	10	61

Key: P—price level; Q—quality of goods; R—stock range. The notation $\underset{rsr}{PQR}$ indicates that shop r is preferred to s on the basis of price level, shop s is preferred to r on the basis of quality of goods, and shop r is preferred to s on the basis of stock range.

Generation of the logit vector in [8.51] using the general sequence of matrix operations [8.30] can then be shown to require

$$\underset{1 \times 30}{f'} = [f_{A|AB}, f_{B|AB}, f_{A|AC}, \ldots, f_{E|EF}, f_{F|EF}]$$

$$= [10, 20, 8, \ldots, 18, 11] \tag{8.53}$$

$$\underset{30 \times 30}{A} = I_{30} \quad \text{(i.e. an identity matrix of order 30)} \tag{8.54}$$

$$\underset{15 \times 30}{K} = \begin{bmatrix} 1-1 & 0 \; 0 & \ldots & 0 \; 0 \\ 0 \; 0 & 1-1 & \ldots & 0 \; 0 \\ \cdot & \cdot & & \cdot \\ \cdot & \cdot & & \cdot \\ \cdot & \cdot & & \cdot \\ 0 \; 0 & 0 \; 0 & \ldots & 1-1 \end{bmatrix} = I_{15} \otimes [1-1] \tag{8.55}$$

A second example of a paired comparison experiment is shown in Table 8.4. This presents the hypothetical responses of 363 consumers to an experiment in which four shops were compared on the basis of price level, quality of goods and stock range. The subscript α is now necessary, and $\alpha = 1$, 2 and 3 correspond to the criteria of price level, quality of goods and stock range respectively.

For each pair of shops and each criterion, α, the linear logit model [8.47] now becomes

$$E_A(\bar{L}_{rs;\alpha}) = E_A\left[\log_e \frac{f_{r|rs;\alpha}}{f_{s|rs;\alpha}} \right] = \log_e \pi_{r;\alpha} - \log_e \pi_{s;\alpha}$$

$$= \beta_{rR;\alpha} - \beta_{sR;\alpha} \tag{8.56}$$

(see Wrigley 1980a and Imrey *et al.* 1976 for further details). In matrix terms the model takes the form [8.50], where

$$\underset{1 \times 18}{L'} = [\bar{L}_{AB;1}, \bar{L}_{AB;2}, \bar{L}_{AB;3}, \bar{L}_{AC;1}, \ldots, \bar{L}_{CD;2}, \bar{L}_{CD;3}] \tag{8.57}$$

$$\underset{1 \times 9}{\beta'} = [\beta_{AD;1}, \beta_{BD;1}, \beta_{CD;1}, \beta_{AD;2}, \ldots, \beta_{BD;3}, \beta_{CD;3}] \tag{8.58}$$

and

$$\mathbf{X} = \begin{bmatrix} 1 & -1 & 0 & 0 & 0 & 0 & 0 & 0 & 0 \\ 0 & 0 & 0 & 1 & -1 & 0 & 0 & 0 & 0 \\ 0 & 0 & 0 & 0 & 0 & 0 & 1 & -1 & 0 \\ 1 & 0 & -1 & 0 & 0 & 0 & 0 & 0 & 0 \\ 0 & 0 & 0 & 1 & 0 & -1 & 0 & 0 & 0 \\ 0 & 0 & 0 & 0 & 0 & 0 & 1 & 0 & -1 \\ 1 & 0 & 0 & 0 & 0 & 0 & 0 & 0 & 0 \\ 0 & 0 & 0 & 1 & 0 & 0 & 0 & 0 & 0 \\ 0 & 0 & 0 & 0 & 0 & 0 & 1 & 0 & 0 \\ 0 & 1 & -1 & 0 & 0 & 0 & 0 & 0 & 0 \\ 0 & 0 & 0 & 0 & 1 & -1 & 0 & 0 & 0 \\ 0 & 0 & 0 & 0 & 0 & 0 & 0 & 1 & -1 \\ 0 & 1 & 0 & 0 & 0 & 0 & 0 & 0 & 0 \\ 0 & 0 & 0 & 0 & 1 & 0 & 0 & 0 & 0 \\ 0 & 0 & 0 & 0 & 0 & 0 & 0 & 1 & 0 \\ 0 & 0 & 1 & 0 & 0 & 0 & 0 & 0 & 0 \\ 0 & 0 & 0 & 0 & 0 & 1 & 0 & 0 & 0 \\ 0 & 0 & 0 & 0 & 0 & 0 & 0 & 0 & 1 \end{bmatrix} \qquad [8.59]$$

18×9

Generation of the logit vector in [8.57] using the general sequence of matrix operations [8.30] can then be shown to require a vector of observed proportions of the form

$$\mathbf{f}' = [\mathbf{f}'_{AB}, \mathbf{f}'_{AC}, \mathbf{f}'_{AD}, \mathbf{f}'_{BC}, \mathbf{f}'_{BD}, \mathbf{f}'_{CD}] \qquad [8.60]$$

1×48

a typical element of which is

$$\mathbf{f}'_{BC} = (1/61)\,[8,\ 6,\ 10,\ 7,\ 6,\ 10,\ 6,\ 8]; \qquad [8.61]$$

an operator matrix $\mathbf{A}$ of the form

$$\mathbf{A} = \mathbf{I}_6 \otimes \begin{bmatrix} 1 & 0 & 1 & 0 & 1 & 0 & 1 & 0 \\ 0 & 1 & 0 & 1 & 0 & 1 & 0 & 1 \\ 1 & 1 & 0 & 0 & 1 & 1 & 0 & 0 \\ 0 & 0 & 1 & 1 & 0 & 0 & 1 & 1 \\ 1 & 1 & 1 & 1 & 0 & 0 & 0 & 0 \\ 0 & 0 & 0 & 0 & 1 & 1 & 1 & 1 \end{bmatrix} \qquad [8.62]$$

36×48

and an operator matrix **K** of the form

$$\mathbf{K} = \mathbf{I}_{18} \bigotimes [1 - 1]$$

[8.63]

18 × 36

EXAMPLE 8.3   *Residential preferences of schoolchildren in Southend-on-Sea*

Longley and Wrigley (1984) have recently used a paired comparison procedure of the type considered in Section 8.4.3 to study the residential preferences of 2,246 schoolchildren attending the ten secondary schools in Southend-on-Sea shown Fig. 8.9. The form of paired comparison procedure used was a generalization of the Bradley–Terry model which allowed for 'ties' in preference. This generalization was suggested by Rao and Kupper (1967) and was first analysed using the weighted least squares logit model approach by Beaver (1977). Its advantage is that it allows respondents to express 'no preference' when evaluating two alternatives. That is to say, respondents are allowed to be indifferent between alternatives through ignorance of the alternatives or because of genuine 'ties' in preference, and they are not forced, as in some earlier residential preference studies, to rank all choice alternatives and maintain consistent preference transitivities across all alternatives. Longley and Wrigley contend that this procedure results in more accurate scaling of environmental images, and a feature of their study is a comparison of the results of the modified paired comparison procedure with those of a more traditional 'ranking of alternatives/ principal components' procedure.

The Rao–Kupper generalization of the Bradley–Terry model accounts for the occurrence of ties in preferences by postulating the existence of a threshold parameter, θ, which in this case reflects the respondents's inability to discriminate between two residential districts

Fig. 8.9
Schools and residential districts used in analysis of preference

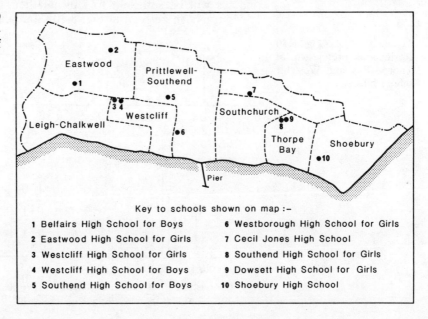

Key to schools shown on map :–

1 Belfairs High School for Boys   6 Westborough High School for Girls

2 Eastwood High School for Girls   7 Cecil Jones High School

3 Westcliff High School for Girls   8 Southend High School for Girls

4 Westcliff High School for Boys   9 Dowsett High School for Girls

5 Southend High School for Boys   10 Shoebury High School

when, in fact, some sort of 'objective' difference between the districts does exist. Instead of the conventional Bradley–Terry form [8.44], the probabilities associated with the comparison of choice alternatives A_r and A_s are now written as

$$\text{Prob}\,(A_r \rightarrow A_s) = \frac{\pi_r}{\pi_r + \theta\pi_s} \qquad \text{Prob}\,(A_s \rightarrow A_r) = \frac{\pi_s}{\theta\pi_r + \pi_s}$$

[8.64]

$$\text{Prob}\,(A_r = A_s) = \frac{(\theta^2 - 1)\,\pi_r\pi_s}{(\pi_r + \theta\pi_s)(\theta\pi_r + \pi_s)}$$

For each choice pair (r, s) there are three observed proportions, $f_{r|rs}$, $f_{s|rs}$, and $f_{0|rs}$, where the latter indicates the proportion of respondents who have 'no preference' between r and s. As a result, two logits

$$\bar{L}_{rs} = \log_e \frac{f_{r|rs}}{1 - f_{r|rs}} \qquad \text{and} \quad \bar{L}_{sr} = \log_e \frac{f_{s|rs}}{1 - f_{s|rs}} \qquad \textbf{[8.65]}$$

can be defined for each choice pair, and the usual one-equation linear logit model for the simplest paired comparison experiment [8.47] becomes a two-equation model of the form

$$E_A(\bar{L}_{rs}) = -\log_e \theta + \log_e \pi_r - \log_e \pi_s = \tau + \beta_{rR} - \beta_{sR}$$

$$E_A(\bar{L}_{sr}) = -\log_e \theta - \log_e \pi_r + \log_e \pi_s = \tau - \beta_{rR} + \beta_{sR}$$

[8.66]

The β and τ parameters can then be estimated using the usual weighted least squares procedure and the required preference and threshold parameters (π_r and θ) can be recovered from the β and τ parameters. (For further details and an illustrative example, see Wrigley 1980a.)

The detailed results of the study need not concern us here. Just two findings will be sufficient to indicate their nature.
1. First, as can be seen in Fig. 8.10, Southend schoolchildren show a

Fig. 8.10
Residential preference of Thorpe Bay and Westcliff schoolchildren

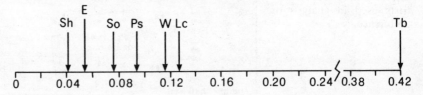

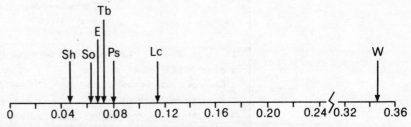

Key: E – Eastwood; Lc – Leigh-Chalkwell; Ps – Prittlewell-Southend;
Sh – Shoebury; So – Southchurch; Tb – Thorpe Bay; W – Westcliff.

consistent preference in favour of their home area. This confirms the 'local dome of desirability' effect identified in all previous residential preference studies.

2. Second, there is a limited amount of evidence in Fig. 8.11, in the consistent decline in the proportion of the sample who select the 'no-preference' option, and in associated pairwise significance tests, to suggest an increasing spread of preferences and improved differentiation amongst individual pairs of districts as secondary schoolchildren become older (Year 1, 11–12 years; Year 2, 12–13 years; Year 3, 13–14 years). As such, the results conform with the general evolution and sequential refinement of residential preferences discerned in Gould's (1975) Swedish national level study. In that study, Gould concluded that mental maps of residential preference 'crystallize out of a mental flux of indifference and uncertainty at an early age.'

Fig. 8.11
Residential preference of Year 1, Year 2 and Year 3 schoolchildren

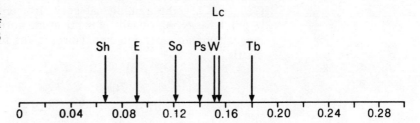

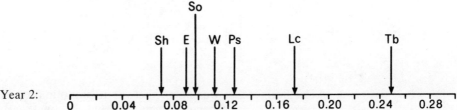

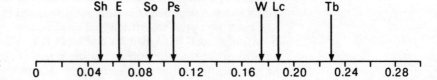

8.5 An alternative treatment of error structures in cell (f) linear logit models

In Sections 2.4.1 and 4.1 it was shown how the dichotomous linear logit model expressed in terms of unobservable probabilities, i.e.

$$L_{12|g} = \log_e \frac{P_{1|g}}{P_{2|g}} = x'_g \beta$$

[8.67]

must be approximated using the observed proportions $f_{1|g}$ and $f_{2|g}$ as estimates and written in the form

$$\bar{L}_{12|g} = \log_e \frac{f_{1|g}}{f_{2|g}} = x'_g \boldsymbol{\beta} + (\bar{L}_{12|g} - L_{12|g}) \qquad \text{[8.68]}$$

In [8.68] the term $(\bar{L}_{12|g} - L_{12|g})$ is an error term introduced to take account of the fact that the left-hand side now contains only an estimate $(\bar{L}_{12|g})$ of the true logit $(L_{12|g})$, and it was shown to have asymptotic properties

$$E_A(\bar{L}_{12|g} - L_{12|g}) = 0, \quad \text{Var}_A(\bar{L}_{12|g} - L_{12|g}) = 1/(n_{+|g} P_{1|g}(1 - P_{1|g})) \text{ [8.69]}$$

The generalization of these results to the multiple response category case was considered in Sections 4.1 and 4.2, where it was shown that $(\bar{L}_{rR|g} - L_{rR|g})$, the multiple-category error term equivalents to that in [8.68], have asymptotically, a multivariate normal distribution with zero means and variances and covariances in the form of a $(R-1) \times (R-1)$ matrix

$$\mathbf{V_{L_g}} = \frac{1}{n_{+|g}} \begin{bmatrix} 1/P_{R|g} + 1/P_{1|g} & 1/P_{R|g} & \cdots & 1/P_{R|g} \\ 1/P_{R|g} & 1/P_{R|g} + 1/P_{2|g} & \cdots & 1/P_{R|g} \\ \cdot & \cdot & & \cdot \\ \cdot & \cdot & & \cdot \\ \cdot & \cdot & & \cdot \\ 1/P_{R|g} & 1/P_{R|g} & \cdots & 1/P_{R|g} + 1/P_{R-1|g} \end{bmatrix}$$

$$\text{[8.70]}$$

Furthermore, it was noted that the asymptotic properties of the error terms are not affected when $P_{1|g}$ and $P_{r|g}$ in [8.69] and [8.70] are replaced by the observed proportions $f_{1|g}$ and $f_{r|g}$. As a result $\mathbf{V_{L_g}}$ is a consistent estimator of $\mathbf{V_{L_g}}$.

Although this assumption of the structure of the error terms in linear logit models has underpinned all our discussion in Chapters 4 and 8, and has been accepted in virtually all applications of such models, it has been questioned by Amemiya and Nold (1975) and Parks (1980). These authors argue that the approximation errors $(\bar{L}_{rR|g} - L_{rR|g})$ are only one of the possible sources of error, and that there are also what can be termed 'specification errors' which represent the effects of variables inappropriately excluded or included in the vector of explanatory variables x'_g (see also Section 10.5.4). If the approximation errors are denoted u_{rg} and the specification errors are denoted v_{rg}, then a multiple response category linear logit model can in this case be written (following equation [4.11]) as:

$$\log_e \frac{f_{r|g}}{f_{R|g}} = x'_g \boldsymbol{\beta}_r + v_{rg} + u_{rg} \qquad r = 1, \ldots, R-1 \qquad \text{[8.71]}$$

Parks (1980) argues that the specification errors can be asumed to have zero means, and a variance-covariance matrix which is made up of the elements

$$E(v_{rg}v_{sg^*}) = \begin{cases} \sigma_{rs} \text{ for } g = g^* \quad \text{for all } r \text{ and } s \\ 0 \quad \text{for } g \neq g^* \quad \text{for all } r \text{ and } s \end{cases} \quad [8.72]$$

where $r = 1, \ldots, R-1$, and $s = 1, \ldots, R-1$. This is the so-called contemporaneously correlated error structure that is often assumed for systems of 'seemingly unrelated' equations (see Theil 1971, 297). It assumes that the specification errors within any single group may be correlated but that such errors are uncorrelated across groups.[†] This appears to be a reasonable assumption because, for any group, an error made in specifying the odds of selecting one particular response category will be offset among the remaining categories.

Both types of error in model [8.71] can be combined to give $\varepsilon_{rg} = v_{rg} + u_{rg}$. Assuming that the approximation errors are independent of the specification errors, the combined error terms for each sub-population g can be shown to have, asymptotically, zero means and variances and covariances in the form of an $(R-1) \times (R-1)$ matrix

$$\mathbf{Q}_g = \mathbf{V}_{\mathbf{L}_g} + \Delta \quad [8.73]$$

where Δ is the contemporaneous cross-category variance-covariance matrix derived from [8.72]. Across all sub-populations, $g = 1, \ldots, G$, the combined variance-covariance matrix will take the block diagonal form

$$\mathbf{Q} = \begin{bmatrix} \mathbf{V}_{\mathbf{L}_1}+\Delta & 0 & \cdots & 0 \\ 0 & \mathbf{V}_{\mathbf{L}_2}+\Delta & \cdots & 0 \\ \cdot & \cdot & & \cdot \\ \cdot & \cdot & & \cdot \\ \cdot & \cdot & & \cdot \\ 0 & 0 & \cdots & \mathbf{V}_{\mathbf{L}_G}+\Delta \end{bmatrix} \quad [8.74]$$

rather than the simple block diagonal form assumed in Chapters 4 and 8

$$\mathbf{V_L} = \begin{bmatrix} \mathbf{V}_{\mathbf{L}_1} & 0 & \cdots & 0 \\ 0 & \mathbf{V}_{\mathbf{L}_2} & \cdots & 0 \\ \cdot & \cdot & & \cdot \\ \cdot & \cdot & & \cdot \\ \cdot & \cdot & & \cdot \\ 0 & 0 & \cdots & \mathbf{V}_{\mathbf{L}_G} \end{bmatrix} \quad [8.75]$$

[†] This has interesting connections to our discussion of the IIA property of multiple-category logit models in Chapter 10, and to the search for less restrictive discrete choice models which permit correlated random components. Parks (1980, 297) notes that by including a specification error term with a variance-covariance structure [8.72] in model [8.71], the odds of choosing choice alternative r over alternative R may be affected by the presence of other choice alternatives. In other words, the model appears to be free from the IIA property.

When the elements $\mathbf{V}_{L_g}$ and Δ in [8.74] are consistently estimated and the inverse of $\mathbf{Q}$ is substituted in place of $\mathbf{V}_L^{-1}$ in the generalized least squares estimator [4.36], the modified estimator will still be consistent. However, if the specification error formulation of the logit model is correct, then the modified estimator (in which $\mathbf{Q}^{-1}$ appears) can be expected to be asymptotically more efficient than the conventional estimator [4.36]. Furthermore, in this case, it follows that the conventionally computed estimates of the standard errors of the parameters derived from [4.36] will be downwardly biased. Parks (1980) provides a simple example where standard error estimates in a conventionally structured three-category linear logit model are biased downwards by as much as half compared with the estimates computed using $\mathbf{Q}^{-1}$. However, as yet, there appears to have been little adoption of the specification-error modification in empirical applications of linear logit models.

CHAPTER 9
Special topics in log-linear modelling

In Chapters 5 and 6 an introduction to the form, estimation and testing of log-linear models was provided. The aim of those chapters was to communicate certain key features of the log-linear modelling approach to the analysis of contingency tables, and particular stress was placed upon the differences in the form of log-linear models for cell (f) and cell (g) problems (i.e. where we either use, or make no use of, any response/explanatory distinction which might exist among the variables). As a result, discussion of many other topics in log-linear modelling (e.g. how to handle zero cell frequencies, when to consider collapsing contingency tables, how to handle outliers, how to deal with contingency tables which have certain 'special' forms, etc.) was deferred to a later stage. In this chapter we return to the discussion of log-linear models and consider these previously omitted topics. It should be noted that, in general, the issues raised in the discussion of these topics apply with equal validity to either cell (g) or cell (f) type log-linear models. As such, our discussion can be simplified by presenting, in the main, only the form of the cell (g) type models. The equivalent cell (f) type models can then be derived, in a straightforward manner using the principles discussed in Chapter 6.

9.1 Combining categories and collapsing tables

In Section 5.4 we discussed the fact that there are considerable advantages to be gained from analysing multidimensional contingency tables directly rather than as a series of two-dimensional tables formed by splitting or collapsing a multidimensional table. We noted that the analysis of two-dimensional tables formed by such a process of collapsing can give rise to fallacious or paradoxical results of the type revealed in the so-called Simpson's paradox. That is to say, we noted that the estimates of the λ parameters representing interactions between variables will often differ considerably in both value and direction as between the collapsed two-dimensional tables and the original uncollapsed table. Nevertheless, we did not rule out all forms of table collapsing and, indeed, there are circumstances in which complex multidimensional tables can usefully be simplified by either combining two or more adjacent categories of a variable, or by collapsing over one or more variables.

In the case of the simplest multidimensional contingency table, the three-dimensional table, Bishop, Fienberg and Holland (1975, 39) demonstrate that we can collapse over a variable (and expect to derive the same measure of the interaction between the other two variables from the collapsed table as we would from the uncollapsed table) if the variable we collapse over is independent of *at least one* of the other two variables. For example, we can collapse over variable C and get estimates of the interaction between variables A and B (i.e. estimates of λ_{ij}^{AB} for all i and j) from the collapsed table which are identical to those we would derive from the uncollapsed table, if variable C is independent of either A or B or both. In other words, the estimates of interaction between variables A and B derived from the collapsed table are consistent with those which could be derived from the uncollapsed table if $\lambda_{ik}^{AC} = 0$ or $\lambda_{jk}^{BC} = 0$, or both, for all i, j and k (and by hierarchical principles if $\lambda_{ijk}^{ABC} = 0$ also). For a good illustration of such a 'collapsible' table, the reader should consult Example 9.1.

It should be noted, however, that the converse of Bishop, Fienberg and Holland's theorem is not true (see Fienberg 1980, 49). Whittemore (1978) has been able to demonstrate that cases exist where $\lambda_{ijk}^{ABC} = 0$, $\lambda_{ik}^{AC} \neq 0$ and $\lambda_{jk}^{BC} \neq 0$, and yet the estimates of λ_{ij}^{AB} for the two-dimensional table (collapsed over C) are the same as for the original uncollapsed table. However, such cases are not common and Fienberg (1980, 50) suggests it is reasonable to consider working with collapsed tables only when the conditions of Bishop, Fienberg and Holland's original theorem are met.

In general, we can say that the conditions which apply to the collapsing of variables in a three-dimensional contingency table also apply to the combining of two or more adjacent categories of any variable within the table. If the structure of the table is such that we can safely collapse a variable, then we can also combine any subset of the categories of that variable without changing the structure by introducing spurious interaction effects or masking effects which are present. However, if the structure of the table is such that we cannot safely collapse a particular variable, then it follows that if we combine some of the categories of that variable we may risk finding relationships between the other two variables which are not present in the full table. For example, in the case where variable C is not independent of either A or B (i.e. $\lambda_{ik}^{AC} \neq 0$ and $\lambda_{jk}^{BC} \neq 0$) it may be unwise to combine adjacent categories of variable C for we may find a relationship between A and B in the reduced $I \times J \times K^*$ table (where $K^* < K$) even though A and B may be independent in the full $I \times J \times K$ table.

In the case of multidimensional tables with more than three categories, Bishop, Fienberg and Holland's collapsibility theorem for three-dimensional tables can be generalized in a straightforward manner. They demonstrate (Bishop, Fienberg and Holland 1975, 47) that if the variables in a multidimensional table are divided into three mutually exclusive groups:

Group 1: the variable or variables to be collapsed.
Group 2: variables independent of those in the first group (i.e. having λ-terms linking them which are zero).

Group 3: the remaining variables which are not independent of those in the first group;

then variables in the first group are collapsible with respect to the λ-terms involving the second group, but are not collapsible with respect to the λ-terms involving only those variables in the third group. (It should be noted that the first parameter, λ, in the log-linear model will always change when one or more variables are collapsed.)

As an illustration of the use of this collapsibility theorem, let us reconsider Table 5.28; Griffin's four-dimensional table of cases of non-fatal deliberate self-harm admitted to Bristol hospitals in 1972–73. The model which we fitted to that table in Chapter 5 took the form

$$\log_e m_{ijkl} = \lambda + \lambda_i^A + \lambda_j^B + \lambda_k^C + \lambda_l^D + \lambda_{ij}^{AB} + \lambda_{ik}^{AC} + \lambda_{il}^{AD} + \lambda_{kl}^{CD} + \lambda_{ikl}^{ACD}$$

[9.1]

Would it now be safe to consider collapsing Table 5.28 over variable B (sex of the individual) and treating male and female cases together in a reduced three-dimensional table?

Following Bishop, Fienberg and Holland's collapsibility theorem, we can divide the variables in Table 5.28 into the three mutually exclusive groups described above. In group 1 we have variable B, in group 2 we have variables C and D, and in group 3 we have variable A. It follows, therefore, that variable B is collapsible with respect to the λ-terms involving C and D, i.e. we could obtain valid estimates of λ_k^C, λ_l^D, λ_{kl}^{CD}, λ_{ik}^{AC}, λ_{il}^{AD}, and λ_{ikl}^{ACD} from the collapsed three-dimensional table. However, variable B is not collapsible with respect to the λ-terms involving only those variables in the third group (in our case just variable A), i.e. we cannot obtain valid estimates of λ_i^A from the collapsed three-dimensional table.

As in the previous case of three-dimensional tables we can say that the conditions which apply to the collapsing of variables in tables with more than three dimensions also apply to the combining of adjacent categories of any variable within such tables. If the structure of the table is such that we can safely collapse a variable, then we can also combine a subset of the categories of that variable without changing the structure of the table or the λ-terms which describe that structure. In other words, we can combine one or more categories of the variable and expect to find that the same log-linear model describes the structure of the reduced table. Conversely, if we cannot safely collapse a particular variable, then it follows that combining some of the categories of that variable may give a reduced table described by λ-terms which differ in value from those in the unreduced original table.

EXAMPLE 9.1   *Oak, hickory and maple distributions in Lansing Woods, Michigan*

Diggle (1979) presents maps of the distributions of 929 oak, 703 hickory and 514 maple trees in a square 19.6 acre sample area of Lansing Woods, Clinton Country, Michigan. Fingleton (1983) has subsequently presented this information in the form of the three-dimensional $2 \times 2 \times 2$ table shown in Table 9.1 where the frequencies in the table are based on the presence or absence of species in 576 grid cells superimposed on the original maps.

Table 9.1
The distribution of oaks, hickories, and maples in 576 grid cells superimposed on the maps of Lansing Woods (cell values from Fingleton 1982)

Table 9.1 The distribution of oaks, hickories, and maples in 576 grid cells superimposed on the maps of Lansing Woods (cell values from Fingleton 1982)

	Maples (M)			
	Present		Absent	
	Hickories (H)		Hickories (H)	
Oaks (O)	Present	Absent	Present	Absent
Present	84	91	177	86
Absent	32	30	61	15

Assuming that the observations in Table 9.1 are spatially independent (see also the further analysis in Section 9.6.4), Fingleton found that the best fitting log-linear for Table 9.1 takes the form

$$\log_e m_{ijk} = \lambda + \lambda_i^O + \lambda_j^H + \lambda_k^M + \lambda_{ij}^{OH} + \lambda_{jk}^{HM} \qquad [9.2]$$

According to Bishop, Fienberg and Holland's collapsibility theorem, we should, therefore, be able to collapse over variable O and derive the same value for the interaction between the other two variables (HM) from the collapsed table as we are able to derive from model [9.2] for the uncollapsed table. This can be confirmed from Table 9.2, which shows that the estimate of λ_{11}^{HM} obtained from the two-way hickory-maple marginal table (-0.225) is the same as the estimate of λ_{11}^{HM} obtained from model [9.2]. Similarly, we should be able to collapse over variable M and derive the same value for the interaction between variables O and H from the collapsed table as we are able to derive from the uncollapsed table. This also can be confirmed from Table 9.2, which shows that the estimate of λ_{11}^{OH} obtained from the two-way oak-hickory marginal table (-0.084) is identical to the estimate of λ_{11}^{OH} obtained from model [9.2].

Table 9.2 Interaction estimates from model [9.2] and from collapsed tables

Model 9.2:		$\hat{\lambda}_{11}^{OH} = -0.084,$		$\hat{\lambda}_{11}^{HM} = -0.225$

| | | | Hickories | |
			P	A	
$\hat{\lambda}_{11}^{OH}$ from 2×2		P	261	177	
	Oaks				table = -0.084
marginal (collapsed)		A	93	45	

| | | | Maples | |
			P	A	
$\hat{\lambda}_{11}^{HM}$ from 2×2		P	116	238	
	Hickories				table = -0.225
marginal (collapsed)		A	121	101	

9.2 Sampling zeros

In Chapters 5 and 6 we saw that multidimensional contingency tables will often have a large number of cells. For example, the simplest four-dimensional table (a 2×2×2×2 table) has sixteen cells and we are often faced with much larger four-dimensional tables such as that in Example 6.3, which has 384 cells. Similarly, the simplest five-dimensional table has thirty-two cells and commonly encountered five-dimensional tables have hundreds of cells. In these situations, it is clear that total sample size must also be large if certain cells in the table are not to be empty, i.e. if certain cells are not to have zero observed frequencies.

However, there are clearly practical limits on the size of samples used in research. As a result, it may be difficult, in many cases, to obtain a sufficiently large and well-structured sample to avoid certain cells having no recorded observations (i.e. to avoid the occurrence of what are termed 'sampling zeros'). In this section we will discuss the problems which are caused by such sampling zeros and consider some of the solutions which have been suggested. The reader should note, however, that not all zero cell frequencies in contingency tables are sampling zeros. They can also be of another type which are termed 'fixed' or 'structural' zeros and which occur when certain cells in a table logically take zero values. Discussion of such structural zeros is deferred until Section 9.3. In this section we confine ourselves to just those zeros which, theoretically, should disappear by increasing the sample size.

9.2.1 Sampling zeros and saturated log-linear models

Saturated log-linear models have as many parameters as there are cells in the contingency table and such models fit the observed cell frequencies perfectly, i.e. the estimates of the expected cell frequencies are identical to the observed cell frequencies. To obtain the parameter estimates of saturated models we can, therefore, simply substitute observed cell frequencies in place of the expected cell frequencies in expressions similar to those in [5.12] to [5.14] and [5.61] (see Goodman 1971a, 38). Alternatively, we can allow one of the iterative estimation procedures (see Section 5.6) to perform essentially the same task. Unfortunately, this procedure breaks down if the contingency table contains zeros. The reason for this is that the natural logarithm of a zero cell takes the value $-\infty$, and it thus becomes impossible to calculate the parameter estimates of the saturated model.

Several procedures have been suggested to overcome this problem. 1. The first suggestion often proposed is to consider reducing the size of the table in some way, usually by combining one or more adjacent categories of a variable. This practice is widespread in the social and environmental sciences and, in addition, it is often adopted when there are non-zero cells with observed frequencies of less than five.[†]

[†] The reason for this stems from advice offered by statisticians such as Cochran and Fisher in the pre-computer era. This advice, based upon practical experience and intuition, led to practical rules such as 'the minimum expected cell frequency should exceeed five' being suggested and perpetuated in introductory statistics texts. However, as Fienberg (1980, 172–76) demonstrates, such rules are somewhat conservative and, today, we have much more insight into the problems of dealing with large sparse multidimensional tables.

However, as we saw in Section 9.1, any such combining of categories must be undertaken with considerable care if it is not to change the structure of the table and the λ-terms which describe that structure. In this respect, it is also important to assess the extent to which the original categories of a particular variable are 'meaningful' in terms of the substantive problem under investigation. As Fienberg (1977, 110) has noted: 'when the categories for a given variable are truly meaningful, collapsing of categories is not necessarily a good procedure'.

2. The second, and widely used, procedure is to 'smooth' the contingency table and eliminate the zero cell frequencies by the addition of a small constant, generally 0.5, to each cell in the table. Goodman (1970, 1971a) was responsible for the adoption of this procedure in the context of log-linear models and it is analogous to similar suggestions by Berkson (1953, 1955) and Grizzle, Starmer and Koch (1969) in the context of linear logit models (see Sections 2.4.1 and 4.2 and also Gart and Zweifel 1967). Indeed, Goodman recommends the addition of 0.5 as a standard practice irrespective of whether there are zero cell frequencies in the table. As we saw in Example 5.5, many users of log-linear models adopt Goodman's advice on this matter and standard computer programs such as the BMDP log-linear modelling routines facilitate that advice by including a single command which adds the required constant to all cells in the table.[†]

3. Despite the widespread use of the practice of adding 0.5 to each cell, the justification for the procedure and the specific choice of 0.5 as the constant is subject to debate. In particular, Fienberg and Holland (1970, 1973) and Bishop, Fienberg and Holland (1975, 401–33) have discussed the issue and have proposed a more general method of smoothing contingency tables and eliminating sampling zeros based on the use of *pseudo-Bayes* cell estimates.

In essence, the pseudo-Bayes procedure involves the calculation of an alternative constant $\hat{K}$ based upon an intelligent estimate of the underlying cell probabilities from which we have obtained our observed cell frequencies. This then replaces the widely used but somewhat arbitrary constant of 0.5. In the case of a two-dimensional table, for example, it is computed (see Bishop, Fienberg and Holland 1975, 401) as

$$\hat{K} = \frac{N^2 - \sum_i \sum_j n_{ij}^2}{\sum_i \sum_j (n_{ij} - NP_{ij}^*)^2} \qquad [9.3]$$

where P_{ij}^* for all i and j, represent the estimates of the cell probabilities based on external information or on the data themselves. Having derived the value of $\hat{K}$, the pseudo-Bayes estimates of the cell frequencies are then computed using the expression

$$\hat{m}_{ij}^* = \frac{N}{N + \hat{K}} (n_{ij} + \hat{K}P_{ij}^*) \qquad [9.4]$$

---

† Users of the GLIM package should recall that in Section 5.6.2 we noted how Nelder recommends the substitution of 0.5 in place of any zero observed cell frequencies in the iterative weighted least squares procedure.

In this way, we obtain a table in which all zero cells have been eliminated and which theoretically combines our actual observed data with our 'previous knowledge' which is represented by the estimated probability values.

Bishop, Fienberg and Holland (1975, 401) claim that the pseudo-Bayes method is 'clearly superior to the generally accepted practice of adding $^1/_2$ to the count in each cell of a large sparse table'. However, the precise use of the method depends upon whether the researcher has reasonable *a priori* grounds for estimating the probabilities P^*_{ij}, etc. or whether he must use the structure of the observed data in some way to build data-dependent estimates of the probabilities. These issues are complex and demand detailed appraisal which is beyond the scope of this book. The interested reader should consult Bishop, Fienberg and Holland (1975).

9.2.2 *Sampling zeros and unsaturated log-linear models*

The procedures for the elimination of sampling zeros which we have just discussed can be applied with equal validity in the case of unsaturated log-linear models. However, an important additional element enters the picture in the case of unsaturated models. This additional element concerns the fact that the iterative maximum likelihood procedures (see Section 5.6) have the capacity to produce non-zero estimated expected frequencies (i.e. non-zero values of $\hat{m}_{ijk}$) for cells with zero observed frequencies (i.e. zero n_{ijk}). In other words, in the case of unsaturated models, if an observed cell frequency is zero it does not necessarily follow that the estimate of the expected frequency for that cell also equals zero. As log-linear models and their parameters are defined in terms of the logarithms of expected cell frequencies, it follows that it is only when the estimates of the expected frequencies equal zero that any difficulties will occur.

If there is only one sampling zero in the contingency table, the estimates of the expected cell frequencies under any of the non-saturated models in the hierarchical set will always be positive and will, therefore, create no problems. Similarly, even when the observed table contains several sampling zeros the iterative estimation procedures will often eliminate them all. Nevertheless, sometimes there will be just 'too many' sampling zeros in the table or they may be distributed in such a way that one or more marginal totals may also be zero. In these circumstances, the iterative estimation procedures will converge to a table which has estimated cell values of zero in some of the cells that originally had sampling zeros, and this will create two types of problem. First, as in the case of the saturated model above, it may become impossible to calculate the estimates of the λ parameters. Second, in order to test the goodness-of-fit of the model it will be necessary to adjust the degrees of freedom.

In the case of tables which have marginal totals which are zero, both the observed and expected frequencies for all cells included in that total must be zero. Consequently, once it is known that a marginal total is zero, that fit of a model for those cells must be perfect. As a result, we must delete those degrees of freedom associated with the fit of the zero cell frequencies. Bishop, Fienberg and Holland (1975, 115) and

Fienberg (1977, 110) present general formula for computing the degrees of freedom in such cases. It takes the form

$$\text{d.f.} = (T_e - Z_e) - (T_K - Z_K) \qquad\qquad [9.5]$$

where

T_e = total number of cells in table

T_K = total number of parameters that require estimating in the model

Z_e = number of cells with zero estimated expected frequencies

Z_K = number of parameters which cannot be estimated because of zero marginal totals.

In most cases in geography and environmental science, the application of this formula will be reasonably straightforward (see Example 9.3) However, certain more complicated cases can arise where zero cell frequencies can be distributed in such a way that they give rise to multiple zero marginal totals (e.g. where zero marginal totals persist in lower order, i.e. collapsed, tables which underlie the original table). In such cases great care must be taken when counting the parameters which cannot be estimated because of the zero marginal totals (i.e. when calculating the figure Z_K to be used in expression [9.5]). In these circumstances, the loss of independent parameters will be less than the number of marginal zeros (see Reynolds 1977, 162; Bishop, Fienberg and Holland 1975, 116 for further details).

EXAMPLE 9.2   *Non-fatal deliberate self-harm in Bristol (continued)*

In Example 5.5 we considered Griffin's (1984) analysis of the four-dimensional $3\times2\times2\times2$ table shown in Table 5.28. The table classifies 1,443 individual cases of non-fatal deliberate self-harm admitted to Bristol's teaching hospitals in 1972–73. In Table 9.3 a reformulation of this data is presented as a $5\times2\times3\times2$ table.[†] The four variables in the table remain the same (i.e. *A,* age of individual; *B,* sex of individual; *C,* time of admission to hospital; *D,* location of home of individual) but now the age variable has five categories (12–19, 20–29, 30–47, 48–65; 66 and over) and the time of admission to hospital variable has three categories ('day-time hours': 9.00 am–9.00 pm; 'evening hours': 9.00 pm–midnight; 'night-time hours': midnight–9.00 am).

It can be seen that in this sixty-cell reformulated table, two cells are empty, i.e. we have two sampling zeros. However, these are not distributed in such a way that any marginal totals are also zero. As such, the table presents no difficulties to the analyst. The iterative estimation procedures of the standard computer programs will provide non-zero estimates of the expected cell frequencies under any of the non-saturated models in the hierarchical set. To illustrate this, consider Table 9.4. This gives the estimates of the expected cell frequencies

[†] This is done simply to illustrate a point concerning sampling zeros. It is *not* intended to give the impression that the number of categories does not matter or that the researcher is permitted to manipulate the original table in any way he pleases.

Table 9.3
An alternative classification of cases of non-fatal deliberate self-harm admitted to Bristol hospitals in 1972–73

		Variable D											
		Inner city						Outer city					
		Night		Day		Evening		Night		Day		Evening	
		Variable B		Variable B		Variable B		Variable B		Variable B		Variable B	
		M	F	M	F	M	F	M	F	M	F	M	F	Total
Variable A	12–19	5	25	13	68	8	26	9	22	12	53	4	21	266
	20–29	34	56	51	108	20	43	32	24	36	76	15	29	524
	30–47	28	32	37	65	16	24	20	31	40	85	10	30	418
	48–65	7	14	17	43	2	5	3	8	21	33	5	9	167
	66+	1	8	11	20	0	2	1	0	4	16	2	3	68
	Total:	75	135	129	304	46	100	65	85	113	263	36	92	1,443

Table 9.4
Estimates of expected cell frequencies under model [9.6]

		Variable D											
		Inner city						Outer city					
		Night		Day		Evening		Night		Day		Evening	
		Variable B		Variable B		Variable B		Variable B		Variable B		Variable B	
		M	F	M	F	M	F	M	F	M	F	M	F
Variable A	12–29	7.2	22.8	14.3	66.7	6.1	27.9	7.4	23.6	11.5	53.5	4.5	20.5
	20–29	38.1	51.9	53.0	106.0	21.1	41.9	23.7	32.3	37.4	74.6	14.8	29.2
	30–47	25.6	34.4	34.4	67.6	13.6	26.4	21.8	29.2	42.1	82.9	13.6	26.4
	48–65	8.4	12.6	18.7	41.3	2.2	4.8	4.4	6.6	16.9	37.1	4.4	9.6
	66+	3.1	5.9	8.3	22.7	0.5	1.5	0.3	0.7	5.3	14.7	1.3	3.7

under the model

$$\log_e m_{ijkl} = \lambda + \lambda_i^A + \lambda_j^B + \lambda_k^C + \lambda_l^D + \lambda_{ij}^{AB} + \lambda_{ik}^{AC}$$
$$+ \lambda_{il}^{AD} + \lambda_{jk}^{BC} + \lambda_{kl}^{CD} + \lambda_{ikl}^{ACD} \qquad \textbf{[9.6]}$$

which has a satisfactory fit to the data of Table 9.3 ($G^2 = 22.62$ with twenty three degrees of freedom – no adjustment in the degrees of freedom being necessary as no zero estimated cell frequencies are present in Table 9.4).

EXAMPLE 9.3

Relationships between tree species and tree height in the forests of South Island, New Zealand.

Table 9.5 presents a set of data abstracted for the purposes of this chapter from a much larger study by Veblen and Stewart (1982) of species regeneration in the evergreen conifer-broadleaved forests of South Island, New Zealand. In Table 9.5 the numbers of trees (>1.4 m tall) of five different species (Dacrydium biforme, Metrosideros umbellata, Phyllocladus alpinus, Quintinia acutifola, Weinmannia racemosa) are recorded within three different height categories (within or above the main canopy, below main canopy but greater than half the height of the main canopy, 'over-topped'–less than half the height of the main canopy) for two different forest stands on the northern face of the Kelly Range in South Island, New Zealand. For our purposes, the important feature of Table 9.5 is the occurrence of three sampling zeros arranged in such a way that a marginal total (in the species by stand marginal table) is also zero. In this example, we will consider how to handle such a marginal zero in the log-linear modelling of the three-dimensional $5 \times 3 \times 2$ contingency table.

Table 9.5

Tree frequencies within five species and three height classes in two stands in the Kelly Range

		Variable C (*stand*)					
		K–1			K–2		
		Variable B (*height*)			Variable B (*height*)		
		Overtop	*Below M. C.*	*Main Can.*	*Overtop*	*Below M. C.*	*Main Can.*
Variable A (*species*)	Dac.bif	122	24	49	0	0	0
	Met.umb	95	19	39	12	2	17
	Phy.alp	762	62	43	97	2	19
	Qui.acu	499	55	31	812	68	97
	Wei.rac	39	19	14	70	23	52

To the data of Table 9.5 we can fit a hierarchical set of log-linear models. Some members of this set are shown in Table 9.6. On the basis of an inspection of the G^2 statistics it is clear that, other than the saturated model, the only model with an acceptable goodness-of-fit is the model in which the three-dimensional interaction parameters are set to zero. This model is the type-2 model of Table 5.11, i.e. the pairwise association model. It implies that each pair of variables is associated but that the association of each pair is unaffected by the level of the third variable. That is to say, tree height is related to tree species but this relationship is unaffected by which forest stand in the Kelly Range is being studied. Similarly, tree species is related to forest stand but this relationship is unaffected by tree height, and so on.[†]

Although Table 9.6 reports the G^2 statistics for the log-linear models, it does not report the associated degrees of freedom. The reason for this

---

[†] The reader should note that these conclusions are limited in their generality to the specific context of Table 9.5. When other tree species are added to Table 9.5, the pairwise association model no longer provides an adequate fit and it is necessary to include a set of three-variable interaction parameters in the model.

Model specification	G^2
$\log_e m_{ijk} = \lambda + \lambda_i^A + \lambda_j^B + \lambda_k^C + \lambda_{ij}^{AB} + \lambda_{ik}^{AC} + \lambda_{jk}^{BC} + \lambda_{ijk}^{ABC}$	0.00
$\log_e m_{ijk} = \lambda + \lambda_i^A + \lambda_j^B + \lambda_k^C + \lambda_{ij}^{AB} + \lambda_{ik}^{AC} + \lambda_{jk}^{BC}$	6.69
$\log_e m_{ijk} = \lambda + \lambda_i^A + \lambda_j^B + \lambda_k^C + \lambda_{ij}^{AB} + \lambda_{ik}^{AC}$	52.58
$\log_e m_{ijk} = \lambda + \lambda_i^A + \lambda_j^B + \lambda_k^C + \lambda_{ij}^{AB} + \lambda_{jk}^{BC}$	1,041.51
$\log_e m_{ijk} = \lambda + \lambda_i^A + \lambda_j^B + \lambda_k^C + \lambda_{ij}^{AB}$	1,063.41
$\log_e m_{ijk} = \lambda + \lambda_i^A + \lambda_j^B + \lambda_k^C + \lambda_{jk}^{BC}$	1,248.17
$\log_e m_{ijk} = \lambda + \lambda_i^A + \lambda_j^B + \lambda_k^C$	1,306.08
$\log_e m_{ijk} = \lambda + \lambda_i^A + \lambda_j^B$	1,421.71
$\log_e m_{ijk} = \lambda + \lambda_i^A$	4,296.05

Table 9.6 Some members of the hierarchical set of log-linear models fitted to Table 9.5

is that for a number of the models it is necessary to adjust the degrees of freedom to take account of the effects of the marginal zero. To illustrate how this adjustment is made we will consider the case of the selected 'acceptable' model; the pairwise association model.

To adjust the degrees of freedom in this context, we substitute into the general formula [9.3]. In the case of the pairwise association model, we know that $T_e = 30$, and we also know that $T_K = 22$, i.e. there are twenty-two parameters that require estimation in the pairwise association model (under the 'centred effect' coding constraints these are λ; $\lambda_1^A \ldots \lambda_4^A$; λ_1^B, λ_2^B; λ_1^C; $\lambda_{11}^{AB} \ldots \lambda_{41}^{AB}$; $\lambda_{12}^{AB} \ldots \lambda_{42}^{AB}$; $\lambda_{11}^{AC} \ldots \lambda_{41}^{AC}$; $\lambda_{11}^{BC}, \lambda_{21}^{BC}$). Furthermore, we know that the two-dimensional marginal table $[AC]$ (i.e. tree species by forest stand) has a zero cell (i.e. $n_{1+2} = 0$). Any estimated cell frequency based on this zero cell will also be zero and therefore under the pairwise association model there will be three cells with zero estimated expected frequencies[†] (i.e. $Z_e = 3$). Finally, since the marginal table $[AC]$ contains one zero, the number of parameters which require estimation in the set $\lambda_{11}^{AC} \ldots \lambda_{41}^{AC}$ is reduced by one (i.e. $Z_K = 1$), since $\lambda_{11}^{AC} = -\lambda_{12}^{AC}$ and λ_{12}^{AC} is known *a priori* to be zero. As a result, the degrees of freedom for the pairwise association model are given by the expression

$$\text{d.f.} = (30-3) - (22-1) = 6$$

With a G^2 value of 6.99 and six degrees of freedom it can readily be seen that the pairwise association model has an acceptable fit to the data of Table 9.5.

9.3 Structural zeros and incomplete contingency tables

Theoretically, at least, sampling zeros can be made to disappear by simply increasing the sample size sufficiently. Structural or fixed zeros,

[†] Many of the standard computer programs for log-linear modelling (e.g. BMDP) will automatically list the number of zero estimated expected frequencies under any model but will not give a value for Z_K, i.e. the number of parameters which do not require estimating because of the zero marginal totals.

on the other hand, will not disappear as sample size increases. They represent cells in the contingency table which are empty *a priori* because certain combinations of the variables are impossible (a classical example is that in a hospital study one cannot expect to find male obstetrical or hysterectomy patients or female prostatectomy patients).

Contingency tables with structural zeros are generally termed *incomplete* tables and it is important that these zero cells must remain empty in any subsequent analysis, i.e. under any model which is fitted to the table. As such it is not appropriate to use the methods of Section 9.2.1 to 'fill in' the cells containing the structural zeros. Neither is it appropriate to combine categories of variables or collapse variables to eliminate the structural zeros. It is important that structural zeros are recognized as such (and not mistaken as sampling zeros) and that appropriate methods are adopted for analysing structurally incomplete tables.

When considering methods for handling structural zeros and incomplete tables, it is useful to distinguish between what are termed *separable* incomplete tables and *inseparable* incomplete tables. A separable incomplete table is one which can be divided into a set of non-interconnected complete tables. For example, Figure 9.1 shows a two-dimensional table with a certain number of structural zeros (indicated by 0). This table can simply be divided into two non-interconnecting complete tables as shown, and each sub-table can be analysed in the usual log-linear manner. The composite model for the incomplete table is then the union of the models for the complete sub-tables. The overall goodness-of-fit of the composite model is assessed using the sum of the separate G^2 values and the sum of the separate degrees of freedom for the two separate models.

Fig. 9.1
A separable incomplete table

Variable A \ Variable B	j=1	2	3	4
i=1	n_{11}	0	0	n_{14}
i=2	0	n_{22}	n_{23}	0
i=3	0	n_{32}	n_{33}	0
i=4	n_{41}	0	0	n_{44}
i=5	n_{51}	0	0	n_{54}

Var. A \ Var. B	j=1	4
i=1	n_{11}	n_{14}
i=4	n_{41}	n_{44}
i=5	n_{51}	n_{54}

Var. A \ Var. B	j=2	3
i=2	n_{22}	n_{23}
i=3	n_{32}	n_{33}

Most incomplete tables, however, are not separable. In geography they are often the type of inseparable tables shown in Tables 9.7 and 9.8 The first of these is an inter-regional migration table which by definition must have zeros down its principal diagonal. The second is a triangular table which by definition must have structural zeros in the upper triangle as the tax revenue of a CBD cannot exceed or equal the tax revenue of the SMSA of which it forms part. In these circumstances we must analyse the tables using what are termed *quasi* log-linear models.

Table 9.7
Zero diagonal inter-regional migration table

		Destination region			
		$j = 1$	2	. . .	J
Origin regin	$i = 1$	0	n_{12}	. . .	n_{1J}
	$i = 2$	n_{21}	0	. . .	n_{2J}
	.	.	.		.
	.	.	.		.
	.	.	.		.
	$i = I$	n_{I1}	n_{I2}	. . .	0

Table 9.8
Triangular table

		Total tax revenue of CBD			
		<$5 m.	*$5–14.99 m.*	*$15–24.99 m.*	*≥$25 m.*
Total tax revenue of SMSA	<$5 m.	n_{11}	0	0	0
	$5–14.99 m.	n_{21}	n_{22}	0	0
	$15–24.99 m.	n_{31}	n_{32}	n_{33}	0
	≥$25 m.	n_{41}	n_{42}	n_{43}	n_{44}

To illustrate the concept of quasi log-linear models we can first consider the situation in which we wish to fit a log-linear model of independence to an inseparable incomplete two-dimensional contingency table. In this case, instead of testing the usual null hypothesis (see Section 5.1) which states that

$$H_0 : P_{ij} = P_{i+}P_{+j} \qquad\qquad [9.7]$$

we test the natural extension of this hypothesis to the incomplete table; in other words, we test a null hypothesis which states that

$$H_0 : P_{ij} = P_{i+}P_{+j} \qquad \text{for all cells which are not} \qquad [9.8]$$
$$\text{structural zeros}$$

Following Goodman (1968), the null hypothesis [9.8] is termed the hypothesis of *quasi-independence*. In other words, it is a form of independence conditional on the restriction of our attention to part of the table only. The log-linear model which expresses this hypothesis can be written in the form

$$\log_e m_{ij} = \lambda + \lambda_i^A + \lambda_j^B \qquad \text{for all } ij \text{ cells which are not} \qquad [9.9]$$
$$\text{structural zeros}$$

In other words, it is a log-linear model of independence applied only to the non-structural-zero cells.

Model [9.9] is, of course only a special form of the most general log-linear model for a two-dimensional incomplete contingency table. This most general model takes the form

$$\log_e m_{ij} = \lambda + \lambda_i^A + \lambda_j^B + \lambda_{ij}^{AB} \qquad \text{for all cells which are not} \qquad \textbf{[9.10]}$$
$$\text{structural zeros}$$

where, assuming a 'centred effect' coding system, the parameter constraints associated with this model take the form

$$\sum_{i=1}^{I} \delta_i^{(2)} \lambda_i^A = \sum_{j=1}^{J} \delta_j^{(1)} \lambda_j^B = 0 \qquad\qquad \textbf{[9.11]}$$

$$\sum_{i=1}^{I} \delta_{ij} \lambda_{ij}^{AB} = \sum_{j=1}^{J} \delta_{ij} \lambda_{ij}^{AB} = 0 \qquad\qquad \textbf{[9.12]}$$

$$\delta_{ij} = 1 \text{ for all } ij \text{ cells which are not structural zeros} \qquad \textbf{[9.13]}$$
$$= 0 \text{ otherwise}$$

$$\delta_i^{(2)} = 1 \quad \text{if } \delta_{ij} = 1 \quad \text{for some } j \qquad\qquad \textbf{[9.14]}$$
$$= 0 \quad \text{otherwise}$$

$$\delta_j^{(1)} = 1 \quad \text{if } \delta_{ij} = 1 \text{ for some } i \qquad\qquad \textbf{[9.15]}$$
$$= 0 \quad \text{otherwise}$$

The model of quasi-independence can be seen, therefore, to be the model obtained from [9.10] by setting

$$\lambda_{ij}^{AB} = 0 \quad \text{for all } ij \text{ cells which are not} \qquad\qquad \textbf{[9.16]}$$
$$\text{structural zeros}$$

Clearly, by setting other terms in [9.10] to zero we can specify a range of quasi log-linear models for a two-dimensional incomplete table.

The parameters and expected cell frequencies of such quasi log-linear models can be estimated without difficulty using the iterative maximum likelihood estimation procedures described in Section 5.6. The only requirement is to ensure that the initial estimates of the expected cell frequencies (the $\hat{m}_{ij}^{(0)}$) are set to zero in the case of the structural zeros (in the case of the iterative proportional fitting procedure this can be achieved by setting $\hat{m}_{ij}^{(0)} = \delta_{ij}$, as defined in expression [9.13]). However, when the goodness-of-fit of any quasi log-linear model is assessed, we must remember to adjust the degrees of freedom to take account of the structural zero cells. Instead of using the standard expression [5.32] for the degrees of freedom, we must now use the expression

d.f. = number of cells in table − number of structural
zero cells − number of parameters that require   **[9.17]**
estimating

Although, in most circumstances, use of this expression will create no difficulties, care is needed in determining the number of parameters that need to be estimated. It must be remembered that those parameters referring to the empty cells are known *a priori* to be zero and, therefore, do not require estimation.

The quasi log-linear models which are used in the case of incomplete two-dimensional tables can readily be extended to handle incomplete multidimensional tables. For example, the most general quasi log-linear model for a three-dimensional table takes the form (see Bishop, Fienberg and Holland 1975, 210)

$$\log_e m_{ijk} = \lambda + \lambda_i^A + \lambda_j^B + \lambda_k^C + \lambda_{ij}^{AB} + \lambda_{ik}^{AC} + \lambda_{jk}^{BC} \qquad [9.18]$$

$$+ \lambda_{ijk}^{ABC} \quad \text{for all } ijk \text{ cells which are not structural zeros.}$$

Assuming a 'centred effect' coding system, the parameter constraints associated with this model take the form

$$\sum_{i=1}^I \delta_i^{(23)} \lambda_i^A = \sum_{j=1}^J \delta_j^{(13)} \lambda_j^B = \sum_{k=1}^K \delta_k^{(12)} \lambda_k^C = 0 \qquad [9.19]$$

$$\sum_{i=1}^I \delta_{ij}^{(3)} \lambda_{ij}^{AB} = \sum_{j=1}^J \delta_{ij}^{(3)} \lambda_{ij}^{AB} = \sum_{i=1}^I \delta_{ik}^{(2)} \lambda_{ik}^{AC} = \ldots = \sum_{k=1}^K \delta_{jk}^{(1)} \lambda_{jk}^{BC} = 0 \quad [9.20]$$

$$\sum_{i=1}^I \delta_{ijk} \lambda_{ijk}^{ABC} = \sum_{j=1}^J \delta_{ijk} \lambda_{ijk}^{ABC} = \sum_{k=1}^K \delta_{ijk} \lambda_{ijk}^{ABC} = 0 \qquad [9.21]$$

where

δ_{ijk} = 1 if the *ijk* cell is not a structural zero
        = 0 otherwise   [9.22]

$\delta_{ij}^{(3)}$ = 1 if δ_{ijk} = 1 for some *k*
        = 0 otherwise   [9.23]

$\delta_i^{(23)}$ = 1 if δ_{ijk} = 1 for some *jk*
        = otherwise   [9.24]

and similar definitions for $\delta_{ik}^{(2)}$, $\delta_{jk}^{(1)}$, $\delta_j^{(13)}$ and $\delta_k^{(12)}$. The equivalent parameter constraints assuming a 'cornered effect' coding system are straightforward extensions of [9.19] to [9.21] and [9.22] to [9.24] and their definition is left as an exercise for the reader.

As in the case of incomplete two-dimensional tables, we can define a range of unsaturated quasi log-linear models for incomplete multi-dimensional tables by setting λ-terms in [9.18] to zero. As usual, we normally restrict our attention to the hierarchical set of models. Principles of estimation and adjustment of degrees of freedom are identical to those discussed in the case of two-dimensional tables.

EXAMPLE 9.4   *Filtering in the housing market of Kingston, Ontario*

Table 9.9 presents data on the occupancy of 2,495 (approximately 15 per cent) of the homes in Kingston, Ontario, Canada in 1958 and 1963. The data are reported in Yeates (1974, 178) and derive from a study by Kirkland (1969) of the filtering process in the housing market. This is the process by which lower-income families move upwards in the housing market into better quality housing formerly occupied by higher-income families, and/or higher-income families move downwards into lower quality housing formerly occupied by lower-income families (see Robson 1975, 37). In each of the two years, 1958 and 1963, homes are classified into seven groups according to the occupation of the head of the household which occupies the home, and these seven groupings (unskilled, service, craft, clerical, sales, managerial, professional) are ranked according to their average income as indicated by the 1961 census of Canada. A filtering upwards of families into homes previously occupied by families of higher economic status is therefore indicated by entries to the left of the principal diagonal (e.g. of the 138 homes occupied by clerical workers in 1958, eleven were occupied by unskilled workers in 1963, twenty-three were occupied by service workers in 1963, and twelve were occupied by craft workers in 1963). Filtering downwards of families into homes previously occupied by families of lower economic status is indicated by entries to the right of the principal diagonal (e.g. of the 568 homes occupied by craft workers in 1958, thirteen were occupied by clerical workers in 1963, twenty-two by sales personnel, twenty-nine by those in managerial occupations, and eight by those in the professions).

Table 9.9
Occupancy of 2,495 homes in Kingston, Ontario in 1958 and 1963 (observed frequencies from Yeates 1974, 178 and Kirkland 1969). Estimates of expected frequencies under model of independence shown in parentheses below observed frequencies

The large observed frequencies along the principal diagonal of Table 9.9 (179, 359, . . . 207) indicate the considerable degree of stability in the occupant status of homes between the two years. As such, the occupancy classifications of homes in the two years are clearly not independent. This is confirmed by fitting the two-dimensional log-linear model of independence [5.11] which has a G^2 goodness-of-fit value of 2327.0 with thirty-six degrees of freedom. This is well within the conventional upper 5 per cent tail of the χ^2 distribution, and an

	1963 occupancy							
	1	2	3	4	5	6	7	Total
$i = 1$: Unskilled	179 (53.0)	62 (80.5)	39 (71.3)	7 (19.6)	18 (28.0)	10 (30.3)	5 (37.3)	320
2: Service	116 (111.1)	359 (169.0)	103 (149.5)	31 (41.1)	32 (58.6)	12 (63.5)	18 (78.3)	671
3: Craft	64 (94.0)	95 (143.0)	337 (126.6)	13 (34.8)	22 (49.6)	29 (53.7)	8 (66.2)	568
4: Clerical	11 (22.8)	23 (34.7)	12 (30.8)	66 (8.5)	7 (12.1)	6 (13.1)	13 (16.1)	138
5: Sales	27 (40.4)	40 (61.4)	25 (54.4)	15 (15.0)	96 (21.3)	18 (23.1)	23 (28.5)	244
6: Managerial	10 (41.5)	24 (63.2)	19 (55.9)	12 (15.4)	23 (21.9)	146 (23.7)	17 (29.3)	251
7: Professional	6 (50.2)	25 (76.3)	21 (67.5)	9 (18.6)	20 (26.5)	15 (28.7)	207 (35.3)	303
Total	413	628	556	153	218	236	291	2,495

(1958 occupancy)

examination of the estimates of the expected values under this model in Table 9.9 shows just how poor the fit of the model is.

As the major failing of the model of independence concerns its serious underprediction of the frequencies along the principal diagonal of Table 9.9, it is useful to treat these diagonal cells in a different fashion to the rest of the table. One way to do this is to exclude the diagonal cells and to test for independence in the remaining portion of the table. In this case the usual hypothesis of independence becomes one of quasi-independence [9.9]. Fitting the model of quasi-independence [9.9] results in a G^2 goodness-of-fit statistic of 153.75 with twenty-nine degrees of freedom calculated according to expression [9.17] as

$$\text{d.f.} = 49-7-13 = 29.$$

Although the model of quasi-independence represents a marked and significant improvement over the previous model of independence, its G^2 value lies within the conventional upper 5 per cent tail of the χ^2 distribution with twenty-nine degrees of freedom. Thus the quasi-independence model is not an adequate representation of the data in Table 9.9. This implies that even if we consider only those homes which had undergone a change in occupant status between 1958 and 1963 their occupancy classifications are not independent (i.e. there is significant association between the occupant status of homes in 1958 and 1963 even when we exclude from attention those homes which had remained stable in occupant status terms). The nature of the failure of the model can be seen by comparing the observed frequencies shown in Table 9.9 with the estimates of the expected frequencies under the model of quasi-independence shown in Table 9.10. It can be seen that there is a greater amount of transition of homes from certain occupant status categories to others (e.g. from service in 1958 to unskilled in 1963) than the model of quasi-independence predicts. Similarly, there is a smaller amount of transition between other pairs of occupant status categories than the model predicts (e.g. from professional in 1958 to unskilled in 1963).

Table 9.10
Estimates of expected cell frequencies under model of quasi-independence

		1963 occupancy						
	$j =$	1	2	3	4	5	6	7
$i =$ 1: Unskilled		34.6	51.1	35.8	11.9	17.9	12.6	11.7
2: Service		86.7	128.2	89.8	29.8	44.8	31.7	29.3
3: Craft		57.1	84.5	59.2	19.6	29.5	20.9	19.3
4: Clerical		15.2	22.5	15.8	5.2	7.9	5.6	5.1
5: Sales		32.4	48.0	33.6	11.1	16.8	11.9	11.0
6: Managerial		22.3	33.0	23.1	7.7	11.5	8.1	7.5
7: Professional		20.3	30.0	21.0	7.0	10.5	7.4	6.9

(1958 occupancy)

For a further geographically oriented example of the fitting of a quasi-independence model, the reader should consult Haberman (1979, 444–55) who investigates a regional migration table.

<u>EXAMPLE 9.5</u>   *Plant type, soil type and slope aspect*

As we saw in Example 9.3, sampling zeros are regularly encountered in the contingency tables reported in the environmental science research literature. However, tables which include structural zeros are rarely reported. For this reason, we must use in this example a hypothetical environmental science oriented table to illustrate certain aspects of the analysis of structural zeros in contingency tables.

Table 9.11 is a hypothetical three-dimensional table in which the numbers of four different plant types have been recorded in a small random sample of quadrats of varying soil type and slope aspect. For the purposes of this example, let us assume that it is known *a priori* that plant type 2 cannot survive in localities where soil type 1 is present and that, as a result, the zeros in Table 9.11 must be treated as structural zeros.

Table 9.11
A hypothetical table showing numbers of certain plant types recorded in a small random sample of quadrats of varying soil type and slope aspect

	Variable C			
	Soil type 1		*Soil type 2*	
	Variable B		Variable B	
	South-facing slopes	*North-facing slopes*	*South-facing slopes*	*North-facing slopes*
Plant type 1	4	2	9	7
Plant type 2	0	0	4	8
Plant type 3	42	7	19	10
Plant type 4	57	20	71	31

(Variable A)

On the basis of fitting a hierarchical set of quasi log-linear models to Table 9.11, and an assessment of the G^2 statistics of these models, the most 'acceptable' model is found to take the form

$$\log_e m_{ijk} = \lambda + \lambda_i^A + \lambda_j^B + \lambda_k^C + \lambda_{ij}^{AB} + \lambda_{ik}^{AC} \qquad \textbf{[9.25]}$$

for all *ijk* cells which are not structural zeros and to have a G^2 statistic of 4.86.

The question which must now be answered is, What are the appropriate degrees of freedom for this quasi log-linear model? To answer this we use the expression [9.17]. We know that there are sixteen cells in the table, two structural zeros, but how many parameters require estimation? To answer this, it is useful to set out these parameters explicitly and to recall the constraints [9.19] and [9.20] imposed on the parameters.

Overall mean	λ	
Main effect, plant type	$\lambda_1^A,\ \lambda_2^A,\ \lambda_3^A$	
Main effect, aspect	λ_1^B	**[9.26]**
Main effect, soil type	λ_1^C	
Interaction effect, plant type × aspect	$\lambda_{11}^{AB},\ \lambda_{21}^{AB},\ \lambda_{31}^{AB}$	
Interaction effect, plant type × soil type	$\lambda_{11}^{AC},\ -,\ \lambda_{31}^{AC}$	

It can be seen that the interaction effect λ_{21}^{AC} is missing since it is known *a priori* to be zero because of the zero which appears in the plant type by soil type two-dimensional marginal table (confirm this by collapsing Table 9.11 over variable B). The number of parameters which require estimating is therefore eleven, and the degrees of freedom for model [9.25] are given by the expression

$$d{\cdot}f{\cdot} = 16-2-11 = 3.$$

In this particular case, model [9.25] is a quasi-conditional - independence model. It implies that for a given plant type (other than plant type 2) there is no interactive relationship between the soil type and slope aspect of the localities in which that plant type is to be found. A useful exercise for the reader would be to reformulate this analysis as a cell (f) type problem in which variable A (plant type) is regarded as the response variable and variables B and C (soil type and slope aspect) as explanatory variables.

9.4   Outliers or rogue cells

In Section 5.8 we saw that the analysis of residuals provides a useful aid in the process of selecting 'acceptable' log-linear models. Residuals of the types [5.91] to [5.95] are useful in indicating systematic patterns in the data which might otherwise be overlooked, and they also enable us to pick out the isolated large anomalies which we term 'outlier' or 'rogue cells'.

Having identified such 'outliers' on the basis of large residual values, the question which must then be answered is whether these cells represent statistically significant anomalies. One way we can attempt to determine this is by exploiting the quasi log-linear model and incomplete table methodology of the previous section. To do this, we fit the same log-linear model as used to compute the residuals for the complete table to the incomplete table formed by excluding the 'outlier' cells. We then compute the difference between the goodness-of-fit G^2 statistics of the incomplete table model and the full table model, and this difference is assessed for statistical significance in the light of the difference in degrees of freedom between the complete and incomplete table models. If the 'outlier' cells are found to be significantly different using this procedure, we might then proceed to model the non-outlier cells separately from the outlier cells, using some appropriate quasi log-linear model to fit the non-outlier cells.

EXAMPLE 9.6   *Opinions about a television series in urban and rural areas (continued)*

In Example 5.6 we considered Lee's (1978) pedagogic illustration of the potential value of residual analysis in log-linear modelling. The example concerned the analysis of a three-dimensional contingency table (Table 5.33) in which sixty six randomly selected adults are cross-classified according to their sex (variable A), their opinion about the content of a television series (variable B – divided into two categories 'approved'

294 Special topics in log-linear modelling

or 'unapproved') and their location (variable C – divided into two categories, 'rural' and 'urban'). In Example 5.6 we considered Lee's treatment of the table as a cell (g) type log-linear modelling problem. However, we noted that it may be more appropriate to regard variable B (opinion about the television series) as a response variable and variables A and C as explanatory variables. Both Mantel (1979) and Upton (1980) have adopted a similar argument and have provided re-analyses of Lee's table as a cell (f) type problem. In this example, we will consider some aspects of Upton's reappraisal of Table 5.33.

In line with the principles which we established in Chapter 6, Upton treats the structural relationships between the explanatory variables as fixed or given 'facts of life' and he considers only those models in which the interaction between the explanatory variables (A and C) and the lower-order relatives of this interaction are included. The simplest model of this type takes the form

$$\log_e m_{ijk} = \lambda + \lambda_i^A + \lambda_k^C + \lambda_{ik}^{AC} + \lambda_j^B \qquad [9.27]$$

However, the fit of this model to Table 5.33 is poor. It has a G^2 value of 25.90 with three degrees of freedom. Despite this, the standardized residuals from this model are suggestive. They indicate that urban males in Table 5.33 have a very different pattern of approval than the other groups. Indeed, it appears that the urban male cell frequency of seventeen in Table 5.33 may well be a significant outlier.

To check this possibility, Upton excludes the outlier cell, and fits a quasi log-linear version of model [9.27] to the incomplete table which remains after the exclusion of the outlier cell. The fit of this quasi log-linear model to the incomplete table is very good. It has a G^2 value of 0.32 with two degrees of freedom (the degrees of freedom having been reduced by one because of the excluded cell). The change in G^2 associated with excluding the outlier cell is therefore ($25.90 - 0.32 = 25.58$) and the difference in degrees of freedom between the complete and incomplete table models ($3 - 2 = 1$). There can be little doubt, therefore, that the excluded cell represents a significant anomaly, differing significantly from the pattern in the rest of the table. An appropriate treatment of Table 5.33 is, therefore, to treat the outlier cell as a separate entity and to model the non-outlier cells using [9.27]. It will be recognized that in these circumstances model [9.27] is a quasi-independence model. As a result, we can say that Table

Table 9.12
Hypothetical panel survey data on main mode of travel to work

1982 main mode of travel	1983 main mode of travel					
	Car	Bus	Train	Cycle	Walk	Total
Car	563	149	62	13	15	802
Bus	96	401	75	10	16	598
Train	73	82	350	4	9	518
Cycle	4	10	8	75	6	103
Walk	6	11	3	14	109	143
Total	742	653	498	116	155	2,164

5.33 shows that, in general, opinion about the television series is independent of sex or location with the single exception of males living in urban areas. Elsewhere, about thirty per cent of respondents are in favour of the television series, but more than 90 per cent of males living in urban areas are in favour of the series.

For another example of the assessment of outlier cells in a geographically oriented contingency table the reader should consult Upton (1981).

9.5 Square tables, symmetry and marginal homogeneity

Two-dimensional contingency tables in which both the row and column variables have the same number of categories and same category definitions occur fairly frequently in geography and environmental science. We have already encountered one example of such a square table in Table 9.9 and another simple example would be that shown in Table 9.12 in which a panel survey of journey to work behaviour in a city allows us to determine, at two points in time, the main mode of travel to work of the same sample of individuals. In this type of table (a regional migration table would have the same form), cells along the principal diagonal represent stability or lack of change, and the off-diagonal cells represent change of some form.

In this type of table, it is natural for us to direct our attention to an examination of the shifts which are revealed in the table. In particular, we might question whether the marginal frequencies of the table show evidence of change (in the case of Table 9.12, for example, we might question whether there has been any significant change in the overall proportions of the panel choosing each particular travel mode in the two years, perhaps as a result of known transport policy changes). Alternatively, we might direct our attention to the individual off-diagonal cells, and question to what extent these cells values are symmetric, i.e. to what extent is there a perfect correspondence of shifts around the principal diagonal. To answer these questions we must test the hypotheses of *symmetry*, *marginal homogeneity* and *quasi-symmetry*.

9.5.1 Symmetry

The hypothesis of symmetry in a two-dimensional square table can be stated in terms of probabilities as:

$$H_0 : P_{ij} = P_{ji} \quad \text{for all } i \neq j \qquad\qquad [9.28]$$

or equivalently, in terms of expected cell frequencies as:

$$H_0 : m_{ij} = m_{ji} \quad \text{for all } i \neq j \qquad\qquad [9.29]$$

In other words, it is a hypothesis which states that cells situated symmetrically around the principal diagonal of a square contingency table have the same probability of occurrence or the same expected frequencies. In terms of Table 9.12, this hypothesis implies that within particular pairs of travel modes there is perfect correspondence in the upward and downward shifts in usage levels between 1982 and 1983.

As in the case of the contingency table hypotheses which we have considered in previous sections (e.g. independence, conditional independence, conditional equiprobability, quasi independence, etc.), we can represent the hypothesis of symmetry in the form of a log-linear model (see Bishop, Fienberg and Holland 1975, 282). This model can be expressed as

$$\log_e m_{ij} = \lambda + \lambda_i^A + \lambda_j^A + \lambda_{ij}^{AB} \qquad \text{[9.30]}$$

where

$$\lambda_{ij}^{AB} = \lambda_{ji}^{AB} \qquad \text{[9.31]}$$

and where assuming a 'centred effect' coding system

$$\sum_{i=1}^{I} \lambda_i^A = \sum_{i=1}^{I} \lambda_{ij}^{AB} = 0 \qquad \text{[9.32]}$$

Assuming a multinomial sampling model, it is then a straightforward task to demonstrate that the MLEs of the expected cell frequencies take the simple form

$$\hat{m}_{ij} = (n_{ij} + n_{ji})/2 \quad i \neq j \qquad \text{[9.33]}$$
$$\qquad = n_{ij} \qquad\quad i = j$$

Substituting these expressions for $\hat{m}_{ij}$ into the usual definition of the G^2 goodness-of-fit statistic, it then follows that the G^2 statistic used to test the model of symmetry takes the form

$$G_S^2 = 2 \sum_{i=1}^{I} \sum_{j=1}^{I} n_{ij} \log_e \frac{2n_{ij}}{n_{ij} + n_{ji}} \qquad \text{[9.34]}$$

This statistic has an asymptotic χ^2 distribution with $I(I-1)/2$ degrees of freedom under the null hypothesis of symmetry.

9.5.2   Marginal homogeneity

The hypothesis of symmetry (i.e. perfect correspondence of shifts around the principal diagonal of a square contingency table) is a rather strong one. In many circumstances the less demanding hypothesis of marginal homogeneity (i.e. correspondence of the marginal frequencies; or in terms of Table 9.12, no change in the overall proportions of panel members choosing each particular travel mode between 1982 and 1983) will prove sufficiently discriminating for many of our purposes.

In terms of probabilities and expected cell frequencies, the hypothesis of marginal homogeneity can be stated as:

$$H_0 : P_{i+} = P_{+i} \quad \text{or} \quad H_0 : m_{i+} = m_{+i} \quad i = 1, \dots I \qquad \text{[9.35]}$$

In other words, in a 2×2 table, marginal homogeneity is equivalent to symmetry. However, in larger two-dimensional tables this is no longer the case and complete symmetry is a more restrictive hypothesis than marginal homogeneity. In such tables, if the cell frequencies are symmetric then the marginal frequencies will be homogeneous. However, the reverse is not necessarily true.

Although marginal homogeneity is a very useful hypothesis which we will often wish to examine, a major difficulty arises from the fact that it is not easy to test. Unlike the hypothesis of symmetry we cannot represent it in the form of a log-linear model (for further details see Bishop, Fienberg and Holland 1975, 282). To circumvent this problem we, therefore, examine marginal homogeneity via a different route,[†] and we make use of the fact that

$$\text{Symmetry} = \text{marginal homogeneity plus quasi-symmetry} \qquad \textbf{[9.36]}$$

Since the addition of quasi-symmetry to marginal homogeneity results in complete symmetry, and since G^2 statistics are additive, it follows that if we can calculate the G^2 statistic for a model of quasi-symmetry (a statistic we term G_Q^2) we can then exploit the relationship expressed in [9.36] and develop a test statistic for marginal homogeneity conditional upon the model of quasi-symmetry being approximately true. This test statistic has the form

$$G_{MH|Q}^2 = G_S^2 - G_Q^2 \qquad \textbf{[9.37]}$$

and has

$$\frac{I(I-1)}{2} - \frac{(I-1)(I-2)}{2} = I-1 \qquad \textbf{[9.38]}$$

degrees of freedom.

9.5.3 Quasi-symmetry

The conditional test for marginal homogeneity [9.37] demands, therefore, that we can fit a model of quasi-symmetry (i.e. a model which implies that there is symmetry in the square contingency table after allowing for disparity in the marginal frequencies). Such a model can be expressed in log-linear form as

$$\log_e m_{ij} = \lambda + \lambda_i^A + \lambda_j^B + \lambda_{ij}^{AB} \qquad \textbf{[9.39]}$$

where

$$\lambda_{ij}^{AB} = \lambda_{ji}^{AB} \qquad \textbf{[9.40]}$$

and where (assuming a 'centred effect' coding system)

$$\sum_{i=1}^{I} \lambda_i^A = \sum_{j=1}^{J} \lambda_j^B = \sum_{i=1}^{I} \lambda_{ij}^{AB} = 0 \qquad \textbf{[9.41]}$$

[†] A direct approach to fitting the model of marginal homogeneity is possible, but this involves methods beyond the scope of this chapter. The interested reader should consult Bishop, Fienberg and Holland (1975, 294); Ireland, Ku and Kullback (1969); Fleiss and Everitt (1971); Everitt (1977, 116).

Assuming a multinomial sampling model, it can be demonstrated (see Bishop, Fienberg and Holland 1975, 289) that the MLEs for the expected cell frequencies under the model of quasi-symmetry must satisfy the constraints

$$\hat{m}_{i+} = n_{i+} \qquad i = 1, \dots . I \qquad\qquad \text{[9.42]}$$

$$\hat{m}_{+j} = n_{+j} \qquad j = 1, \dots J \text{ (where } J = I) \qquad \text{[9.43]}$$

$$\hat{m}_{ij} + \hat{m}_{ji} = n_{ij} + n_{ji} \qquad i \neq j \qquad\qquad \text{[9.44]}$$

Fitting these constraints is not a completely straightforward process. However, Bishop, Fienberg and Holland (1975, 289) suggest a procedure which has the advantage of allowing us to use the standard computer programs discussed in Chapter 7. This procedure has the following stages.

1. Transform the basic two-dimensional table into a three-dimensional table by writing out the original table twice. The second time the original table is written out, it should be transposed (i.e. the cells on each side of the principal diagonal should be interchanged). In the case of Table 9.12, this implies two tables of the form

563	149	62	13	15		563	96	73	4	6
96	401	75	10	16		149	401	82	10	11
73	82	350	4	9		62	75	350	8	3
4	10	8	75	6		13	10	4	75	14
6	11	3	14	109		15	16	9	6	109

This transformation yields a three-dimensional table of observed values and an associated three-dimensional table of expected values, i.e.

$$\begin{aligned} n_{ijk} = n_{ij} \quad k = 1 \qquad m_{ijk} = m_{ij} \quad k = 1 \\ = n_{ji} \quad k = 2 \qquad = m_{ji} \quad k = 2 \end{aligned} \qquad \text{[9.45]}$$

2. In the context of this newly created three-dimensional table, equations [9.42] and [9.43] imply

$$\hat{m}_{i+k} = n_{i+k} \qquad\qquad\qquad\qquad \text{[9.46]}$$

$$\hat{m}_{+jk} = n_{+jk} \qquad\qquad\qquad\qquad \text{[9.47]}$$

and equation [9.44] implies

$$\hat{m}_{ij+} = n_{ij+} \qquad\qquad\qquad\qquad \text{[9.48]}$$

These expressions [9.46] to [9.48] are recognizable as the maximum likelihood equations for the 'no three-variable interaction' (or pairwise association) log-linear model for a three-dimensional table. It follows, therefore, that the second stage of the procedure is simply to fit this model

$$\log_e m_{ijk} = \lambda + \lambda_i^A + \lambda_j^B + \lambda_k^C + \lambda_{ij}^{AB} + \lambda_{ik}^{AC} + \lambda_{jk}^{BC} \qquad \text{[9.49]}$$

to the newly created three-dimensional table.

3.   The estimates of the expected cell frequencies derived under model [9.49] will be identical in both parts of the newly created three-dimensional table (subject to a simple reversal in the second part) and they provide the required cell estimates under the model of quasi-symmetry. Taking just those estimates for the first part of the three-dimensional table we therefore have the required estimates of the expected cell frequencies under the quasi-symmetry model for the original two-dimensional table.

4. To derive the required goodness-of-fit statistic for the quasi-symmetry model (G_Q^2) we simply halve the G^2 statistic for the model [9.49] fitted to the newly created three-dimensional table, and G_Q^2 can be shown to have $(I-1)(I-2)/2$ degrees of freedom.

If the model of quasi-symmetry provides an adequate fit to the contingency table data, we can then proceed to use statistic [9.37] to test for marginal homogeneity conditional on quasi-symmetry being approximately true. Example 9.7 provides an illustration of the use of such a test.

9.5.4   *Symmetry and marginal homogeneity in multidimensional tables*

So far, we have concentrated upon models of symmetry and marginal homogeneity for two-dimensional tables. When we move from the two-dimensional case to the multidimensional case, the analysis becomes more complex as there are several possible ways to generalize the models of symmetry, quasi-symmetry and marginal homogeneity.[†] To illustrate this we can consider the case of the simplest multi-dimensional table, a three-dimensional table.

The simplest hypotheses of symmetry and marginal homogeneity for a three-dimensional $(I \times I \times K)$ table are what are termed hypotheses of *conditional symmetry*.

$$H_0 : m_{ijk} = m_{jik} \quad i, j = 1, \ldots I; \ k = 1, \ldots K \qquad \text{[9.50]}$$

and *conditional marginal homogeneity*

$$H_0 : m_{i+k} = m_{+ik} \quad i, j, = 1, \ldots I; \ k = 1, \ldots K \qquad \text{[9.51]}$$

To test these hypotheses, in which we examine the table conditional on the level of variable C, we simply use the models of Sections 9.5.1 to 9.5.3 directly on each of the K layers of the table. The model of conditional symmetry has $KI(I-1)/2$ degrees of freedom and the overall goodness-of-fit statistic can be partitioned into K components of the form [9.34] which can be computed separately for each table, each with $I(I-1)/2$ degrees of freedom. Similarly, we can get a conditional test statistic for marginal homogeneity with $I-1$ degrees of freedom for each of the K layers of the table.

Such conditional models, however, are not the only models we can specify for three-dimensional tables. Instead of the hypothesis of

---

[†] Further useful discussion of the issues covered in Sections 9.5.1 to 9.5.4 is to be found in Bhapkar (1979) and Mantel and Byar (1978).

conditional symmetry we can postulate instead a hypothesis which is termed *complete symmetry*. In the case of an $I \times I \times I$ table (where $K = I$) this hypothesis takes the form

$$H_0 : m_{ijk} = m_{ikj} = m_{jik} = m_{jki} = m_{kji} = m_{kij} \qquad \text{[9.52]}$$

for all i, j and k and the equivalent model (which can be represented in log-linear form) has $[I\,(I-1)\,(5I+2)]/6$ degrees of freedom. In its turn, complete symmetry implies two forms of marginal homogeneity; the first based on one-dimensional marginal totals

$$m_{i++} = m_{+i+} - m_{++i} \qquad i = 1, \ldots I \qquad \text{[9.53]}$$

and the second based on two-dimensional marginal totals

$$m_{ij+} = m_{ji+} = m_{i+j} = m_{+ij} \qquad i, j = 1, \ldots I \qquad \text{[9.54]}$$

By analogy with the two-dimensional procedures in Sections 9.5.2 and 9.5.3 we can then specify two models of quasi-symmetry; the first preserving one-dimensional marginal totals and the second preserving two-dimensional marginal totals. Having fitted these models we can then proceed to construct conditional tests for the homogeneity of one-dimensional marginal tables and the homogeneity of two-dimensional marginal totals. Further details of these issues and their extension to higher-dimensional tables is to be found in Bishop, Fienberg and Holland (1975, 229–309); Kullback (1971).

9.5.5 *Alternative log-linear models for square tables*

Although the models and tests of symmetry, quasi-symmetry and marginal homogeneity provide extremely useful insight into the structure of two-dimensional square tables such as Table 9.12, other log-linear models for square tables can also be specified. These alternative models are particularly appropriate when the categories of the variables are ordered in some way (e.g. Table 9.13 represents a two-period panel survey of residential mobility between social areas of a city, and the social areas are ordered according to some measure of social status). Some particularly interesting models for tables of this type, which attempt to take account of what can be termed 'distance'

Table 9.13
Social area residential mobility table

		1983 residence				
	Social area	1	2	.	. .	*I*
Social area	1	983	83	.	. .	5
1982 residence	2	96	894	.	. .	6
	.	.	.			.
	.	.	.			.
	.	.	.			.
	I	10	9	.	. .	906

and 'loyalty' effects, have been proposed by Goodman (1972c) and Haberman (1974a, 215–27) and further developed is the work of Upton (1977) and Upton and Sarlvik (1981).[†]

To understand the structure of these models it is useful to consider Table 9.13. In this table, eighty-three people are shown as having moved from the first ranking social area in 1982 to the second ranking social area in 1983. Similarly, ninety-six people are shown as having moved in the reverse direction. both these cells lie on the same diagonal of the table, and to model these cells (i.e. these shifts from one social area to another) we can consider building into a basic main-effects log-linear model a type of 'diagonal' effect which takes account of the 'status distance' between the two areas. Such an effect can be introduced in a number of ways to produce what Haberman (1979, 501) terms either a 'fixed-distance' model or a 'variable-distance' model. In the latter case, the log-linear model takes form

$$\log_e m_{ij} = \begin{cases} \lambda + \lambda_i^A + \lambda_j^B + \sum_{k=i}^{j-1} \lambda_k^S & i < j \\ \lambda + \lambda_i^A + \lambda_j^B + \sum_{k=j}^{i-1} \lambda_k^S & i > j \\ \lambda + \lambda_i^A + \lambda_j^B & i = j \end{cases}$$ [9.55]

The parameters λ_k^S are the 'status distance' between areas (or 'crossings') parameters, and there are a total of $I - 1$ such parameters.

To the model [9.55] we can also consider adding what is termed a 'loyalty' (or stability) parameter. This takes account of the large cell frequencies typically found along the principal diagonal of such tables, and which in the case of Table 9.13 represent no change in the social area between the two years. A dummy variable can be used to represent such loyalty. A cell of the table can be said to belong to the first category of the dummy variable if the 1982 and 1983 social areas are identical, or the second category of the dummy variable if they are not. Details of the estimation of these models are provided by Goodman, Upton and also Haberman (1979, 500).

EXAMPLE 9.7   *Filtering in the housing market of Kingston, Ontario (continued)*

In Example 9.4 we began an assessment of filtering in the housing market of Kingston, Ontario. We will now re-analyse Table 9.9 using the hypotheses of symmetry, quasi-symmetry and marginal homogeneity.

In the case of Table 9.9, the hypothesis of symmetry can be interpreted as stating that within each pair of occupant categories upward and downward filtering is equally likely to have occurred, e.g. it is as likely for a home occupied by an unskilled worker in 1958 to be occupied by a clerical worker in 1963 as it is for the reverse to occur. Table 9.14 gives the estimates of the expected cell frequencies under the

---

[†] There is a considerable related literature in the sociology journals concerned with the analysis of occupational mobility tables (see Duncan 1979; Goodman 1979a; Hauser 1978). Geographers concerned with residential mobility or regional migration should consult those papers to supplement the basic information on the modelling of square tables provided in this chapter.

Table 9.14
Estimates of expected cell frequencies under model of symmetry

		1963 occupancy						
		$j = 1$	2	3	4	5	6	7
1958 occupancy	$i = 1$: Unskilled	179.0	89.0	51.5	9.0	22.5	10.0	5.5
	2: Service	89.0	359.0	99.0	27.0	36.0	18.0	21.5
	3: Craft	51.5	99.0	337.0	12.5	23.5	24.0	14.5
	4: Clerical	9.0	27.0	12.5	66.0	11.0	9.0	11.0
	5: Sales	22.5	36.0	23.5	11.0	96.0	20.5	21.5
	6: Managerial	10.0	18.0	24.0	9.0	20.5	146.0	16.0
	7: Professional	5.5	21.5	14.5	11.0	21.5	16.0	207.0

model of symmetry, computed using equation [9.33]. Computing the goodness-of-fit statistic G_S^2 from equation [9.34], it can be seen, however, that the model of symmetry does not fit the data of Table 9.9 well. The G_S^2 value is 60.52 with twenty-one degrees of freedom. This is within the upper 5 per cent tail of the χ^2 distribution and thus we can reject the hypothesis of symmetry.

A glance at the marginal totals of Table 9.9 shows that the proportions of homes occupied by the various occupational groups differ in the two years. Some of the differences are small (e.g. 0.117 of homes were occupied by professionals in 1963 compared to 0.121 in 1958), but some are relatively large (e.g. 0.166 of homes were occupied by unskilled workers in 1963 compared to 0.128 in 1958). There are many possible reasons for these differences, e.g. changes in the labour market in Kingston between the two years, differential errors in classifying occupants in the two years, outward migration from the city of certain occupational groups, etc. However, whatever the reasons, we must question to what extent the differences in marginal totals between the two years are statistically significant (i.e. we must assess the hypothesis of marginal homogeneity). We must also ask whether Table 9.9 can be said to be symmetric after allowing for disparity in the marginal frequencies (i.e. we must assess the hypothesis of quasi-symmetry).

As we saw in Sections 9.5.2 and 9.5.3, it is convenient to test for marginal homogeneity conditional upon the model of quasi-symmetry being approximately true. We first, therefore, fit the model of quasi-symmetry using the Bishop, Fienberg and Holland procedure discussed in Section 9.5.3, and Table 9.15 gives the estimates of the

Table 9.15
Estimates of expected cell frequencies under the model of quasi-symmetry

		1963 occupancy						
		$j = 1$	2	3	4	5	6	7
1958 occupancy	$i = 1$: Unskilled	179.0	110.8	63.3	10.3	29.4	13.0	7.2
	2: Service	69.2	359.0	97.2	24.3	38.4	19.1	22.9
	3: Craft	39.7	100.8	337.0	11.5	25.4	25.9	15.7
	4: Clerical	7.7	29.7	13.5	66.0	12.8	10.5	12.8
	5: Sales	15.6	33.7	21.6	9.2	96.0	20.5	21.5
	6: Managerial	7.0	16.9	22.1	7.5	20.5	146.0	16.0
	7: Professional	3.8	20.1	13.3	9.2	21.5	16.0	207.0

expected cell frequencies under this model. The goodness-of-fit statistic for the model (G^2_Q) is 20.57 with fifteen degrees of freedom. Thus the model of quasi-symmetry provides a satisfactory fit to the data of Table 9.9. It provides a significant improvement over the symmetry model and implies that there is a type of symmetry in Table 9.9 which can be discerned after taking into account the disparity between the marginal distributions of occupant categories for 1958 and 1963 (see Haberman 1979, 495 for further discussion of the meaning of quasi-symmetry).

As the quasi-symmetry model provides an adequate fit to the data of Table 9.9, it is reasonable for us to test for marginal homogeneity conditional on quasi-symmetry being true. To do this we simply substitute into statistic [9.37] which in this case gives the value

$$G^2_{MH|Q} = 60.52 - 20.57 = 39.95 \qquad [9.56]$$

with six degrees of freedom. This is well within the upper 5 per cent tail of the χ^2 distribution with six degrees of freedom. Thus we reject the hypothesis of marginal homogeneity and conclude that the marginal distribution of occupant categories (i.e. the proportions of homes occupied by the various occupational groups) in 1963 in Kingston is significantly different from the marginal distribution in 1958.

9.6 Some remaining issues

9.6.1 *The multiplicative form of the log-linear model*

In Chapters 5 and 6 and the previous sections of this chapter, all our log-linear models have been written in additive form, i.e. the natural logarithm of the expected cell frequency has been decomposed into a set of additive components. Although this is by far the most popular way of writing log-linear models, the reader should be aware that he will sometimes encounter (particularly in the work of Goodman) an equivalent multiplicative specification.

In the usual additive form, the saturated log-linear model for a two-dimensional table is written

$$\log_e m_{ij} = \lambda + \lambda_i^A + \lambda_j^B + \lambda_{ij}^{AB} \qquad [9.57]$$

In the equivalent multiplicative form it is written

$$m_{ij} = \gamma \, \gamma_i^A \, \gamma_j^B \, \gamma_{ij}^{AB} \qquad [9.58]$$

Taking natural logarithms of [9.58] and comparing with the equivalent additive model [9.57] we have

$$
\begin{aligned}
\lambda &= \log_e \gamma & \gamma &= e^\lambda \\
\lambda_i^A &= \log_e \gamma_i^A & \gamma_i^A &= e^{\lambda_i^A} \\
\lambda_j^B &= \log_e \gamma_j^B & \gamma_j^B &= e^{\lambda_j^B} \\
\lambda_{ij}^{AB} &= \log_e \gamma_{ij}^{AB} & \gamma_{ij}^{AB} &= e^{\lambda_{ij}^{AB}}
\end{aligned}
\qquad [9.59]
$$

In other words, the λ parameters are simply the natural logarithms of the equivalent γ parameters. Assuming a 'centred effect' coding system, the usual constraints [5.15] and [5.19] which specify that the λ parameters sum to zero are equivalent to stating that the product of the corresponding γ terms is unity. That is to say,

$$\sum_{i=1}^{I} \lambda_i^A = \sum_{j=1}^{J} \lambda_j^B = \sum_{i=1}^{I} \lambda_{ij}^{AB} = \sum_{j=1}^{J} \lambda_{ij}^{AB} = 0 \qquad [9.60]$$

and

$$\prod_{i=1}^{I} \gamma_i^A = \prod_{j=1}^{J} \gamma_j^B = \prod_{i=1}^{I} \gamma_{ij}^{AB} = \prod_{j=1}^{J} \gamma_{ij}^{AB} = 1 \qquad [9.61]$$

are equivalent sets of constraints. When there are only two categories for a variable (e.g. variable A) this implies that $\gamma_1^A \gamma_2^A = 1$ and thus $\gamma_1^A = 1/\gamma_2^A$. This is equivalent to the relationship $\lambda_1^A = -\lambda_2^A$ which we have often used in our previous discussion.

Although the multiplicative specification of log-linear models is much less frequently employed in practice, some authors argue that it permits a somewhat easier interpretation of the parameters of the model. This is a matter of taste, and the issue is not discussed here to avoid unnecessarily confusing the discussion of parameter interpretation provided in Chapter 5 (further information is provided by Goodman 1972a; Payne 1977; Upton 1978a, 91, and both λ and γ parameters are automatically provided by standard log-linear modelling computer programs such as BMDP). However, the reader should be prepared to encounter the multiplicative specification of log-linear models in his wider reading.

9.6.2   Log-linear models for tables with ordered categories

With the exception of the 'distance' models in Section 9.5.5, a characteristic of all the log-linear models we have discussed in Chapters 5, 6 and 9 is that they have made no assumptions about the ordering of categories for any of the variables. In many tables, however, it is likely that one or more of the variables will have categories which fall into a natural order, and it is often valuable to attempt to utilize this aspect of the data.

There have been many proposals for dealing with ordered categories in log-linear modelling (e.g. Haberman 1974b; Simon 1974; Clayton 1974; Bock 1975; Williams and Grizzle 1972). For simplicity, however, we will describe here only the most straightforward method which can be adopted if the categories of the ordered variables can be assumed to be equally spaced. In this approach (see Goodman 1971a; Everitt 1977, 100; Upton 1978a, 104 for illustrations) we apply standard *orthogonal polynomial* coefficients to obtain measures of the linear and non-linear trends associated with the ordered variables.

In the case of a three-dimensional $3 \times 2 \times 2$ table with an ordered variable (A) having three equally spaced categories (with category 1 the 'high' rank), the linear and nonlinear (quadratic) main effects associated with the ordered variable can be obtained from the usual λ-parameters as follows;

$$\text{Linear main effect: } \lambda_L^A = [(1)\lambda_1^A + (0)\lambda_2^A + (-1)\lambda_3^A]/2 \tag{9.62}$$
$$= [\lambda_1^A - \lambda_3^A]/2$$

$$\text{Quadratic main effect: } \lambda_Q^A = [(1)\lambda_1^A - (2)\lambda_2^A + (1)\lambda_3^A]/4 \tag{9.63}$$
$$= [\lambda_1^A + \lambda_3^A - 2\lambda_2^A]/4$$

Similarly, linear and nonlinear (quadratic) effects may be found for the interaction between the ordered variable (A) and the unordered variables (B and C). These take the form:

$$\lambda_{L1}^{AB}, \ \lambda_{L2}^{AB}, \ \lambda_{Q1}^{AB}, \ \lambda_{Q2}^{AB}$$
$$\lambda_{L1}^{AC}, \ \lambda_{L2}^{AC}, \ \lambda_{Q1}^{AC}, \ \lambda_{Q2}^{AC} \tag{9.64}$$
$$\lambda_{L11}^{ABC}, \ \lambda_{L12}^{ABC}, \ \lambda_{L21}^{ABC}, \ \lambda_{L22}^{ABC}, \ \lambda_{Q11}^{ABC}, \ \lambda_{Q12}^{ABC}, \ \lambda_{Q21}^{ABC}, \ \lambda_{Q22}^{ABC}$$

The size of these interaction parameters indicate the similarity or otherwise of the trend of the ordered variable (A) in the different categories of the unordered variables (B and C).

The computation of linear and quadratic effects of this type presents no additional problems for the researcher. They can be derived automatically using the standard computer programs for log-linear modelling (e.g. BMDP and ECTA) discussed in Chapter 7. For example, in the widely available BMDP program this can be achieved by using the ORTHogonal statement.

9.6.3   Causal analysis with log-linear models

In a number of social science disciplines (particularly sociology and political science) researchers are often interested in providing a 'causal' interpretation of the statistical relationships they uncover whilst modelling a set of data. To provide such interpretation, causal analysis procedures based initially on partial correlation analysis (Blalock 1964) and subsequently on path analysis (see Duncan 1966, 1975; Asher 1976; Macdonald 1977) have become widely used (despite the criticism which has been directed at the procedures, e.g. Bibby 1977). In geography and environmental science, however, despite some introductory accounts of these procedures (e.g. Johnston 1978, 93–98; Pringle 1980) they have found little favour and have been very rarely applied.

More recently, causal modelling has been extended to categorical data. Goodman (1973a; 1973b; 1979b) has demonstrated how log-linear models can be used to provide an analogue to path analysis for categorical variables. Like conventional path analysis, Goodman's analogue has two key features. First, the causal models make explicit the assumed order of priority of the categorical variables. Second, the relationships between the variables are described in terms of path diagrams.

To provide an informal illustration of Goodman's methods, let us reconsider Example 6.2. In this example, cell (f) type log-linear models were fitted to Fingleton's data on shopping behaviour in Manchester presented in Table 5.25. This table is a four-dimensional 2×2×2×2 table with variables consisting of: A, age of consumer; B, income of consumer; C, car ownership; D, shopping centre patronage. In Example 6.2, variable D (shopping centre patronage) was treated as a response variable and variables A, B and C as explanatory variables.

An alternative way to analyse Table 5.25 is as shown in Fig. 9.2(a). In other words, we can assume that age and income of the consumer are determinants of (i.e. are causal antecedents of) car ownership, and that car ownership then determines shopping centre patronage (i.e. car ownership acts as an intervening variables between age, income and shopping centre patronage). In Fig. 9.2, single-headed arrows point from the assumed cause to assumed effect, whereas double-headed arrows denote relationships which are assumed to be not necessarily causal in nature.

Fig. 9.2
(a) A possible path diagram for Table 5.25; (b) an alternative possible structure

(a)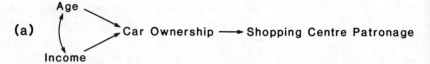

(b)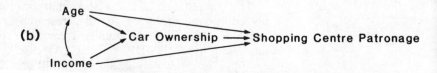

Analysis of Fig. 9.2(a) would begin at the left of the diagram and move progressively to the right. First, two-dimensional log-linear models would be used to analyse the two-dimensional marginal table of age by income, and the assumed relationship between age and income would be assessed by considering how good or poor was the fit of the log-linear model of independence. If the model of independence fitted well then the two variables should not be connected in Fig. 9.2 by the double-headed arrow. If only the saturated two-dimensional model fitted satisfactorily then the double-headed arrow would be required. Following this analysis, the links between the remaining variables would then be assessed using cell (f) type log-linear models or their equivalent logit model formulations. These models would assess: (a) car ownership given age and income, and then (b) shopping centre patronage given age, income and car ownership. Details of the estimation of the component models need not concern us (they are provided by Goodman 1973a; 1973b; 1979b; Fienberg 1977, 91–105; Knoke and Burke 1980, 42–47). All we really need to be aware of is that this procedure will lead us to confirm or modify (perhaps into the form of Fig. 9.2b) the causal structure postulated in Fig. 9.2(a). The resulting causal structure (model) which best represents the observed data will be the sum of the best fitting component models, and the G^2 statistic for testing the goodness-of-fit of this resulting model is, likewise, simply the sum of the G^2 statistics for the component models.

At this stage in the analysis, Goodman suggests (by analogy with conventional path analysis) the assignment of numerical values to the arrows in the resulting causal structure (path) diagram. These numerical values are the parameter estimates of the previously mentioned cell (f) type log-linear models or their logit model equivalents. However, the assignment of numerical values is a controversial issue and Fienberg (1977, 92) suggests a more cautious approach in which we view the assignment of numerical values as problematic, and limit ourselves to an indication of sign for causal relationships, in a fashion similar to that described by Blalock (1964).

Although Goodman has provided an extremely valuable analogy to conventional path analysis which can be applied to systems of categorical variables, the analogy to the conventional path analysis approach breaks down in a number of places (see Fienberg 1977, 91). In particular, the assignment of numerical values to the arrows in the path diagram is controversial because:

(a) there is no 'calculus' of path coefficients and thus no formal way to calculate the magnitude of effects along indirect paths between variables;

(b) multiple categories for a variable lead to multiple numerical values (estimated path coefficients) associated with a given arrow in the path diagram;

(c) the possible existence of three-variable or higher-order interaction terms in log-linear models may result in the need for more complex visual representation of the path diagram (see the suggestions of Knoke and Burke 1980, 47).

In addition, certain commentators (e.g. Birnbaum 1982) have expressed a wider sense of uneasiness with the whole approach. Nevertheless, of more importance in the context of this book may be the fact that geographers and environmental scientists have so far made little use of conventional path analysis, and may thus prove reluctant to use its categorical data analogue.

9.6.4 Log-linear models and spatially dependent data

Quantitative geographers and related research workers in the environmental sciences have given a great deal of attention over the past 15 years to the problems which derive from applying classical statistical models and inferential procedures, which assume independent observations, to geographical data which typically exhibit systematic ordering over space. During this period, the analysis of spatial data has developed from a stage of relatively uncritical application of standard inferential tests and statistical models, to a stage of wide appreciation of the effects of spatially dependent data. Tests for spatial autocorrelation have been developed, classical statistical models and tests have been modified to handle spatially dependent data, and extensive research on spatial process modelling and spatial-time series analysis has been conducted (see Cliff and Ord 1973, 1981; Hepple, 1974; Haggett *et al*. 1977; Bennett 1979; Haining 1980; Griffith 1980). However, as yet there has been little linkage between this research tradition and the new methods for categorical data analysis described in this book.

In the context of log-linear modelling what this implies is that considerable care must be taken when applying the models of Chapters 5, 6 and 9 to spatially dependent data. In these circumstances, the standard model selection procedures of log-linear modelling (see Section 5.7) may erroneously detect interaction effects between variables (or even main effects) which are, in fact, spurious and a consequence of the spatially dependent observations.

At the time of writing this chapter, it is encouraging to note that the first attempts to link the log-linear modelling and spatial dependence literatures have begun to appear. Fingleton (1983) has drawn attention to the growing literature on the question of categorical data and complex sampling schemes (e.g. Altham 1976, 1979; Cohen 1976; Fienberg 1979; Brier 1980; Fellegi 1980; Holt, Scott and Ewings 1980) and has noted that spatial sampling produces similar problems (i.e. positive dependence between observations from similar origins, which in the survey analysis context is said to produce a 'design effect' exceeding 1). As a result, he suggests exploiting some of the central features of this categorical data/complex sampling scheme literature to accommodate spatial dependence effects in log-linear modelling.

Using the Lansing Woods oak-hickory-maple data set reproduced in Table 9.1, and only very minimal assumptions about the nature of the spatial autocorrelation process, Fingleton is able to accommodate spatial dependence effects into the inferential process of selecting a suitable log-linear model. He does this by calculating a lower bound (X^2_L) for the 'true' Pearson chi-square goodness-of-fit statistic (X^2_t), and he uses this in place of the standard calculation of X^2 (see equation [5.2]), which in the presence of positive spatial dependence will assume an inflated value and will effectively become an upper bound (X^2_U) for the 'true' chi-square statistic $(X^2_t)^{\dagger}$.

Assuming that the observations in Table 9.1 are spatially independent, standard model selection techniques (in fact Aitkin's simultaneous test procedure) suggest that the best fitting model is

$$\log_e m_{ijk} = \lambda + \lambda_i^O + \lambda_j^H + \lambda_k^M + \lambda_{ij}^{OH} + \lambda_{jk}^{HM} \qquad [9.65]$$

(where O-oak, H-hickory, M-maple). That is to say, standard model selection procedures imply a conditional independence model in which there is an interaction (negative) between the oak and hickory distributions and the hickory and maple distributions but, conditional on the hickory distribution, the distribution of oaks is independent of the maple distribution. However, assuming a moderate amount of spatial dependence in the data (i.e. dependence over a limited spatial extent), and using the lower-bound (X^2_L) statistic in the inferential process, Fingleton finds that the independence model

$$\log_e m_{ijk} = \lambda + \lambda_i^O + \lambda_j^H + \lambda_k^M \qquad [9.66]$$

---

† This approach parallels that of Altham (1979) who, in the analogous situation of temporally dependent categorical observations, suggests the use of the inequalities $X^2_L \leq X^2_t \leq X^2_U$ as a highly practical method of handling the problems caused by the sequential dependence of observations. In the temporal dependence case, Altham argues that it is a reasonable simplifying assumption to suggest that observations r or more units apart in a time sequence are independent, whilst those less than r units apart are positively related. As a result, she demonstrates that the lower bound chi-square statistic (X^2_L) takes the form $X^2/(2r-1)$, in which the conventional chi-square statistic is adjusted by a deflation factor for the inflation caused by positive temporal dependence. Extending this procedure to the case of spatially dependent categorical observations, Fingleton assumes that on a square lattice a cut-off distance d exists beyond which observations are independent and within which they are positively related. He then demonstrates that the conventional chi-square statistic must be adjusted for the inflation caused by positive spatial dependence to give the lower bound chi-square statistic $X^2_L = X^2/[1+2d(d+1)]$.

and also the even more parsimonious model

$$\log_e m_{ijk} = \lambda + \lambda_i^O + \lambda_j^H \qquad\qquad [9.67]$$

become satisfactory representations of the data of Table 9.1. In other words, the *OH* and *HM* interactions which are found to be necessary when model selection is conducted assuming spatially independent observations are likely to be spurious products of the spatial dependence of the observations in Table 9.1.

Fingleton's approach can be extended in a number of ways. For example, he shows how knowledge of the precise nature of the spatial dependence in the data might ultimately be used to estimate variance-covariance matrices which include terms representing the structure of the spatial dependence. However, such extensions are likely to be difficult and to involve stringent assumptions, such as second-order stationarity. Despite this, the overall spirit and direction of Fingleton's approach (and its roots in the work of Altham and Holt *et al.*) is a promising one. If this initial research can be built upon, the next decade could see an immensely valuable integration of the spatial dependence, categorical data analysis, and sampling survey literatures in which quantitative geographers and environmental scientists can make a distinctive contribution.

PART 4

Discrete choice modelling

In this part of the book we link together developments in the statistical modelling of categorical data with associated developments in microeconomic theory, and show how the statistical models of the earlier parts of the book serve to operationalize an important research field which has come to be known as 'discrete choice modelling'. Many geographers have made important contributions to this area of research, and their students are likely to first encounter elements of the new approaches to categorical data analysis (logit models, probit models, etc.) via this research. Such readers who, in the first instance, are using the book to provide the statistical background to courses in discrete choice modelling, may wish to proceed to this part of the book directly from a reading of Chapters 2 and 3. In contrast, other readers may come to it following a much wider consideration of all the chapters in Parts 2 and 3, and others from a reading of just Part 2. Part 4 has been designed to be sufficiently self-contained to meet all these possible backgrounds.

CHAPTER 10

Statistical models for discrete choice analysis

For many social scientists, much of the utility of the statistical models for categorical data derives from their intimate connection to certain important developments in micro-economic theory. In fact, the linkage of developments in the statistical analysis of categorical data and developments in micro-economic consumer theory has given rise to a major new research field which has come to be known as *discrete choice modelling*. Research in this field is already of considerable importance in transportation science and economics (Hensher and Johnson 1981a; Manski and McFadden, 1981a) and it has major implications for many areas of geographical enquiry, particularly various aspects of urban and behavioural analysis (Wrigley and Longley 1984). In this chapter, an introduction to some of the basic concepts of discrete choice analysis and the statistical models which underpin this research field will be provided.

10.1 Random utility maximization, discrete choice theory and multinomial logit models

Many important decisions which an individual must take in his or her life involve selection from a limited and constrained set of discrete alternatives: e.g. choice of a house, neighbourhood, car, occupation, educational institution, marital status, number of children, the mode of travel on a work trip or shopping trip, the destination of a shopping or recreation trip, and so on. In these circumstances, conventional 'marginalist' micro-economic consumer theory with its assumption that individual demand is the result of utility maximization by a representative consumer whose decision variable is *continuous* (i.e. who is selecting fractional quantities from a continuously divisible choice continuum) is patently unrealistic. In these situations, conventional consumer theory needs modifying to allow the focussing of attention on choices at what economists term the *extensive* margin (i.e. discrete choices), rather than choices at the *intensive* margin which is treated in

nal analysis. Over the past ten years in the work of McFadden,
xiva, Manski, Hensher, Williams, Daly, Daganzo, Heckman,
n, Gaudry, Horowitz and many others, significant progress has
nade in this direction, and it has been shown that a logically
ent discrete choice theory can be developed based upon the
lesis of random utility maximization.

re are two interpretations of random utility maximization which
e adopted. These have been termed the *inter-personal* and
personal interpretations respectively. The first is characteristic of
of the discrete choice modelling literature in economics, trans-
tion science and geography (see McFadden 1981; Manski 1981;
ams 1981; Wrigley 1982; Wrigley and Longley 1984). In this
pretation, the distribution of demands in the population is
eived to be the result of individual preference maximization, but
preferences are viewed as being influenced, in part, by variables which
are unobserved by the analyst/modeller. Because certain choice-
relevant attributes are unobserved, and because the valuation of
observed attributes may vary from individual to individual, a random
element enters an individual's utility function and utility functions are
assumed to vary over the population of decision makers. The second
interpretation (the intra-personal) is that found in the literature of
psychology (see Luce and Suppes 1965). This assumes that each
individual draws a utility function from a random distribution each time
a decision must be made. That is to say, the individual is a classical
utility maximizer given his/her state of mind, but his/her state of mind
varies randomly from one choice situation to the next. The two
interpretations are formally indistinguishable in their effect on the
observed distribution of demand, and the random element in each
implies that discrete choice problems must be handled using some form
of probabilistic model.

If we adopt the first (inter-personal) interpretation of random utility
maximization, we can derive a probabilistic discrete choice model in the
following way (see Manski 1981; Hensher and Louviere 1981 for further
details, and also Williams, 1981 for an alternative derivation written in
terms of 'surplus' i.e. utility minus cost).

First, we assume that each decision-maker i is faced with a set of R
available choice alternatives $\mathbf{A}_i = \{A_{1i}, \ldots A_{ri}, \ldots A_{Ri}\}$. For simplicity
we may wish to make the further assumption that the same choice
alternatives are available to all decision makers, but this is not strictly
necessary (see Manski 1981, 63 and also the discussion of 'unranked'
alternatives in Section 2.11.2).

Second, we assume that an unobservable utility value $U_{ri} = U(x_{ri}^\phi, x_i^\psi)$
is associated with the choice of each alternative r by each individual i
(where using the notation of Ch. 2, x_{ri}^ϕ is a vector of attributes of the
choice alternative r faced by individual i, and x_i^ψ is a vector of
socio-economic characteristics of individual i), and that each individual i
wishes to select an alternative which yields maximum utility. That is to
say, each individual is assumed to be a utility maximizer who will select
choice alternative r if and only if

$$U_{ri} > U_{si} \quad \text{for } s \neq r \quad s = 1, \ldots R \qquad [10.1]$$

Third, we assume that the analyst/modeller knows the structure of the

U function up to a finite parameter vector, has observed specific values of a subset of the many possible attributes x_{ri}^{ϕ} and socio-economic characteristics x_i^{ψ}, and knows up to a finite parameter vector the distribution of unobserved characteristics across the population.

In this context (and letting $x_{ri} = f(x_{ri}^{\phi}, x_i^{\psi})$ denote a vector of utility-relevant functions of the observed values of x_{ri}^{ϕ} and x_i^{ψ}, where x_{ri} is the usual shorthand expression used in Chapters 2 and 3 – see [2.118] and [A2.28]) we next assume that the analyst/modeller imposes a probability distribution on the unobserved utility vector (U_{ri}, $r = 1, \ldots R$) conditioned on the known matrix $X_i = [x_{1i}, \ldots x_{ri}, \ldots x_{Ri}]$ and on an unknown parameter vector θ. The vector θ includes parameters of the utility function U, parameters of the distribution of unobserved socio-economic characteristics, and unknown attributes of the choice alternatives. Choice probabilities can then be derived as

$$P(r|X_i, \theta) = \text{Prob}(U_{ri} \geqslant U_{si}, s = 1, \ldots R|X_i, \theta) \qquad [10.2]$$

In this expression $P(r|X_i, \theta)$ denotes the probability that decision-maker i faced with choice set A_i will select alternative r or, more formally, that choice alternative r will be selected by individual i conditional upon X_i and the parameters θ.

In most practical applications, we can assume that the random utility function has a linear-in-parameters additive form (though Koppelman 1981, Louviere 1981a and others have utilized non-linear forms). This can be written as

$$U_{ri} = x'_{ri}\beta_r + (x'_{ri}\gamma_{ri} + \eta_{ri})$$
$$= V_{ri}^* + \varepsilon_{ri} \qquad [10.3]$$

The first term V_{ri}^* on the right-hand side of [10.3] is referred to as the systematic or 'representative' component of utility. The second term, ε_{ri}, is the random component and it can be see to be composed of two parts. One part, η_{ri}, is a random disturbance which captures the effects of unobserved attributes of the choice alternative and unobserved socio-economic characteristics of the decision-maker. The other part represents the idiosyncratic tastes of individual i, i.e. the difference between the tastes of individual i and the average tastes of individuals with identical observed characteristics. (Notice that the 'deviation' or taste variation parameters vector γ_{ri} has an individual, i, subscript to allow this).

The vector of random components $\varepsilon_i = [\varepsilon_{1i}, \ldots \varepsilon_{ri}, \ldots \varepsilon_{Ri}]$ has a distribution, conditioned on X_i, which lies within the parametric class $F(\varepsilon_i|X_i, \omega)$. Thus, given the expression [10.3] and the fact that $\theta = (\beta, \omega)$, the form of the choice probabilities specified in [10.2] depends on the distribution F chosen for the random components. In the early theoretical development and empirical application of discrete choice modelling (e.g. McFadden 1974; Ben-Akiva 1973; Domencich and McFadden 1975; Richards and Ben-Akiva 1975) it proved convenient to assume that F was the independent and identically distributed type I extreme-value (double exponential) distribution

$$F(\varepsilon_i|X_i, \omega) = \prod_{r=1}^{R} \exp\left[-\exp(-\varepsilon_{ri})\right] \qquad [10.4]$$

and that there was no random taste variation across individuals (i.e. the elements of $\boldsymbol{\gamma}_{ri}$ in equation [10.3] all equal zero and thus individuals with identical observed characteristics must have identical tastes). Under these assumptions the choice probabilities have the form

$$P(r|\mathbf{X}_i, \boldsymbol{\beta}_r) = \frac{e^{x'_{ri}\boldsymbol{\beta}_r}}{1 + \sum\limits_{s=1}^{R-1} e^{x'_{si}\boldsymbol{\beta}_s}} = \frac{e^{V_{ri}^*}}{1 + \sum\limits_{s=1}^{R-1} e^{V_{si}^*}} \qquad r = 1, \ldots R \qquad [10.5]$$

This is the familiar multiple response category logistic model which has played such an important role in the earlier chapters of the book. As shown in Chapter 2 (equations [2.102], [2.101]) it can readily be converted to a linear equation in which the left-hand side is the log-odds, i.e. to a linear logit model. However, it has become the practice in discrete choice modelling to refer to both the logistic expression of [10.5] and its linear logit re-expression as the *multinomial logit* (MNL) model and, in this chapter, we will similarly adopt this practice.

Before proceeding to examine the properties and applications of the MNL discrete choice model, it will be useful, at this stage, if we pause for a moment and take the opportunity to re-examine the form of the model [10.5] and clarify a number of issues which the reader is likely to encounter (and perhaps be confused by) in his wider reading.

First, the reader should note that in [10.5] the logistic model is written in the standard fashion we adopted in Chapter 2, and that it uses the shorthand expression $x'_{ri}\boldsymbol{\beta}_r$ (see equations [2.118] and [A2.28]) for the 'representative' component of utility V_{ri}^*. Readers who have consulted Appendix 2.1 will note, however, that this is the 'asymmetric' manner of writing the logistic model. We can also present an equivalent 'symmetric' form of the model

$$P(r|\mathbf{X}_i, \boldsymbol{\beta}_r) = \frac{e^{V_{ri}}}{\sum\limits_{s=1}^{R} e^{V_{si}}} \qquad r = 1, \ldots R \qquad [10.6]$$

in which the 'representative' component of utility is now written as V_{ri} to indicate that it takes a slightly different form.[†] This is the more usual

---

[†] To understand the difference between V_{ri}^* in the 'asymmetric' form of the MNL model, and V_{ri} in the 'symmetric' form, it is necessary to consult the final section of Appendix 2.1. In equations [A2.23] and [A2.24] it was shown how dividing through the numerator and denominator of the 'symmetric' multiple-category logistic model

$$P_{r|i} = \frac{e^{z'_{ri}\boldsymbol{\beta} + z'_{ri}\boldsymbol{\beta}_r + z'_i\boldsymbol{\beta}_r}}{\sum\limits_{s=1}^{R} e^{z'_{si}\boldsymbol{\beta} + z'_{si}\boldsymbol{\beta}_s + z'_i\boldsymbol{\beta}_s}} \qquad r = 1, \ldots R$$

by $e^{z'_{Ri}\boldsymbol{\beta} + z'_{Ri}\boldsymbol{\beta}_R + z'_i\boldsymbol{\beta}_R}$ and imposing the constraints $\boldsymbol{\beta}_R = \mathbf{0}$ produces the 'asymmetric' multiple-category logistic model.

$$P_{r|i} = \frac{e^{(z_{ri} - z_{Ri})'\boldsymbol{\beta} + z'_{ri}\boldsymbol{\beta}_r + z'_i\boldsymbol{\beta}_r}}{1 + \sum\limits_{s=1}^{R-1} e^{(z_{si} - z_{Ri})'\boldsymbol{\beta} + z'_{si}\boldsymbol{\beta}_s + z'_i\boldsymbol{\beta}_s}} \qquad r = 1, \ldots R$$

where the numerator equals 1 when $r = R$. (In these equations we use, as in Appendix 2.1, z to represent the vector of explanatory variables, and add $z'_{ri}\boldsymbol{\beta}_r$ to equations [A2.23] to represent type ϕ

way of expressing the MNL model in the discrete choice modelling literature and, in subsequent sections of this chapter, we will often adopt this 'symmetric' form and use V_{ri} as the shorthand expression for the representative component of utility.

Second, the reader should note that, following our previous practice, it will prove convenient in subsequent sections of this chapter to simplify the left-hand side of equations [10.5] and [10.6] to the form $P_{r|i}$. The more formal expressions used in equations [10.5], [10.6] and [10.2] are useful as unambiguous initial statments of the fact that choice probabilities are conditioned on a known matrix of explanatory variables and an unknown parameter vector. However, once this is understood by the reader, we can adopt the less cumbersome expression $P_{r|i}$ which we have used elsewhere in the book (see footnote, Section 2.2).

Third, the reader should note that he will occasionally encounter (see Williams 1981; Williams and Ortuzar 1982; Hensher and Manefield 1981; Hensher and Louviere 1981) the MNL model written in the form

$$P_{r|i} = \frac{e^{\S V_{ri}}}{\sum\limits_{s=1}^{R} e^{\S V_{si}}} \qquad r = 1, \dots R \qquad [10.7]$$

where § (sometimes written as λ or θ) is the dispersion parameter of the extreme value distribution and which takes the form

$$\S = \pi / \sigma \sqrt{6} \qquad [10.8]$$

In most cases, it is not necessary to make this parameter explicit and it is normally absorbed into the utility function (see Williams 1981, 57; Williams and Ortuzar 1982, 174). However, the reader should note the important fact that σ in [10.8] is the common standard deviation of the random components ε_{ri} $(r = 1, \dots R)$. It follows, therefore, that in the MNL model the variance of each of the random components is the same and equal to

$$\sigma^2 = \pi^2 / 6\S^2 \qquad [10.9]$$

---

variables entered in alternative specific form). It can be seen, therefore, that in the 'asymmetric' form of the MNL model

$$V^*_{ri} = (z_{ri} - z_{Ri})'\boldsymbol{\beta} + z'_{ri}\boldsymbol{\beta}_r + z'_i\boldsymbol{\beta}_r$$

or in the more usual terms of Chapters 2 and 3

$$V^*_{ri} = x'_{ri}\boldsymbol{\beta}_r = \beta_{1r} + \sum_{k=2}^{K_1} \beta_k(X^\phi_{rik} - X^\phi_{Rik}) + \sum_{k=K_1+1}^{K_2} \beta_{kr}X^\phi_{rik} + \sum_{k=K_2+1}^{K} \beta_{kr}X^\psi_{ik}$$

   (ASC)   (Generic-type ϕ)   (ASV-type ϕ)   (ASV-type ψ)

On the other hand, in the 'symmetric' form of the MNL model

$$V_{ri} = z'_{ri}\boldsymbol{\beta} + z'_{ri}\boldsymbol{\beta}_r + z'_i\boldsymbol{\beta}_r$$
$$= \beta_{1r} + \sum_{k=2}^{K_1} \beta_k X^\phi_{rik} + \sum_{k=K_1+1}^{K_2} \beta_{kr}X^\phi_{rik} + \sum_{k=K_2+1}^{K} \beta_{kr}X^\psi_{ik}$$
   (ASC) (Generic-type ϕ). (ASV-type ϕ)   (ASV-type ψ)

and where necessary we use the shorthand expression

$$V_{ri} = \check{x}'_{ri}\boldsymbol{\beta}_r$$

Although we will not concern ourselves with the dispersion parameter §
in this chapter, we will make use of the fact that in the MNL model each
of the random components ε_{ri} ($r = 1, \ldots R$) has the same variance σ^2.

The MNL discrete choice model can be seen, therefore, to be a multiple
category logistic/logit model with a form identical to the models
discussed in Chapters 2 and 3. The principles of parameter estimation,
hypothesis testing, goodness-of-fit measurement, etc which we established in Chapter 2 continue to hold, as does the typology of explanatory
variables and response categories.[†] It follows, therefore, that the MNL
model has significant advantages for empirical research (notably its
computational tractability) and, aided by the diffusion of readily
accessible package programs (see Ch. 7), it has become very widely
used. Nevertheless, despite its wide use, it has long been recognized that
the MNL discrete choice model is based upon restrictive assumptions
and that it has properties which are not always desirable.

Most important in this respect is the property of the MNL model
known from the work of Luce (1959) as 'independence from irrelevant
alternatives' (IIA); a property which implies that the ratio of choice
probabilities for any two alternatives r and s (i.e. the odds of choosing r
over s) should not depend on what other alternatives are available to the
decision-maker. The IIA property is both a great strength and also a
potential weakness of the MNL discrete choice model. It is a strength in
the sense that it: (a) underpins the simple form and computational
tractability of the MNL model, and (b) allows the analyst/modeller to
introduce new choice alternatives into the choice set (or to eliminate
existing choice alternatives) and to derive the resulting choice probabilities without the need to re-estimate the parameters of the MNL model.
However, it is a weakness in the sense that it implies highly restrictive
patterns of cross-substitution between choice alternatives, and this can
result in the MNL model producing counter-intuitive predictions when
there is distinct 'similarity' between certain alternatives in the choice
set.

At this stage it is appropriate to defer further discussion of the IIA
property and the problems which accompany it to Section 10.2. Instead,
we will now turn to an illustration of the empirical application of a MNL
discrete choice model; an application which exploits the IIA property of
the MNL model to introduce a new alternative, post-estimation, into
the model.[‡]

EXAMPLE 10.1   *The collapse and re-opening of the Tasman Bridge*

On Sunday 5 January 1975 a bulk ore carrier (the *Lake Illawarra*)
loaded with zinc concentrate collided with the Tasman Bridge. Two piers
of the bridge were demolished by the accident and three of the bridge's
spans collapsed into the River Derwent. The collapse of the Tasman

---

[†] Consequently, these issues will not be reconsidered in this chapter, and the reader is referred to
Chapter 2 for the necessary details.

[‡] There are literally hundreds of empirical applications of MNL discrete choice models. Some of
these have already been reported in Chapters 2 and 3 as empirical examples of straightforward multiple
category logistic/logit models (without at that stage any discussions of the discrete choice theory
underpinnings). In this chapter we therefore, present just one example of the application of a MNL
discrete choice model, and refer the reader to Examples 2.8–2.11, 3.3 and 3.4 for other illustrations.

Bridge was a major catastrophe for the transport system of Tasmania. As Fig. 10.1 shows, the bridge has a vital role to play in the Hobart urban area. Since its opening in 1964 it has facilitated major shifts in the residential location and commuting patterns in the area, opening up the eastern side of the river as a dormitory area for Hobart's CBD on the western side of the river. By 1974, 28.2 per cent of the population in the Hobart urban area were living on the eastern side of the river but only 5.3 per cent of the employment was located there. As a result, 43,930 vehicle crossings of the bridge were made per weekday in 1974.

With the collapse of the bridge, emergency river crossing points were established. These are shown on Fig. 10.1 and are respectively: the Lindisfarne and Bellerive ferries (for foot passengers only and operating in much the same location as the collapsed bridge); a Bailey bridge (which could take vehicles but was much further upstream); the Risdon punt (which could take both vehicles and foot passengers). Although these emergency facilities were adequate, they significantly increased the travel time of commuter journeys.

It took until October 1977 to rebuild and re-open the bridge. In the meantime, a unique opportunity was provided to estimate a discrete choice model of the selection of transport route and mode by commuters; a model which would have the advantage of permitting a

Fig. 10.1
Local area and cross-river transport links in Tasman Bridge study

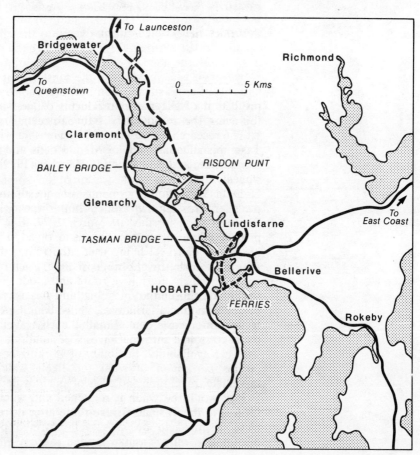

Table 10.1
Hensher's finally selected
specification of the representative component in
model [10.10]

Explanatory variables (alternative specific to)		Parameter estimate
FPDUMY	alternative specific constant (ferry passenger)	1.6136 (0.2787)
CDPDMY	alternative specific constant (car driver, Risdon punt)	−1.6138 (0.3300)
WLK	walking time – generic variable	−0.03885 (0.0125)
WT	waiting time – generic variable	−0.12843 (0.0234)
IVT	in-vehicle time – generic variable	−0.01481 (0.0066)
IVC/I	total in-vehicle cost divided by personal gross income – generic variable	−0.03693 (0.0120)
OVC/I	total out-of-vehicle cost divided by personal gross income – generic variable	−0.07050 (0.0526)
CDTMDY	travel during peak hours (car driver)	−0.24428 (0.0484)
CDPURP	multiple purpose journey (car driver)	0.18194 (0.0725)
CDAVAL	car availability (car driver)	0.88766 (0.3579)
CDEMPL	cars supplied by employer (car driver)	1.2131 (0.3270)
CPTMDY	travel during peak hours (car passenger)	−0.13802 (0.0469)
FPBHERS	inflexibility of working hours (ferry passenger, bus passenger)	−0.34069 (0.2241)

$$\rho^2 = 0.48$$

prediction of the patronage levels of the various routes and modes following the re-opening of the Tasman Bridge.

The research was conducted by Hensher with the cooperation of the Tasmanian Department of Main Roads and the Traffic Management Committee, and is reported in Hensher (1979). It involved three major stages:

1. Data collection by means of questionnaires distributed to all travellers crossing the river (in both directions) between 6.15 a.m. and 12 noon on Thursday 10 March 1977. Each traveller was asked to provide details of the trip he was involved in at the time of the survey, i.e. mode of travel being used, the route, the walking time, waiting time, in-vehicle travel time, costs, etc. Each traveller was also asked to consider which alternative route and mode combinations he might use to the one chosen that morning, and was asked to supply the same type of details on the alternatives, i.e. walking time, waiting time, in-vehicle travel time, costs, etc. Finally, each traveller was asked to supply socio-economic information on occupation, age, income, household size, car availability, flexibility of working hours, etc. and information on his *perception* of what the likely walking time, waiting time, in-vehicle travel time and costs of an equivalent journey via the Tasman Bridge would be when it reopened and whether he would switch (or return) to the Tasman Bridge route after its reopening. Most travellers were assumed to be in a well-informed position to provide such information on a Tasman Bridge journey when it reopened as most

individuals had experienced the Tasman Bridge route on repeated occasions up to 1975.

2. From a subset of 1,324 individuals making home-based work trips a set of MNL models of the form [10.10] was estimated, in each of which there were $R = 5$ route and mode combination choice alternatives ($BB(CD)$ = car driver crossing by the Bailey bridge; $BB(CP)$ = car passenger crossing by the Bailey bridge, FP = ferry passenger; $P(CD)$ = car driver crossing by the Risdon punt; $BB(BP)$ = bus passenger crossing by the Bailey bridge).

$$P_{r|i} = \frac{e^{V_{ri}}}{\sum\limits_{s=1}^{5} e^{V_{si}}} \qquad r = BB(CD), BB(CP), FP, P(CD), BB(BP)$$

[10.10]

In other words, a set of MNL models was estimated for the emergency route and mode of travel combinations only.

Each of the MNL models had a different specification of V_{ri} (the representative utility component) and, after a process of incremental assessment and modification, the one chosen by Hensher as the most suitable specification took the form shown in Table 10.1. The interpretation of the parameter estimates of this specification are discussed by Hensher (1979) and the details need not concern us here. All we need to note is that all the parameter estimates have the expected sign; that the five generic variables have negative parameter estimates; that there is a negative relationship between the likelihood of choosing the car as driver or passenger and travel during the morning peak period ($CDTMDY$, $CPTMDY$); that the existence of multiple purpose journeys ($CDPURP$), availability of car, ($CDAVAL$) and employer-provided cars all increase the likelihood of travelling to work by car as driver; and that persons with fixed working hours ($FPBHRS$) appear to have a lower likelihood of using scheduled services. In addition, Hensher went on to compute direct and cross elasticities from the parameter estimates of Table 10.1. These elasticities summarize the effects on route and mode choice caused by changing the values of the explanatory variables. However, in the case of the model [10.10], only ($CDTMDY$, $CPTMDY$) the 'travel during peak hours' variables had elasticities greater than unity. This suggests that in the emergency period, prior to the reopening of the Tasman Bridge, the scope for significant changes in the shares of traffic held by the various emergency route and mode travel combinations was extremely limited. That is to say, there was very little the transport managers in the area could have done in the emergency period to alter the travel patterns across the river through simple manipulation of the levels of service and costs of the various emergency route and mode travel combinations.

3. An attempt was made by Hensher to provide a short-term forecast of what the traffic shares of the various route and mode combinations would be following the reopening of the Tasman Bridge. To do this he: (a) used the information collected on each traveller's perception of the likely levels of services and costs which would be offered by the Tasman Bridge route when the bridge reopened; (b) exploited the IIA property of the MNL model to introduce the new Tasman Bridge choice alternative into the existing model [10.10] without the need to re-estimate its

parameters. To see how this was done, note that model [10.10] can be written in a longer, more explicit, form as[†]

$$P_{BB(CD)} = \frac{e^{V_{BB(CP)}}}{e^{V_{BB(CD)}} + e^{V_{BB(CP)}} + e^{V_{FP}} + e^{V_{P(CD)}} + e^{V_{BB(BP)}}}$$

.   .

.   .   **[10.11]**

.   .

$$P_{BB(BP)} = \frac{e^{V_{BB(BP)}}}{e^{V_{BB(CD)}} + e^{V_{BB(CP)}} + e^{V_{FP}} + e^{V_{P(CD)}} + e^{V_{BB(BP)}}}$$

Because of the IIA property, the new choice alternative (TAS)[‡] can be introduced into [10.11] without the need to re-estimate its parameters, i.e. model [10.11] becomes

$$P_{BB(CD)} = \frac{e^{V_{BB(CP)}}}{e^{V_{BB(CD)}} + e^{V_{BB(CP)}} + e^{V_{FP}} + e^{V_{P(CD)}} + e^{V_{BB(BP)}} + e^{V_{TAS}}}$$

.   .

.   .   **[10.12]**

.   .

$$P_{TAS} = \frac{e^{V_{TAS}}}{e^{V_{BB(CD)}} + e^{V_{BB(CP)}} + e^{V_{FP}} + e^{V_{P(CD)}} + e^{V_{BB(BP)}} + e^{V_{TAS}}}$$

From [10.12] it is then a simple matter to calculate the predicted choice probabilities for the various route and mode combinations (i.e. forecasts of traffic shares). The forecasts Hensher derived from this procedure are shown in the second column of Table 10.2. Assuming that all the emergency routes are maintained with the same levels of service as in the emergency period, the Tasman Bridge is shown to attract 47.7 per cent of the traffic after its reopening. However, this is a very unlikely scenario. Figure 10.1 shows that the ferry services cross the river in the same location as the bridge and Hensher anticipated that the ferry services would, therefore, be eliminated once the Tasman Bridge was reopened. Column three of Table 10.2 shows the revised traffic forecasts assuming the elimination of the ferry services. The Tasman Bridge route is now shown to attract 85.5 per cent of the traffic.

The forecast of traffic shares produced by Hensher's MNL discrete

---

[†] $V_{BB(CD)}$, $V_{BB(CP)}$, etc. are the representative components of utility for that particular alternative. They can be obtained from the general representative utility expression V_{ri} (as given in Table 10.1) in a straightforward manner, e.g.

$$\hat{V}_{BB(CP)} = \begin{aligned}&-0.03885WLK -0.12843\ WT -0.01481\ IVT -0.03693\ IVC/I \\ &-0.07050\ OVC/I -0.13802\ CPTMDY\end{aligned}$$

[‡] No details are available on which mode an individual was likely to use if the Tasman Bridge route were chosen. Hence it was necessary to treat $TAS(CD)$, $TAS(CP)$ and $TAS(BP)$ as a single alternative TAS.

Table 10.2

Predicted traffic shares in emergency period and following reopening of Tasman Bridge (data from Hensher 1979)

choice model can, of course, be compared with the traffic share figures obtained from traffic monitoring *after* the reopening of the bridge in late 1977. Hensher (1979, 257) notes that following the reopening of the bridge 85.18 per cent of peak hour traffic flows for *all* journey purposes were observed to use the Tasman Bridge route.[†] It appears, therefore, that the MNL discrete choice model has produced a remarkably strong aggregrate prediction though the reader must, of course, exercise due caution when interpreting this result.

	Emergency period (March 1977) predictions from model [10.10] (%)	With Tasman Bridge reopened, and no other changes (%)	With Tasman Bridge, no ferries (%)
Car driver via Bailey bridge	18.3	9.61	10.75
Car passenger via Bailey bridge	3.6	1.87	2.11
Ferry passenger	75.3	39.40	—
Car driver via Risdon punt	1.2	0.62	0.70
Bus passenger via Bailey bridge	1.5	0.79	0.89
Car driver, car passenger or bus passenger via Tasman Bridge	—	47.70	85.50

Before leaving this example the reader should note one further important point. When a new choice alternative (such as the Tasman Bridge) is introduced, post-estimation, into a MNL model, no satisfactory method exists for assigning a value to the new alternative's ASC (alternative specific constant). In general, the representative component of the new alternative is often restricted to the set of generic variables only, e.g. in the case of the Tasman Bridge in model [10.12]

$$\hat{V}_{TAS} = -0.03885\, WLK - 0.12843\, WT - 0.01481\, IVT$$
$$\qquad\quad -0.03693\, IVC/I - 0.07050\, OVC/I \qquad\qquad [10.13]$$

As alternative specific constants are normally included in discrete choice models to capture the mean effect of unobserved factors which influence the selection of choice alternatives, the omission of an ASC for the new alternative is potentially a serious problem. Hensher (1981) has provided an empirical exploration of this issue in the context of the Tasman Bridge data set. His results demonstrate that the omission of an ASC for the new alternative is likely to have an important effect on the predictions obtained from the model, and that some procedure for assigning a value to, and including, an ASC for the new alternative should be sought.

[†] Hensher (1981) has subsequently reported an observed value of '70%+' for the appropriate home-based work trip journeys.

10.2   The IIA property and its implications

In Section 10.1 we noted that IIA (independence from irrelevant alternatives) is a property of the MNL model which implies that the ratio of choice probabilities for any two alternatives r and s (i.e. the odds of choosing r over s) should not depend on what other alternatives are available to the decision maker. Figure 10.2 shows an illustration of this property for a reasonably homogeneous population of consumers in one small suburban area of a single city. In Fig. 10.2(a) the consumers are shown as having the option of shopping at two shopping centres, A and B, and the market shares of the two centres are shown to be 0.4 and 0.6 respectively. The ratio of the choice probabilities is therefore $0.4/0.6 = 0.66$. Now, say (as shown in Fig. 10.2b) a third shopping centre is built to serve this suburban area, i.e. a third alternative enters the choice set of each individual. To be consistent with the IIA property the ratio of the probability of choosing shopping centre A over shopping centre B should remain the same even though an additional alternative is now available to each decision-maker. In this particular illustration, this is in fact the case. The new shopping centre C gains a 0.17 market share, reducing the share of centres A and B to 0.33 and 0.50 respectively, but the ratio of the probabilities of choosing A over B remains the same ($0.33/0.50 = 0.66$) even though an additional alternative has entered the choice set.

Fig. 10.2
An illustration of the IIA property

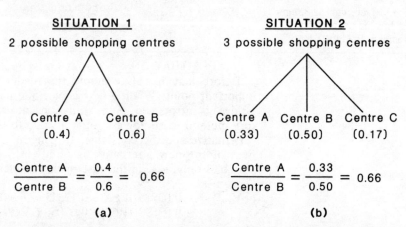

It can be seen in the illustration that the new alternative (centre C), entering the choice set has obtained a share of the market by drawing from the two existing alternatives in the choice set in direct proportion to the original shares of the market held by these existing alternatives. That is to say, the IIA property implies that the new alternative competes equally with each existing alternative.

In many cases, this implication is reasonable, for in many choice sets the alternatives are distinctly different options competing equally with all other alternatives in the choice set, and in these circumstances the IIA property imposes no unrealistic restrictions on the structure of the choice probabilities. However, this is not always the case. In some choice sets there are certain alternatives which are perceived by the choice makers to be similar and, as such, the alternatives in the choice set do not compete equally with each other. In such cases, the IIA property can give rise to unrealistic predictions and, for a considerable time, beginning as far back as Debreu's (1960) review of Luce's (1959) book, examples of situations in which the IIA property of the MNL model will yield counter-intuitive behavioural forecasts have been known.

An example of the type of case where the IIA property of the MNL model gives rise to counter-intuitive predictions is shown in Fig. 10.3.[†] In Fig. 10.3(a) a reasonably homogeneous population of consumers in one small inner-city area is shown as having the option of shopping in the traditional central business district (CBD) of the city or in a purpose-built suburban shopping centre (A). Fifty per cent of the consumers are observed to choose each alternative and the ratio of the choice probabilities is CBD/A = 0.5/0.5 = 1. Next, we imagine that a second purpose-built suburban shopping centre (B) is completed with attributes similar to those of centre A, and that it is located in such a position that the travel times and costs for each consumer in our small

Fig. 10.3
An example in which the IIA property of the MNL model gives rise to counter-intuitive predictions

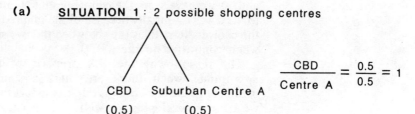

(a) **SITUATION 1** : 2 possible shopping centres

$$\frac{CBD}{Centre\ A} = \frac{0.5}{0.5} = 1$$

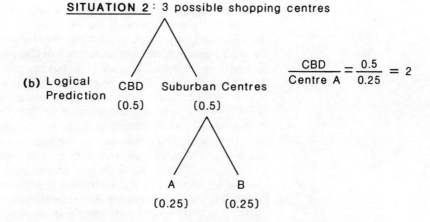

SITUATION 2 : 3 possible shopping centres

(b) Logical Prediction

$$\frac{CBD}{Centre\ A} = \frac{0.5}{0.25} = 2$$

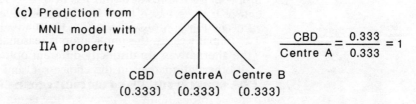

(c) Prediction from MNL model with IIA property

$$\frac{CBD}{Centre\ A} = \frac{0.333}{0.333} = 1$$

[†] This is an alternative version of the classical 'red bus/blue bus' conundrum discussed in the transportion science literature.

inner-city study area are the same to both suburban shopping centres. Logically, we would assume that the resulting market shares of the shopping centres would be as shown in Fig 10.3(b). In other words, we would logically assume that the 50 per cent of shoppers who preferred to shop in the CBD in the first situation would continue to prefer to go to the CBD. However, those who preferred the suburban centre in the first situation would now split evenly between the two suburban centres. That is to say we would logically predict CBD = 0.5, A = 0.25, B = 0.25. Unfortunately, this is not the prediction which an MNL model with the IIA property will give. As shown in Fig 10.3(c), the MNL model will predict, counter-intuitively, that the new suburban shopping centre (B) will draw its market share equally from both the CBD and the existing suburban shopping centre (i.e. it will predict CBD = 0.333, A = 0.333, B = 0.333). It will do this in order that the ratio of the probabilities of choosing the CBD over shopping centre A in the new situation (CBD/A = 0.333/0.333 = 1) will remain the same as in the first situation (CBD/A = 0.5/0.5 = 1). This is clearly unrealistic and, as we have seen, the counter-intuitive predictions produced by the MNL occur essentially because of the marked similarity between two of the choice alternatives (i.e. between the two suburban shopping centres when contrasted to the CBD).

The reason why the IIA property of the MNL model is not appropriate when there are marked similarities between certain members of the choice set is due to the fact that one of the assumptions we made in developing the MNL discrete choice model is not satisfied in such circumstances. In equation [10.4] we assumed that the random components of utility are *independent and identically* distributed with a type *I* extreme-value distribution and that there was no random taste variation across individuals. By the term 'independent' we imply that the random component associated with a particular choice alternative should be uncorrelated with the random components of any of the other alternatives in the choice set (i.e. the random components should be pairwise uncorrelated) and by the term 'identically distributed' we imply that random components should have equal variance σ^2 (see equation [10.9]). When there is distinct similarity between certain alternatives in the choice set the random components of such alternatives become correlated and, as a result, one of the crucial assumptions in developing the MNL discrete choice model is not satisfied. Clearly, in these circumstances, what we need instead are more flexible discrete choice models which do not incorporate this assumption, which at the very least allow the random components to be correlated, and which perhaps also permit unequal variances and random taste variation across individuals. The result of this has been a search for alternative discrete choice models based upon less demanding assumptions and with less restrictive properties of cross-substitution embodied in their structure.

10.3   The search for less restrictive discrete choice models

It is now realized that the counter-intuitive predictions of the MNL discrete choice model produced in cases such as that shown in Fig. 10.3 are not a function of the IIA property *per se*. Instead, they are common to any discrete choice model in which we assume independent and identically distributed random utility components, i.e. models in which we assume that the variance-covariance matrix of the random components takes the very simple but restrictive form

$$\Sigma^I = \begin{bmatrix} \sigma^2 & 0 & .. & & 0 \\ 0 & \sigma^2 & \cdots & & 0 \\ . & . & & & . \\ . & . & & & . \\ . & . & & & . \\ 0 & 0 & \cdots & & \sigma^2 \end{bmatrix} = \sigma^2 \, \mathbf{I} \qquad [10.14]$$

As we have seen, such models produce suspect predictions when there is distinct similarity between choice alternatives because in this case random components are correlated but the model assumes that they are independent. To overcome such problems, models which are based upon less restrictive assumptions about the distribution of the error components are required.

10.3.1 *The multinomial probit model*

Perhaps the most general of the alternative discrete choice models is that produced when it is assumed that the vector of random components ε_i has a multivariate normal distribution with mean vector zero and a general variance-covariance matrix Σ (rather than the restrictive from shown in [10.14]). This assumption produces a model known as the *multinomial probit* (MNP) model which takes the form

$$P_{r|i} = \int_{-\infty}^{\infty} d\varepsilon_r \left[\prod_{s \neq r} \int_{-\infty}^{V_{ri} - V_{si} + \varepsilon_r} d\varepsilon_s \right] \Phi_R\big(\varepsilon_1, \, \ldots \, \varepsilon_R; \, \Sigma\big) \qquad [10.15]$$

where Φ_R is a R-dimensional normal density function with mean vector zero and variance-covariance matrix Σ (see Horowitz 1981b; Johnson and Hensher 1982 and our discussion of the dichotomous probit model in Section 2.3.3).

The MNP model allows the random components of the choice alternatives to be correlated and to have unequal variances. It also permits random taste variation across individuals. To achieve this, it assumes that in the random component

$$\varepsilon_{ri} = x'_{ri} \, \gamma_{ri} + \eta_{ri} \qquad [10.16]$$

(see equation [10.3]), the deviation or taste variation parameters γ_{ri} are drawn from a multivariate normal distribution with mean vector zero and a $K \times K$ variance-covariance matrix Ω, and that the random disturbance η_{ri} is drawn from a multivariate normal distribution with zero means and a $R \times R$ variance-covariance matrix Δ. As a result, the general variance-covariance matrix Σ, in the MNP model is a $R \times R$

matrix which takes the form

$$\Sigma = \mathbf{X}'_i \Omega \mathbf{X}_i + \Delta \qquad\qquad [10.17]$$

where $\mathbf{X}_i$ is the $K \times R$ matrix of explanatory variables used in equation [10.2], whose rth column is x_{ri}. The reader should contrast [10.17] to the restrictive form [10.14] used in the MNL model, and note how [10.14] incorporates the assumptions of no random taste variation across individuals, and independent and identically distributed random disturbances.

Clearly, the MNP model is free from the restrictive assumptions and properties of the MNL model and these useful aspects of the MNP have been known for a considerable time. However, the MNP is extremely intractable in computational terms, and it was not until the late 1970s in the work of Hausman and Wise (1978), Albright, Lerman and Manski (1977) and Daganzo, Bouthelier and Sheffi (1977) that any significant progress was made in developing practical estimation procedures. The most popular and cost-effective of these procedures is that suggested by Daganzo *et al.* (1977) which uses the Clark approximation (Clark 1961) to reduce the estimation problem to one of sequential univariate integration. Daganzo (1979) has distributed a computer program called CHOMP which facilitates the application of the MNP model. However, the accuracy of the estimation procedure when the Clark approximation is adopted is still controversial, and for most users of discrete choice models the MNP model remains conceptually complex and computationally unwieldy. As a result, there has been a search for what might be termed 'half-way house' models which lie somewhere between the generality and complexity of the MNP model and the restrictiveness but tractability of the MNL model.

10.3.2   The dogit model

Many possible 'half-way house' models have been suggested in recent years. The first we will consider is the *dogit* model proposed by Gaudry and Dagenais (1979) which takes the form

$$P_{r|i} = \frac{e^{V_{ri}} + \mu_r \sum_{s=1}^{R} e^{V_{si}}}{\left(1 + \sum_{s=1}^{R} \mu_s\right) \sum_{s=1}^{R} e^{V_{si}}} \qquad \text{where } \mu_r \text{ and } \mu_s \geqslant 0 \qquad [10.18]$$

It gets its name because it attempts to avoid or 'dodge' the researcher's dilemma of having to choose *a priori* between a simple computationally tractable discrete choice model (such as MNL) which commits him to restrictive assumptions and the IIA property, and a model (such as MNP) which is free from such restrictions but is conceptually and computationally complex. Essentially, the dogit model avoids the issue by allowing some pairs of choice alternatives to exhibit the IIA property but, simultaneously, allowing other pairs of choice alternatives to be free from the IIA property (i.e. the dogit model allows the MNL format to hold for some pairs of choice alternatives, but a more general format to hold for other choice pairs). It can be seen from [10.18] that if the

parameters μ_s and μ_r all equal zero then the dogit mo
basic MNL form. If, on the other hand, the μ param
pairs of choice alternatives equal zero (or if $\mu_r/\mu_s =$
those pairs of alternatives) then the IIA property will I
particular pairs, whilst the other pairs of alternative
strained by the IIA property.

Following the work of Ben-Akiva (1977), the dogit n
given a very useful behavioural interpretation. It can b
model allows for a certain degree of 'captivity' of deᴄιᴗᴜn-makers to
particular choice alternatives. In particular, if the two elements in the
numerator of [10.18] are reversed and [10.18] is re-expressed as:

$$P_{r|i} = \frac{\mu_r}{1 + \sum_s \mu_s} + \left(\frac{1}{1 + \sum_s \mu_s} \cdot \frac{e^{V_{ri}}}{\sum_s e^{V_{si}}} \right) \qquad [10.19]$$

then it can be shown that the first component on the right-hand side of
[10.19] is that part of the probability of choosing alternative r which
results from the basic need for or 'captivity' to alternative r, whilst the
second component (shown in brackets) is that part which results from
the 'discretionary' choice of r out of all alternatives in the choice set. In
this way, the dogit model reconciles elements of constrained choice with
elements of free choice, and is compatible with Stone's (1954) approach
to consumer demand theory.

10.3.3 The nested logit model

Although the dogit model has attracted a considerable amount of
attention, by far the most promising and widely adopted of these
'half-way house' models is the *nested* (structured or hierarchical) *logit*
model (see Sobel 1980 for an alternative introductory review). The
nested logit is a special case of a model known as the *generalized
extreme-value* (GEV) model[†]; the GEV being the model derived when
the vector of random components is assumed to have the multivariate
extreme-value distribution

$$F(\varepsilon_i | \mathbf{X}_i, \, \boldsymbol{\omega}) = \exp\left\{ -H[(e^{-\varepsilon_{1i}}, \, \dots, \, e^{-\varepsilon_{Ri}}), \, \mathbf{X}_i] \right\} \qquad [10.20]$$

rather than the type I extreme-value distribution shown in equation
[10.4].[‡]

To illustrate the nested logit model let us consider the simple
journey-to-work mode-choice situation displayed in Fig. 10.4 in which
the choice alternatives are car, bus and metro. Using the MNL model
we would treat this choice problem as shown in Fig. 10.4(a), i.e. we
would treat the modes as distinctly independent alternatives and assume
that each decision-maker chooses one particular mode following a

---

[†] McFadden (1981, 238) calls nested logit models 'tree extreme-value' models to emphasize this
point. We will not consider the GEV model in detail in this chapter as the reader will in almost all cases
encounter its special form, the nested logit model. (For further details on the GEV model see
McFadden 1979, 1981; Manski 1981)

[‡] H in [10.20] is a non-negative, linear homogeneous function.

Fig. 10.4
Alternative decision and model structures for a simple travel-to-work mode-choice example

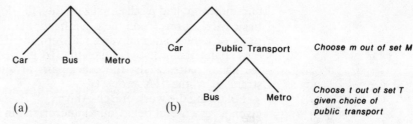

(a)　　　　　　　　(b)

simultaneous evaluation of all three. In contrast, using the nested logit model we would treat the choice problem in the sequential fashion shown in Fig. 10.4 (b) and would group together the modes which show similarities. In this particular case this implies a grouping together of the two public transport modes for, when compared to the private transport (car) alternative, both are 'inflexible' rather than 'flexible' modes and both imply a certain waiting and/or transfer time on any jouney. In a situation such as that shown in Fig. 10.4, where there are distinct similarities between choice alternatives, the MNL model with its restrictive properties of cross-substitution is likely to provide unrealistic predictions. In contrast, the nested logit model permits correlation between the random components of the choice alternatives, is not constrained by the IIA property, and is thus able to embody more general properties of cross-substitution than the MNL model.

The nested logit model for the simple example of Fig. 10.4 is a set of two linked equations which can be written[†]

$$P_{t|\text{PUBTRN}} = \frac{e^{V_t + V_{mt}}}{\sum\limits_{t^*=1}^{T} e^{V_{t^*} + V_{mt^*}}} \qquad t = \text{bus or metro}$$

$$\text{[10.21]}$$

$$P_m = \frac{e^{V_m + \delta \tilde{U}_m}}{\sum\limits_{m^*=1}^{M} e^{V_{m^*} + \delta \tilde{U}_{m^*}}} \qquad m = \text{car or public transport}$$

The first equation in [10.21] gives the conditional probabilities of the choice of bus or metro by individual i given that choice is constrained to public transport. The second equation gives the marginal probabilities of the choice of car (private transport) or public transport.[‡]

In [10.21] $\tilde{U}_m$ is the so-called 'composite utility' term which in this case can be defined as

$$\tilde{U}_m = \log_e \sum_{t^*=1}^{T} e^{V_{t^*} + V_{mt^*}} \qquad \text{[10.22]}$$

This term is a vital component of the nested logit model and it acts as a hierarchical linking mechanism in Fig. 10.4(b). It allows us to assume

---

[†] In equation [10.21] the reader should note that we drop the subscript i for simplicity. In the equation, V_t and V_m denote representative components of utility which are specific to the lower and higher nests in the hierarchy of Fig. 10.4(b) respectively, V_{mt} denotes components which relate to both nests in the hierarchy, and V_m also includes any attributes whose parameters cannot be identified in the lower nest.

[‡] Equivalent marginal probabilities for bus and metro are:

$P_{\text{Bus}} = P_{\text{Bus|Public Trans}} \cdot P_{\text{Public Trans}}$

$P_{\text{Metro}} = P_{\text{Metro|Public Trans}} \cdot P_{\text{Public Trans}}$

that individuals taking decisions at the higher nest
take account of the 'expected maximum utility' of dec
nest. To be consistent with random utility maximization,
δ of the composite utility term must lie within the range 0
where there are more than two levels in the hierarchy, yieldir
composite utility terms and associated parameters δ, the value
δ's should not decline when proceeding from lower to higher lev
the hierarchy.[†]

When the parameter δ of the nested logit model equals 1, it can be
shown that the model will yield equivalent results to a simple MNL
model applied to the simultaneous structure in Fig. 10.4(a), i.e.
equivalent results to a model which assumes uncorrelated random
components. However, this is merely a special case and, in general, the
permitted range of variation of the parameter δ ($0 < \delta \leq 1$) allows the
nested logit model to handle correlated random components. As a
result, the nested logit model embodies more general properties of
cross-substitution than the MNL model, and effectively overcomes the
'similarity of choice alternatives' problem. For this reason empirical
applications of nested logit models are now multiplying rapidly and we
will consider one illustration of such applications in Example 10.4.[‡]

Although examples such as that in Fig. 10.4 provide the clearest
illustration of how the nested logit model is able to overcome the
'similarity of choice alternatives/correlated random components' problem, the reader is also likely to encounter nested logit models in the
context of what in Example 2.10 we termed 'joint' choice models. For
example, in a study of residential location and accommodation in the
private housing sector of a single city, there are likely to be three or
more interlinked decisions.

Decision 1:   where to live (i.e. individual must choose a particular residential location l out of the set of possibilities L).

Decision 2:   what type of dwelling (i.e. the individual must choose a particular dwelling type d out of the set of possibilities D).

Decision 3:   what type of occupant status (i.e. the individual must choose a particular occupant status o out of the set of possibilities O).

In behavioural terms, the most realistic and appealing way to model
such a choice problem is in the form of an interdependent simultaneous
decision structure like that shown in Fig. 10.5(a), where a particular
combination (ldo) of location, dwelling type and occupant status is
selected from the set of possibilities LDO. Unfortunately, a discrete
choice model which assumes such a simultaneous decision structure may
be very difficult and costly to estimate because it demands that all

---

[†] Although the application of nested-type logit models dates back to the work of Ben-Akiva (1973), it was not until the work of Williams (1977), Daly and Zachary (1978), Ben-Akiva and Lerman (1979) and McFadden (1979) that the consistency of such nested logit models with random utility maximization and the correct form and significance of the composite utility term and its associated parameter was established.

[‡] An extension of the nested logit known as the *cross-correlated logit* model has been proposed by Williams and Ortuzar (1982). This has a symmetric form for the variance-covariance matrix Σ of random components rather than the asymmetric form employed by the nested logit model. However, its estimation is complex and its theoretical advantages over the nested logit model are not particularly great. Hence it is unlikely to be adopted in empirical research.

choice models   331

in the hierarchy will
sions in the lower
the parameter,
δ ≤ 1, and
g multiple
s of the
els of

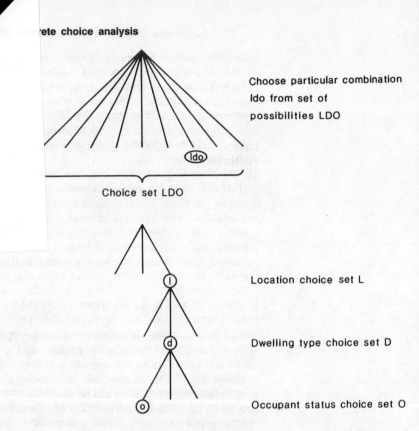

Choose particular combination
ldo from set of
possibilities LDO

Choice set LDO

Location choice set L

Dwelling type choice set D

Occupant status choice set O

possible combinations are considered simultaneously in one model (e.g. if there were thirty possible residential locations, six dwelling types and four occupant status types there would be 720 possible (*ldo*) combinations, all of which would have to be considered simultaneously in one model). This is often completely impractical and, a result, discrete choice models will often assume some form of sequential decision structure such as that shown in Fig. 10.5(b). As we have seen above, nested logit models are the appropriate model forms for such sequential decision structures. Hence, nested logit models not only overcome the 'similarity of choice alternatives/correlated random components' problem but also provide a practical (sometimes the only practical) means of handling 'joint' choice structures.

10.3.4   Elimination-by-aspects models

All the discrete choice models which we have considered so far can be classified as belonging to a general set of choice models which are termed *compensatory* models. In such models we assume that individuals 'trade off' attributes of the choice alternatives in the choice process. That is to say, when evaluating a choice alternative, a low (poor) value on one attribute can be compensated by a high (good) value on another (e.g. the greater distance to one particular shopping centre may be compensated by the lower prices that centre offers). However, we can also recognize another general set of choice models which we termed *non-compensatory* models. Such models are based upon dominance, lexicographic, conjunctive, disjunctive, lexicographic semiorder, maximin, minimax regret, Satislex, and similar decision rules in which we assume there is no 'trade off' of attributes in the choice process, i.e. no compensation of a low (poor) value on one attribute by a high (good) value on another (see Timmermans 1984). Although there are many possible non-compensatory choice models the only ones which have, so far, assumed significance in discrete choice modelling are the elimination-by-aspects (EBA) models proposed by Tversky (1972a, 1972b).[†]

In EBA models, choice is viewed as a process of elimination. At each stage in the process the individual selects a certain attribute (aspect) of the choice alternatives and eliminates all alternatives which do not possess this attribute. This process of selection and elimination then continues until only a single choice alternative remains. Formally this process of aspect selection and elimination implies that EBA models can be expressed in terms of a recursive equation

$$P(r|\mathbf{A}) = \sum_B Q(\mathbf{B}|\mathbf{A}) \, P(r|\mathbf{B}) \qquad\qquad [10.23]$$

where **B** represents a non-empty subset of the original set of choice alternatives **A**. This equation expresses the probability $P(r|\mathbf{A})$ of choosing alternative r from choice set **A** as weighted sum of the probabilities $P(r|\mathbf{B})$ of choosing alternative r from the various subsets **B** of **A**. The weights $Q(\mathbf{B}|\mathbf{A})$ are referred to as 'transition probabilities' and are interpreted as the probability of eliminating from **A** all alternatives that are not included in the subset **B**.

The transition probabilities (which sum to 1 for all subsets **B** in **A**) are given by the expression[‡]

$$Q(\mathbf{B}|\mathbf{A}) = \frac{v_\mathbf{B}}{\sum_{\mathbf{B}*} v_{\mathbf{B}*}} \qquad\qquad [10.24]$$

In the expression $v_\mathbf{B}$ is a non-negative scale value which is a measure of the attributes (aspects) which are shared by all the alternatives in **B** but which are not possessed by any alternative which does not belong to **B**. As such, it can be interpreted as a measure of the unique advantage of the choice alternatives in subset **B**.[§]

The standard MNL model can be shown to be a special case of the EBA model.[‖] However, in general, the EBA model is much less restrictive than the MNL model, and is able to accommodate complex patterns of cross-substitution between choice alternatives. Consequently,

---

[†] EBA models have their roots in the intra-personal rather than inter-personal interpretation of random utility maximization (see Section 10.1). It is probable that EBA models are the only members of the non-compensatory class of models which have, so far, assumed significance in discrete choice modelling because EBA models are compensatory in nature despite the fact that at any given instance in time, choice is assumed to follow a (non-compensatory) conjunctive or lexicographic strategy. Tversky (1972a, 296) notes, therefore, that EBA models can be regarded as being compensatory 'globally' but non-compensatory 'locally'.

[‡] Further details concerning the exact nature of the sumations in [10.23] and [10.24] are to be found in Tversky (1972a, 287–88; 1972b, 346–50).

[§] The reader should note that $v_\mathbf{B}$ can be related to the representative utility component V used in earlier sections in this chapter (see McFadden 1981, 226).

[‖] This MNL special case occurs when all pairs of choice alternatives are (aspect wise) disjoint, i.e. when $v_\mathbf{B} = 0$ when **B** consists of more than one alternative, and where $\log_e v_\mathbf{B}$ is a linear function in the usual MNL parameters **β**.

like the previous models in this section, it is able to overcome the 'similarity of choice alternatives' problem. Unfortunately, these useful properties of the EBA model are not achieved without cost. In particular, the EBA model in its general form includes a large number of parameters which require estimation. As such there has been a need to develop a more parsimonious form of the general EBA model. Tversky and Sattath (1979) have attempted this and have proposed a hierarchical or preference tree version of the general EBA model. This model, which is known as *Pretree*, is very similar in structure and performance to the nested logit model. Both are special cases of more general formulations and Pretree bears the same relationship to the general EBA model as the nested logit model does to the GEV model.[†]

10.3.5   *Weight shifting models*

In our basic derivation of the MNL discrete choice model in Section 10.1 we made the implicit and unstated assumption that the 'generic' parameters of the representative component of utility are 'context free', i.e. we assume that the 'generic' parameters remain stable as the choice set changes in composition. For prediction/forecasting purposes this was a particularly useful implicit assumption to make for the alternative assumption that 'generic' parameters are 'context dependent' and change with choice set composition would have implied a need to model the covariation of parameters with changes in choice set composition.

Although this was a useful simplifying assumption to make, it was not, however, a particularly realistic one. To illustrate this, consider an individual making a three-alternative choice between shopping in the CBD or two similar suburban centres. In this case factors such as relative distances to the shopping centres, price levels, variety of goods available, cost and ease of car parking, etc. are likely to be the important determinants of the individual's choice. However, if the CBD is removed from the choice set, and both suburban centres which remain in the choice set are identical in terms of distance, price level, variety of goods and cost and ease of car parking, then it is reasonable to assume that the individual will tend to 'shift' his attention to those attributes on which the suburban centres do differ, e.g. factors such as opening hours, attractiveness of the shopping environment, store-to-store accessibility within the shopping centre, etc. Clearly, in this case, the parameters of the utility function, which represent the importance of attributes in the choice process, do not remain stable as the choice set changes. The importance or 'weight' given to an attribute 'shifts' as the choice set changes, and the importance of an attribute in influencing choices covaries with the amount of variation which exists across 'other' attributes.

Although the assumption that 'generic' parameters of the utility function are 'context free' and that there is no shifting in the valuation (weight) given to an attribute as the choice set changes is clearly

---

[†] Although no example of the geographical application of an EBA model will be discussed in the chapter the reader will find one in Smith and Slater (1981). These authors propose a particularly interesting new family of migration models belonging to the EBA family in which the traditional spatial interaction model is seen as a special case.

unrealistic, it is an assumption which is not confined to the basic MNL model. It is also a property of other 'simultaneous' choice models, e.g. the MNP and dogit models. In contrast, the 'sequential' choice models (nested logit, Pretree) explicitly acknowledge that attributes used to discriminate between choice alternatives will change in different nests (levels) of the decision hierarchy (though weight shifting is assumed not to occur within particular levels, and the need to prespecify a particular decision tree structure can make such models, in practice, rather insensitive to the 'context dependence/weight shifting' problem).

In an attempt to incorporate context induced parameter instability into discrete choice models Meyer and Eagle (1981, 1982) have proposed and tested a set of adjusted logit models. In the dichotomous case, instead of the usual 'generic variables only' linear logit model

$$\log_e \frac{P_{1|i}}{P_{2|i}} = \sum_{k=1}^{K} \beta_k (X_{1ik}^{\phi} - X_{2ik}^{\phi}) \qquad [10.25]$$

(see equations [2.118], [A2.28]) in which the parameters β_k are assumed to be stable, they propose the *bilinear differences* model

$$\log_e \frac{P_{1|i}}{P_{2|i}} = \sum_{k=1}^{K} \left[\beta_k (X_{1ik}^{\phi} - X_{2ik}^{\phi}) - \sum_{\substack{l=1 \\ l \neq k}} \beta_{kl} (X_{1ik}^{\phi} - X_{2ik}^{\phi}) \, | \, (X_{1il}^{\phi} - X_{2il}^{\phi}) \, | \right] \qquad [10.26]$$

In model [10.26] the influence of an attribute in determining choice is related to: (a) the extent to which choice alternatives differ on *that* attribute, and (b) the extent to which choice alternatives differ on *other* attributes. The model includes the usual K main effects (the parameters β_k)[†] and also $(K^2 - K)$ negative interactions introduced to capture the hypothesis that the marginal effect or importance of a given attribute decreases with increases in variability on 'other' attributes. It should be noted that the traditional model [10.25] assumes that all these interaction (context dependence) terms should be zero. The obvious limitation of model [10.26] is that it has a large number of parameters (K^2). As such it is only parsimonious for choice problems in which a small number of attributes are being examined. However, this limitation can be overcome, and Meyer and Eagle (1981) have proposed an approximation to [10.26] which can be used for problems with larger numbers of attributes.

The bilinear differences model can be generalized to handle multiple choice alternative problems, but the generalization is not without complications. In particular, it is necessary to develop an adjusted MNL model which *simultaneously* overcomes the 'similarity of choice alternatives' problem (and thus represents an alternative to the models discussed in Sections 10.3.1 to 10.3.4) and also the 'context dependence/ weight shifting' problem. Meyer and Eagle (1981, 1982) have proposed such a model which in the 'asymmetric' logistic model form (see

---

[†] Note that in the bilinear differences model the interpretation of these main effect parameters differs slightly because of the inclusion of the interactions. The β_k parameters may now be viewed as 'limiting' rather than average values for the marginal effects of various attributes.

equation [10.5]) is written

$$P_{r|i} \cong \frac{e^{V^*_{ri}}\left(\bar{\alpha}_r^{\bar{\beta}}/\bar{\alpha}_R^{\bar{\beta}}\right)}{1+\sum\limits_{s=1}^{R-1} e^{V^*_{si}}\left(\bar{\alpha}_s^{\bar{\beta}}/\bar{\alpha}_R^{\bar{\beta}}\right)}$$

[10.27]

In this equation

$$V^*_{ri} = \sum_{k=1}^{K}\left[\beta_k(X^{\phi}_{rik}-X^{\phi}_{Rik}) - \sum_{\substack{l=1\\l\neq k}}^{K}\beta_{kl}(X^{\phi}_{rik}-X^{\phi}_{Rik})\,|\,(X^{\phi}_{rik}-X^{\phi}_{Rik})|\right]$$

[10.28]

from model [10.26] or the approximation to this (Meyer and Eagle 1981). The term $\bar{\alpha}_r$ represents a measure of the average degree of perceived dissimilarity between alternative r and all other choice alternatives[†] and $\bar{\beta}$ is an empirical 'similarity sensitivity' parameter. The dissimilarity measures $\bar{\alpha}_r$ are constrained to vary between zero and one, with one reflecting perfect dissimilarity, and their purpose is to overcome the 'similarity of choice alternatives' problem.[‡] It should be noted that, because of 'context dependence/weight shifting', equation [10.27] only provides approximations to the required probabilities. However, this approximation is likely to be weak only in the extreme case where context effects on parameter values are large and the number of choice alternatives is small.

In terms of other discrete choice models discussed in this section, Meyer and Eagle's weight shifting' model lies somewhere between the 'fully simultaneous' choice models (MNL, MNP, dogit) and the 'sequential' hierarchical models (nested logit, Pretree). Like the hierarchical models it recognizes that attributes used to discriminate between choice alternatives will change with variations in choice context but it generalizes that perspective and does not require the prespecification of a particular decision tree structure. Like all models except the MNL it also takes account of the 'similarities of choice alternatives' problem. However, as yet, the model is still in a development stage. Its multiple alternative form provides only an approximation to the required probabilities, its consistency with random utility maximization has not been explored to the same extent as the nested logit model, and there is as yet little empirical experience relating to the estimation of the model. Moreover, the actual proportion of variance explained by adding 'weight shifting' interactions to the utility functions may be very small in relation to the usual main effects. Nevertheless, the model does serve to focus attention on an important characteristic of decision-making and choice behaviour.

---

† One way to derive $\bar{\alpha}_r$ is from the expression

$$\bar{\alpha}_r = \frac{1}{R}\sum_{\substack{s=1\\s\neq r}}^{R}\left|\frac{r^*_{rs}-1}{2}\right|$$

where r^*_{rs} is the observed Pearson correlation between alternatives r and s across their observed attributes.

‡ Meyer and Eagle believe that this method of allowing for differential patterns of similarity between choice alternatives is preferable to that in the dogit model. In particular, similarly effects can be estimated *a priori* rather than being simply corrected for *post-hoc*.

EXAMPLE 10.2   *Location decisions of clothing retailers in Boston*

In Section 10.3.1 we considered the multinomial probit (MNP) model. We saw that the MNP represents perhaps the most general of the alternative discrete choice models and that it allows the random components of the choice alternatives to be correlated and to have unequal variances. It also permits random taste variation across individuals. To illustrate the application of the MNP we will now consider a useful, geographically relevant, application of the MNP provided by Miller and Lerman (1981) as part of a study of the location decisions of clothing retailers in Boston, U.S.A. (for an alternative example see O'Brien 1982).

In their study, Miller and Lerman assume that each clothing retailer will wish to choose the particular store location (l) and store size (s) combination which maximizes the marginal profitability function.

$$U^*_{ls} = x'_{ls}\boldsymbol{\beta} + \delta_1 w_l s + \delta_2 w_l s \bar{I}_l + \delta_3 w_l s \ \log_e(s) + \delta_4 w_l s \ \log_e(w_l)$$
$$+ \delta_5 w_l s \ \log_e(S_l) + \delta_6 w_l s \ \log_e(S_{tl}) + \delta_7 w_l s \ \log_e(M_{tl})$$
$$+ \delta_8 w_l s \gamma_{ls} + \eta_{ls} \qquad [10.29]$$

where x_{ls} is a vector of observable exogenous variables, w_l is the average wage rate paid by stores at location l, $\bar{I}_l$ is the average 'employment intensity' (number of employees per square foot of floor space) at l, S_l and S_{tl} are the total floor space of all stores and type t stores at location l, M_{tl} is the total expected sales of stores of type t at location l and γ_{ls} and η_{ls} are error terms. In [10.29] $\delta_1, \ldots \delta_8$ are 'reduced form' parameters and they can be used to recover the original 'structural' revenue parameters (α and $a_0, a_1, \ldots a_5$). In addition, $\delta_8 \gamma_{ls}$ in the penultimate part of [10.29] can be interpreted as a type of 'taste variation' component as it describes a distribution of δ across the population of stores.

If we assume that the composite error terms

$$\varepsilon_{ls} = (\delta_8 w_l s \gamma_{ls} + \eta_{ls}) \qquad [10.30]$$

(see equation [10.16]) are independent and identically distributed with a type-*I* extreme-value distribution then the choice probabilities of selecting a particular store location and size combination are given by the MNL model. Alternatively, if we assume that both γ_{ls} and η_{ls} are drawn from a multivariate normal distribution with the variance–covariance structures outlined in Section 10.3.1, then the choice probabilities are given by the MNP model.

As part of their empirical research Miller and Lerman fitted both MNL and MNP models to a sample of 161 clothing retailers whose choice set consisted of seven possible locations (a mix of traditional downtown centres, planned downtown centers and suburban centres) and two possible store size categories (under 5,000 square feet). The function $x'_{ls}\boldsymbol{\beta}$ in [10.29] was specified·as

$$x'_{ls}\boldsymbol{\beta} = \mathbf{L}_l^{\text{dum}}\boldsymbol{\beta}_l + \mathbf{S}_s^{\text{dum}}\boldsymbol{\beta}_s + \beta_c \, \text{COST} \qquad [10.31]$$

where $\mathbf{L}_l^{\text{dum}}$ is a vector of location specific dummy variables, $\mathbf{S}_s^{\text{dum}}$ is a vector of size category dummy variables, and the COST term is simply rent at location l times store size. Table 10.4 shows the results of fitting the MNL and three variations of the MNP model.

The three versions of the MNP can be understood from an inspection of Table 10.3 which shows the variance–covariance structures assumed for the three models. Model C represents the closest version to the general MNP form outlined in Section 10.3.1. It permits random taste variation (i.e. Var $(\gamma_{ls}) \neq 0$), permits each location to have a different variance (i.e. $\sigma_l^2 \neq \sigma_{l^*}^2$) and permits correlation between store sizes within a given location (i.e. $\zeta \neq 0$). Models A and B represent special forms of model C. In model B, taste variation is allowed but each location is assumed to have the same variance and no correlation is permitted between store sizes within a given location. In model A no taste variation is allowed, no correlation is permitted and each location is assumed to have the same variance with the value $\pi^2/6$ (see equation [10.9], i.e. the composite random component ε_{ls} in [10.30] is assumed to have a constant variance $\pi^2/6$. Model A is, therefore, almost identical to the MNL model and, as such, it is often given the special name *identity probit*. (We will see in Section 10.4 that the identity probit model can play a useful function in the comparison of MNL and MNP models.)

For the purposes of this example, the substantive interpretation of the parameters in Table 10.4 need not concern us unduly (full details are provided by Miller and Lerman). We will simply note that the size and significance of locational variances in model C suggests that the assumption of independent and identically distributed random components contained in MNP models A and B and the MNL model appears

Table 10.3
The variance-covariance structures assumed in the MNP models of Table 10.4

	Location 1		Location 2			Location 7	
	Size 1	Size 2	Size 1	Size 2	. . .	Size 1	Size 2
Location 1							
Size 1	σ_1^2	$\zeta\sigma_1^2$					
			0		. . .	0	
Size 2	$\zeta\sigma_1^2$	σ_1^2					
Location 2							
Size 1			σ_2^2	$\zeta\sigma_2^2$			
	0				. . .	0	
Size 2			$\zeta\sigma_2^2$	σ_2^2			
.							
.							
.							
Location 7							
Size 1						σ_7^2	$\zeta\sigma_7^2$
	0		0				
Size 2						$\zeta\sigma_7^2$	σ_7^2

MNP model A: $\zeta = 0$; $\sigma_l^2 = \sigma_{l^*}^2$ for all l and l^*; Var$(\gamma_{ls}) = 0$

MNP model B: $\zeta = 0$; $\sigma_l^2 = \sigma_{l^*}^2$ for all l and l^*; Var$(\gamma_{ls}) \neq 0$

MNP model C: $\zeta \neq 0$; $\sigma_l^2 \neq \sigma_{l^*}^2$ for all l and l^*; Var$(\gamma_{ls}) \neq 0$

Table 10.4
Parameter estimates for MNL and MNP models with asymptotic 't statistics' in brackets (adapted from Tables 8 and 9 of Miller and Lerman 1981. Reproduced by permission of Pion Ltd)

	Logit	Probit		
	Model 1	Model A	Model B	Model C
Location specific parameters $(\beta_l)^a$				
Harvard Square	7.264	3.649	4.686	5.036
	(6.10)	(6.36)	(6.55)	(8.54)
Chestnut Hill	4.194	1.961	−1.320	−0.986
	(3.91)	(2.64)	(−1.47)	(−1.78)
Faneuil Hall	7.097	3.336	3.500	3.029
	(5.23)	(4.62)	(3.69)	(4.60)
Boston CBD	10.52	5.461	8.672	9.360
	(6.83)	(5.19)	(7.51)	(8.74)
Newbury/Boylston	8.905	4.862	6.485	6.541
	(6.98)	(6.34)	(7.02)	(7.65)
Prudential Center	7.277	3.480	2.968	3.386
	(4.79)	(5.44)	(3.46)	(3.46)
Small store parameter $(\beta_s)^b$	8.980	3.956	6.663	7.163
	(6.86)	(7.14)	(6.90)	(7.41)
Structural revenue parameters				
a_1	2.254	2.0778	2.4630	2.9382
	(10.97)	(8.57)	(4.08)	(3.59)
a_2	0.1459	0.1164	0.1464	0.2072
	(4.95)	(5.22)	(1.95)	(1.85)
a_3	0.2358	0.1972	0.1163	0.1723
	(5.16)	(4.43)	(1.02)	(1.37)
a_4	−0.5892	−0.4103	−0.6960	−0.9143
	(−4.73)	(−4.67)	(−3.56)	(−2.05)
a_5	0.3301	0.2661	0.7004	0.8727
	(4.01)	(3.50)	(1.65)	(2.08)
α^c	0.1300	0.0585	0.1130	0.1308
	(5.01)	(5.03)	(2.46)	(4.05)
Taste variation variance				
(var γ_{ls})		−	0.07826	0.06805
			(6.90)	(2.04)
Correlation between store sizes within a given location (ζ)		−	−	0.04749
				(0.28)
Location variances (σ_l^2)				
Harvard Square		1.6449^d	1.6449^d	1.3061
				(1.70)
Chestnut Hill				2.0602
				(2.28)
Burlington Mall				0.8290
				(0.97)
Faneuil Hall				1.6255
				(2.20)
Boston CDB				1.9235
				(3.10)
Newbury/Boylston				1.3686
				(2.42)
Prudential Center		↓	↓	1.6449^d

Notes: a. Burlington Mall used as base category for location specific parameters.
b. Store size 2,500–5,000 sq. ft. used as base category for store size parameters.
c. Cost parameter (β_c), see equation [10.31] constrained to value of $-\alpha$.
d. Set at this value ($= 1/6 \; \pi^2$).

not to be a reasonable assumption for this choice problem. More important, however, for our purposes is that Table 10.4 demonstrates that:

(a) the MNP is not simply one model. The reader must understand that there is a family of special cases of the general MNP form discussed in Section 10.3.1. The choice of MNP form simply depends upon what particular variance–covariance structure we are prepared to assume in any particular case;

(b) the MNP appears to be a feasible model even for choice problems with a reasonably large number of alternatives in the choice set. However, there are certain problems which are likely to be encountered in the estimation of large MNP models and more practical empirical experience is necessary before definitive statements can be made concerning the practicality of estimating large MNP models. (Miller and Lerman used a modified version of the program CHOMP and encountered certain difficulties, e.g. the likelihood function of the model was not completely 'well behaved', and the estimation procedure seemed to be very sensitive to the parameter initialization used.)

EXAMPLE 10.3   *Travel mode choice in Montreal*

Although the dogit model of Section 10.3.2 is a very widely discussed alternative to the MNL and MNP models, there are, as yet, very few empirical applications of the model. One of the few available is a study by Gaudry (1980) of travel mode choice in Montreal. For our purposes this is not the most satisfactory of illustrations as it concerns only two travel modes (public transportation and car); and is estimated using zonal market share data rather than individual choice data. Nevertheless, the study does have some useful and interesting features and will serve to reinforce some aspects of the theoretical discussion of Section 10.3.2.

The first of these interesting features concerns the form of the V_{ri} (representative utility) components used in the model. Essentially, Gaudry wished to compare the fit of logit and dogit models to travel mode market share data for Montreal. Rather than simply confine this comparison to the usual linear-in-parameters form of V_{ri}, Gaudry used a Box-Cox transformation to specify the more general functional form of the representative utility component.[†]

$$ V_{ri} = \beta_{1r} + \sum_{k=2}^{K} \beta_k \left(\frac{X_{rik}^{\xi_k} - 1}{\xi_k} \right) \qquad [10.32] $$

It should be noted that the usual linear form of V_{ri} is a special case of [10.32] where $\xi_2, \ldots \xi_K = 1$ and that there are many other special forms which can be derived including a multiplicative version. (For further discussion of this Box-Cox transformation see Hensher and Johnson 1981a, 186, 1981b; Koppelman 1981).

By substituting [10.32] into the dogit model expression [10.18] and by using three generic explanatory variables (OPTC, out of pocket fare;

[†] Note that in this particular case all the explanatory variables are assumed to be generic, and that an ASC is included.

IVT, in-vehicle travel time; WT, waiting time), Gaudry was able to specify a range of variations of the basic dogit model by imposing various restrictions on the nine possible parameters (β_1, β_2, β_3, β_4, μ_1, μ_2, ξ_2, ξ_3, ξ_4). Associated with each of these variants of the dogit model was a comparable logit model (i.e. the equivalent logit model could be obtained from the dogit model [10.18] by simply setting μ_1 and μ_2 to zero and retaining all other features). He then conducted a pairwise comparison of equivalent dogit and logit models. The parameter estimates from one of the best fitting of these pairs of models are shown in Table 10.5. The particular dogit model shown in Table 10.5 is that which is obtained when the parameters ξ_2, ξ_3, ξ_4 of [10.32] are constrained to be equal but the μ and β parameters are unconstrained.

Table 10.5
Parameter estimates from one matched pair of dogit and logit models

Parameter (plus variable or alternative to which it refers)		Dogit	Logit
μ_1	pub. trans	0.0414	0
μ_2	car	0.0468	0
ξ_2	OPTC	3.855	3.570
ξ_3	IVT	3.855	3.570
ξ_4	WT	3.855	3.570
β_2	OPTC	−11.489	−9.43
		(6.880)	(5.30)
β_3	IVT	−0.000004	−0.00001
		(0.000005)	(0.00001)
β_4	WT	(−0.002	−0.004
		(0.0004)	(0.0009)
β_1	ASC	0.38	0.35
		(0.10)	(0.09)

On the basis of certain test statistics and also (most interestingly) an inspection of the log-likelihood surface, Gaudry concluded that in this particular case the dogit model (with its two extra parameters μ_1 and μ_2) does not represent a significant improvement over the basic logit model. Not surprisingly, he also concluded that in this particular case there is no evidence of significant captivity to either public transport or car.

EXAMPLE 10.4 *Travel mode choice in the Rotterdam/Hague metropolitan area*

As part of a valuable introduction to nested logit models, Sobel (1980) presents an application of a nested logit model of travel mode choice extracted from a much larger regional planning study of the Rotterdam/ Hague metropolitan area in the Netherlands. The nested logit model is used to model the decision structure shown in Figure 10.6 and was estimated using observations on the work travel mode choices of 765 commuters.

Estimation of the nested logit model began in the lower nests (levels) of Fig. 10.6. For example, the car sub-model (i.e. whether to choose the car as a driver or passenger for commuting to work) takes the form shown in Table 10.6, and the 'slow' modes sub-model has a similar structure but includes twenty-one rather than nine variables. The parameter estimates from these lower nest sub-models were then substituted into equations of type [10.22] and were used to compute estimates of the required 'composite utility' terms. These estimated 'composite utility' terms were then included with twenty-four other variables in the upper nest mode choice model (i.e. choice between car, public transport or 'slow' modes). The parameter estimates for this 'main' model are shown in Table 10.7, and it is clear that the 'composite utility' parameter associated with 'car' sub-choices is significant and indicates a lack of independence between the car driver and car passenger modes. This lack of independence ('similarity of choice alternatives' problem) would be likely to result in unrealistic forecasts of mode share if a basic MNL model was used to model this mode choice problem.

Table 10.6
Car (lower nest) sub-model for Fig 10.6 (D indicates car driver mode, P indicates car passenger mode)

Variable (alternative specific to)	Parameter estimate	Standard error
ASC (D)	0.0995	0.5622
Number of cars (D)	1.29	0.389
Male commuter dummy (D)	0.930	0.335
Parking costs > 0 and arrival > 9.00 a.m. dummy (D)	2.65	1.244
Distance (P)	0.00931	0.00232
Parking cost (P)	0.390	0.130
Housing density, origin (P)	−0.000413	0.00015
Employment density, destination (P)	−0.000251	0.00013
Unfixed destination dummy (P)	−0.653	0.408

Fig. 10.6
Decision structure modelled in Tables 10.6 and 10.7

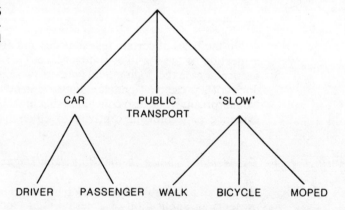

Finally, the reader should note that the standard errors in the upper levels models (e.g. Table 10.7) of a nested logit structure may be biased if standard MNL computer programs are used to compute the parameter estimates. The reason for this is that standard errors of the lower nest parameter estimates are 'passed through' and embedded in the estimated value of the composite utility term. They then have the effect of measurement errors of the composite utility term in the context of the higher nest model estimation. If there are more than two nests, such problems tend to accumulate in successively higher level composite

Table 10.7
Upper nest mode choice model for choice problem of Fig. 10.6 (*C* indicates car mode, *T* indicates public transport mode, *S* indicates 'slow' mode)

Variable (alternative specific to)	Parameter estimate	Standard error
ASC (*C*)	−1.61	0.548
ASC (*T*)	−0.222	0.590
Composite utility term (*C*)	0.477	0.177
Composite utility term (*S*)	0.384	0.282
In vehicle time (*C*, *T*)	−0.0069	0.009
Cost (*C*, *T*)	−0.108	0.089
Final walk time (*T*)	−0.0048	0.009
Outbound headway (*T*)	−0.0313	0.014
Return headway (*T*)	−0.0977	0.024
Rotterdam destination dummy (*T*)	0.826	0.355
Distance if < 10 km (*S*)	−0.363	0.047
Distance if 10–25 km (*S*)	0.326	0.084
Distance if > 25 km (*S*)	−0.780	0.456
Departure > 5.30 p.m. dummy (*S*)	−0.464	0.329
Lunch time trip dummy (*S*)	0.365	0.237
Age < 30 years dummy (*S*)	0.270	0.233
Male dummy (*S*)	0.822	0.337
The Hague destination dummy (*S*)	0.885	0.342
Male dummy (*C*)	0.346	0.398
Age > 55 years dummy (*C*)	−0.987	0.374
Number of cars per driving license if 1 car (*C*)	1.81	0.364
Number of cars per driving license if 2+ cars (*C*)	1.07	0.359
Parking cost > 0 and arrival > 9.00 a.m. dummy (*C*)	−1.00	0.704
No driving license dummy (*C*)	−1.97	0.568
White collar dummy (*C*)	−0.454	0.222
Peak period trip dummy (*C*)	−0.755	0.296

$$\rho^2 = 0.493$$

utility terms. A correction of the standard errors in the upper level models of a nested logit structure is, therefore, often desirable. McFadden (1981, 244, 252–260) illustrates the effect of such a correction and gives the necessary technical details.

10.4 Assessing and comparing the performance of alternative discrete choice models

Given the now wide range of alternative discrete choice models, each having associated advantages and limitations, there is clearly a need for formal procedures which enable us to assess and compare the performance of the various models. Simple measures of the performance of a single model includes the ρ^2 statistic, prediction success table, randomization inference goodness-of-fit test, residual plots, etc. discussed in Section 2.7. However, there are in addition a number of more specialized comparative tests which, for convenience, we can divide into three groups.

10.4.1   Tests of the IIA property of the MNL

A number of tests of this type have been suggested by McFadden, Train and Tye (1976, 1977). The rationale of these tests is that users should first consider whether or not the choice problem they are investigating is consistent with the IIA property of the MNL before considering the adoption of any of the less restrictive but more complex models discussed in Section 10.3

The first type of tests suggested by McFadden, Train and Tye are based upon transformed residuals

$$d_{ri} = \hat{\varepsilon}^*_{ri} - \hat{\varepsilon}^*_{Ri} \sqrt{\hat{P}_{r|i}} \; (1 - \sqrt{\hat{P}_{R|i}} \,)/(1 - \hat{P}_{R|i}) \qquad \text{[10.33]}$$

where

$$\hat{\varepsilon}^*_{ri} = (Y_{ri} - \hat{P}_{r|i})/\sqrt{\hat{P}_{r|i}} \qquad \text{[10.34]}$$

and

Y_{ri} are the observed choices of alternatives (i.e. 1 if r chosen, 0 otherwise).

Under the null hypothesis that the estimated MNL model is a correct specification, the standardized residuals $\hat{\varepsilon}^*_{ri}$ have, asymptotically, zero, mean, unit variance and covariance $E(\hat{\varepsilon}^*_{ri}\, \hat{\varepsilon}^*_{si}) = - \sqrt{\hat{P}_{r|i}\, \hat{P}_{s|i}}$. On the other hand, McFadden, Train and Tye suggest that the transformed residuals, d_{ri} are asymptotically independent, with zero mean and unit variance.

Using the transformed and standardized residuals, McFadden, Train and Tye suggest that three different tests can be performed.

1.   A test of the residual means. Under the null hypothesis that the estimated MNL model is a correct specification, the means of d_{ri} for each choice alternative will be zero. Since the variance of these residuals is asymptotically one, the statistics $\sum_{i=1}^{N} d_{ri}/N$ should be, asymptotically, standard normal. In reasonably large samples, McFadden, Train and Tye suggest that this normality assumption is usually valid, but in small samples they suggest it may be useful to use a non-parametric method (e.g. Wilcoxian sign test) to perform this test.

2.   A test of the residual variances. Unfortunately the statistic suggested by McFadden, Train and Tye is extremely unreliable when there are few exact repetitions of values of the explanatory variables (see Section 2.4.1) and consequently is of little value in the context of this chapter.

3.   A test of the association of residuals. In this test the estimated probabilities $\hat{P}_{r|i}$ for alternative r are ranked (highest to lowest) and the sign of the corresponding standardized residuals, $\hat{\varepsilon}^*_{ri}$ is associated with each $\hat{P}_{r|i}$. A table of the form 10.8 is then derived, in which the ranked estimated probabilities are divided into M cells with an average estimated probability $\bar{\hat{P}}_{r|m}$ in cell m.

In each cell, m, the number of associated positive $(RPOS_m)$ and negative $(N_m - RPOS_m)$ residuals are listed. Under the null hypothesis that the estimated MNL model is a correct specification, McFadden, Train and Tye claim that asymptotically $E(RPOS_m/N_m) = \hat{P}_{r|m}$ (i.e. the number of positive residuals is expected to be higher for cells where m is a low number than for cells where m is a high number). To test this

Table 10.8

Table for test of the association of residuals

	$m = 1$	2	...	M
Positive residuals	$RPOS_1$	$RPOS_2$	...	$RPOS_M$
Negative residuals	$N_1 - RPOS_1$	$N_2 - RPOS_2$	...	$N_M - RPOS_M$
Average estimated probability	$\bar{\hat{P}}_{r\|1}$	$\bar{\hat{P}}_{r\|2}$	...	$\bar{\hat{P}}_{r\|M}$

hypothesis they suggest the test statistic

$$\sum_{m=1}^{M} \frac{(RPOS_m - N_m \bar{\hat{P}}_{r|m})^2}{N_m \bar{\hat{P}}_{r|m}} \qquad [10.35]$$

which they claim has an asymptotic distribution bounded by chi-square distributions with $M-1$ and $M-K-1$ degrees of freedom (where K is the number of parameters estimated in the MNL model). If the computed value of statistic [10.35] lies above the higher of the two bounding chi-square critical values, the null hypothesis that the estimated MNL model is a correct specification is rejected. If the computed value falls between the two bounding critical values, the test is inconclusive. For each choice alternative a new table of positive and negative residuals must be constructed, and the test statistic [10.35] recalculated.

The second type of test suggested by McFadden, Train and Tye (see also Horowitz 1982) is known as the *universal logit* test. To understand this test, the reader must recall that, conventionally, in a MNL discrete choice model the utility of a specific choice alternative (U_{ri}) is considered to depend only on attributes of that particular alternative (r) and characteristics of the choice maker (i). In the universal logit test a MNL model based on this conventional specification of utility is compared with another generalized MNL model in which the utility of a choice alternative (r) is allowed to depend upon the attributes of other alternatives ($s \neq r$) as well. The generalized utility function in this alternative MNL model is constructed so that it includes the utility function of the conventional MNL model as a parametric special case. The conventional MNL special case in then tested against the generalized utility function model by means of a likelihood ratio test. Rejection of the empirical validity of the parametric special case implies that the MNL model being tested is not a correct specification.

In formal terms this test can be expressed in the following way. (a) Replace the expression for the representative component of utility[†] V_{ri}^* in [10.3], by the expanded expression $V^*(x_{ri}, \boldsymbol{\beta}_r)$ and write the representative component of the generalized MNL model as

$$V^*(x_{ri}, \boldsymbol{\beta}_r; \{x_{si} : s \neq r\}, \boldsymbol{\Psi}_r) \qquad [10.36]$$

In this expression $\{x_{si} : s \neq r\}$ represents attributes of alternatives other than r, and $\boldsymbol{\Psi}_r$ is a vector of parameters associated with $\{x_{si} : s \neq r\}$.

[†] It is convenient in terms of the notation adopted in this book to use V_{ri}^* rather than V_{ri} here. If V_{ri} were to be used then x_{ri} in Section 10.4.1 would have to be replaced by $\tilde{x}_{ri}$ for notational consistency (see footnote, page 317).

(b) The null hypothesis that the conventional MNL being tested is a correct specification implies that $H_0 : \boldsymbol{\Psi}_r = \mathbf{0}$ for all r, that is to say:

$$V^*(\boldsymbol{x}_{ri}, \boldsymbol{\beta}_r; \{\boldsymbol{x}_{si} : s \neq r\}, \boldsymbol{\Psi}_r) = V^*(\boldsymbol{x}_{ri}, \boldsymbol{\beta}_r) \qquad [10.37]$$

for all r.

(c) To test this null hypothesis we compare the maximized log-likelihood value of the conventional MNL model (denoted $\log_e \Lambda(\hat{\boldsymbol{\beta}})$ when $\boldsymbol{\Psi}_r = \boldsymbol{0}$ for all r) with the maximized log-likelihood value of the generalized MNL model, denoted $\log_e \Lambda(\hat{\boldsymbol{\beta}}, \hat{\boldsymbol{\Psi}})$. Under the null hypothesis, i.e. if the conventional MNL being tested is a correct specification, then the test statistic

$$-2 \left[\log_e \Lambda\,(\hat{\boldsymbol{\beta}}) - \log_e \Lambda\,(\hat{\boldsymbol{\beta}}, \,\hat{\boldsymbol{\Psi}})\right] \qquad [10.38]$$

is distributed asymptotically as chi-square with degrees of freedom equal to the number of extra parameters contained in the vectors $\boldsymbol{\Psi}_r$.

This test is straightforward to perform, it simply involves fitting two MNL models (the conventional and the generalized) to the same set of data. However, when there are a large number of alternatives in the choice set, the vectors, $\boldsymbol{\Psi}_r$ must together have a large number of parameters if the generalized utility function of each alternative is to include all the attributes of the other alternatives. This can make the estimation of the generalized MNL model computationally cumbersome, and it may cause the test to have low power because of the large number of degrees of freedom associated with extra parameters, $\boldsymbol{\Psi}_r$. These problems can be avoided by allowing only a small subset of the attributes $\{\boldsymbol{x}_{si} : s \neq r\}$ to enter the generalized utility function for alternative r. However, there are no commonly accepted systematic procedures for selecting an appropriate subset.

The third type of test suggested by McFadden, Train and Tye is known as the *conditional choice* test. This test exploits the fact that the IIA property of the MNL model implies that consistent estimates of the utility function parameters can be obtained using data sets which are restricted to a subset of the full set of choice alternatives. On this basis, a logical test of the validity of the MNL model involves dividing the total sample of choice makers into two independent groups. From the first group parameter estimates are obtained using the full set of choice alternatives, and from the second group parameter estimates are obtained using a restricted subset of choice alternatives. The equality of the two sets of parameter estimates is tested using a likelihood ratio test. Rejection of the hypothesis that the utility function parameters are the same for the full and restricted choice sets implies that the MNL model under consideration is not a valid specification.

Under the null hypothesis that the two sets of parameter estimates are the same (i.e. that the MNL model under consideration is a correct specification), the statistic

$$-2 \left[\log_e \Lambda(\hat{\boldsymbol{\beta}}_{FR}) - \{\log_e \Lambda(\hat{\boldsymbol{\beta}}_F) + \log_e \Lambda(\hat{\boldsymbol{\beta}}_R)\}\right] \qquad [10.39]$$

is asymptotically distributed as chi-square with K degrees of freedom, where K is the number of parameters in the model. In this test statistic, $\log_e \Lambda(\hat{\boldsymbol{\beta}}_F)$ is the maximized log likelihood for the full choice set group,

$\log_e \Lambda(\hat{\boldsymbol{\beta}}_R)$ is the maximized log likelihood for the restricted choice set group, and $\log_e \Lambda(\hat{\boldsymbol{\beta}}_{FR})$ is the maximized log-likelihood for the combined groups. For the test to be valid it is important that the two groups of choice makers are independent, otherwise the test statistic [10.39] is not distributed asymptotically as chi-square.

This test is computationally straightforward. However, in practice, two problems occur which make it slightly more difficult to use than it might seem. First, the results of the test depend on which choice alternatives are included in the restricted subset. As a result, it may be desirable to carry out the test several times using a different restricted choice subset each time. Second, the requirement that the two groups of choice makers be independent means that relatively large data sets may be needed to obtain good power with this test.

10.4.2 Tests of the MNL against specific alternative discrete choice models

A number of tests of this type have been suggested. For convenience, we will consider just two: tests of the MNL against MNP and nested logit model specifications.

Horowitz (1981a) has suggested an approximate likelihood ratio test of the MNL model against the MNP model which exploits the fact that the MNL model and the identity probit model (which was introduced in Example 10.2 and which has a variance-covariance matrix of the form $\boldsymbol{\Sigma}^{\mathbf{I}} = (\pi^2/6)\mathbf{I}$) are virtually identical. The test is performed in the following way.

(a) Let $V^*(\boldsymbol{x}_{ri}, \boldsymbol{\beta}_r)$ be the representative component of utility for the MNL model under consideration.

(b) Estimate the parameters $\boldsymbol{\beta}_r$ and $\boldsymbol{\Sigma}$ of a general MNP model which has the same representative component of utility as the MNL model.

(c) Exploit the virtual equivalence between the MNL and identity probit models and test the null hypothesis $H_0 : \boldsymbol{\Sigma} = (\pi^2/6)\,\mathbf{I}$. If this null hypothesis is rejected, then the implication is that the identity probit is not a valid alternative to the general MNP model, and, because of the virtual equivalence between the identity probit and MNL models, we can proceed to infer that the MNL model is not a valid specification for the particular data set under consideration.

(d) Test the null hypothesis by using the statistic

$$-2\,[\log_e \Lambda(\text{MNL}) - \log_e \Lambda(\text{MNP})] \qquad\qquad [10.40]$$

where $\log_e \Lambda(\text{MNL})$ and $\log_e \Lambda(\text{MNP})$ represent the maximized log likelihoods for the MNL and MNP models respectively. Under the null hypothesis, this test statistic has approximately a chi-square distribution with degrees of freedom equal to the number of non-redundant parameters in the variance-covariance matrix $\boldsymbol{\Sigma}$ of the general MNP model that must be constrained to achieve $\boldsymbol{\Sigma}^{\mathbf{I}}$, the identity probit form of the variance–covariance matrix. The distribution is approximate (even asymptotically) because, conventionally, $\log_e \Lambda(\text{MNL})$ is used in [10.40] rather than the maximized log likelihood obtained from the identity probit model.

Although it is often extremely useful, this test has the computational disadvantage of requiring the estimation of the parameters of a general MNP model. This makes the test extremely difficult to perform when there are large numbers of parameters or choice alternatives. There are, however, two possible ways to overcome this problem. The first involves constraining some of the components of Σ, i.e. reducing the number of parameters it is necessary to estimate by testing the MNL against a less general form of the MNP such as that described in model B in Example 10.2. The second possibility involves the development and use of a test statistic which retains the power of the approximate likelihood ratio test[†] but does not require the estimation of the parameters of the general MNP model. Horowitz (1981b, 1982) has suggested a Lagrangian multiplier test which meets these requirements, but discussion of the details of this test lies beyond the scope of this chapter.

Our second example of tests of the MNL against specific alternative discrete choice models involves the comparison of nested logit and simple MNL specifications. To understand this test the reader must recall that in Section 10.3.3 it was established that the parameter, δ, of the composite utility term of a two-level nested logit model must lie within the range $0 < \delta \leqslant 1$, and that when it equals 1 the model will yield equivalent results to a simple MNL model. Consequently, it follows that in a multi-level nested logit model, the MNL model is a special case where all the parameters δ's have the value 1. In addition, it follows that a parameter $\delta_j = 1$ at any particular level j in the hierarchy, implies the possibility of simplifying the nested logit model structure at that level. (For example, in Fig. 10.7a; if the δ at the second level equals 1, then that nested logit model structure is mathematically equivalent to, and can be simplified to, the structure in Fig. 10.7b). Sobel (1980) has suggested utilizing these properties as a means of testing alternative nested logit model structures. He suggests: (a) estimating the parameters of each feasible and reasonable nested logit model structure for a

Fig. 10.7
Simplifying a nested logit structure when composite utility parameter equals one

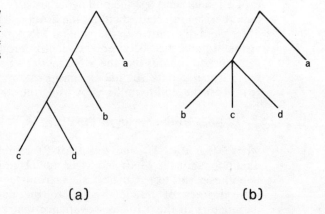

(a)   (b)

[†] Horowitz (1981a, 1982) has provided useful simulation-based evidence concerning the power of this approximate likelihood ratio test and also that of the universal logit and conditional choice tests discussed in Section 10.4.1 In the cases he considered, the approximate likelihood ratio test and the universal logit test were shown to have high power but the power of the conditional choice test was highly variable.

given number of choice alternatives; (b) rejecting a structure if any of the parameters, δ, do not satisfy the constraint $0 < \delta \leqslant 1$ (and also if the δ's do not decline in value when proceeding from lower to higher levels); (c) simplifying a structure if a δ is not significantly different from 1; (d) accepting the simple MNL model as the preferred form if all the δ's in a particular nested structure are approximately equal to 1. Clearly, this is a much more informal test procedure than Horowitz's likelihood ratio or Lagrangian multiplier test of the MNL against the MNP, but it does not preclude the use of more formal test procedures.

10.4.3   A generalized test procedure for comparing the performance of any pair of discrete choice models

One very general procedure for the comparison of competing discrete choice models has recently been suggested by Halperin $et\ al,$. (1984). This procedure is an adaptation of Hubert and Golledge's (1981) heuristic method of comparing related matrices (see also Golledge $et\ al.$ 1981) and it utilizes the correlation statistic

$$r_{\mathbf{A},(\mathbf{B}-\mathbf{C})} = \frac{r_{\mathbf{AB}} - \hat{r}_{\mathbf{AC}}}{\sqrt{2(1 - r_{\mathbf{BC}})}} \qquad [10.41]$$

where $\mathbf{A}$ is a proximity matrix representing actual choice data, and $\mathbf{B}$ and $\mathbf{C}$ are proximity matrices computed from the predicted probabilities for two different discrete choice models. In essence, the correlation statistic [10.41] enables the researcher to test the null hypothesis that neither model (represented by $\mathbf{B}$ or $\mathbf{C}$) is a better representation of the choice data than the other, and the null hypothesis is evaluated using a non-parametric inference strategy in which the observed value of the correlation statistic is compared to a reference distribution constructed using a randomization model. If the observed correlation is at a suitable extreme in the reference distribution, the null hypothesis is rejected. An extreme position in the upper-tail (lower-tail) of the reference distribution implies that the model represented by $\mathbf{B}$ ($\mathbf{C}$) is the better representation of the choice data.

In more detail, the proximity measure b_{ik} in $\mathbf{B}$ (or c_{ik} in $\mathbf{C}$) is defined by Halperin $et\ al.$ (1984) in terms of predicted probability differences for each choice alternative between individual i and individual k

$$b_{ik}\ (\text{or }c_{ik}) = \sqrt{\sum_{r=1}^{R} (\hat{P}_{r|i} - \hat{P}_{r|k})^2} \qquad [10.42]$$

and it is suggested that the proximity measures are then standardized to have a mean of zero and variance of one. The proximity measure a_{ik} in $\mathbf{A}$ is defined by Halperin $et\ al.$ as

$$a_{ik} = \frac{1}{2} \sum_{r=1}^{R} |Y_{ri} - Y_{rk}| \qquad [10.43]$$

where Y_{ri} and Y_{rk} are the observed choices of alternative r by individuals i and k (i.e. 1 if r is chosen, 0 otherwise). Further details of this test procedure and a geographical example of its use are to be found in Halperin $et\ al.$ (1984).

10.5   A brief guide to some remaining statistical issues

The aim of this chapter has been to provide an introduction to some of the basic concepts of discrete choice modelling and to demonstrate how the statistical models which underpin this research field are linked to the wider developments in the analysis of categorical data. A full coverage of discrete choice modelling has not been attempted as this would have required a book length treatment of the type provided by Hensher and Johnson (1981a) and Manski and McFadden (1981a). Nevertheless, some guide, albeit partial, to the remaining statistical issues may prove useful in alerting and directing the reader to the wider literature in the field. The brief summaries in this concluding section attempt to provide such a guide.

10.5.1   *Statistical transformations and the search for appropriate functional form*

In Example 10.3 we saw how Gaudry (1980) used a Box–Cox transformation of the explanatory variables to specify a more general form of the representative utility component, and how this allowed him to test a range of variations of his basic model form. This use of statistical transformations to search for appropriate functional form is now a widely adopted procedure in discrete choice modelling; however, many applications utilize a slightly more general transformation which has been termed the Box–Tukey transformation (see Gaudry and Wills 1978; Hensher and Johnson 1981b; McCarthy 1982; Wrigley and Longley 1984 for examples). This takes the form:

$$V_{ri} = \beta_{1r} + \sum_{k=2}^{K} \beta_{kr} (X_{rik} + \bar{\mu}_k)^{\xi_k} \qquad [10.44]$$

where

$$(X_{rik} + \bar{\mu}_k)^{\xi_k} = \frac{(X_{rik} + \bar{\mu}_k)^{\xi_k} - 1}{\varepsilon_k} \quad \text{if } (X_{rik} + \bar{\mu}_k) > 0, \ \xi_k \neq 0$$
$$= \log_e (X_{rik} + \bar{\mu}_k) \quad \text{if } \xi_k = 0 \qquad [10.45]$$

In [10.44] extra parameters $\bar{\mu}$ (location parameters) have been added to the basic power parameter ξ introduced in the Box–Cox transformation [10.32]. The value of each $\bar{\mu}_k$ is chosen to ensure that $X_{rik} + \bar{\mu}_k$ is greater than zero for all observations. Since it can be shown that as ξ_k approaches zero $((X_{rik} + \bar{\mu}_k)^{\xi_k} - 1)/\xi_k$ approaches the natural logarithm of $X_{rik} + \bar{\mu}_k$, $(X_{rik} + \bar{\mu}_k)^{\xi_k}$ is defined as $\log_e (X_{rik} + \bar{\mu}_k)$ at $\xi_k = 0$ so that the transformation is continuous for all possible values of ξ_k. It should be noted that the usual linear-in-parameters form of V_{ri} is merely a special case of [10.44] where the additional Box–Tukey scaling parameters $\xi_2, \ldots \xi_K$ equal 1 and $\bar{\mu}_2, \ldots \bar{\mu}_K$ equal 0.

Using the linear-in-parameters form as a reference point, a range of alternative functional forms can be assessed by successively setting the ξ and $\bar{\mu}$ parameters in [10.44] to particular values and maximizing the log-likelihood function of the model with respect to the β's for that particular combination of values of the Box–Tukey ξ and $\bar{\mu}$ parameters.

Using this search procedure, an 'optimal' combination of values of the ξ and $\bar{\mu}$ parameters can be found. In this way an 'optimal' functional form can be selected or, perhaps more importantly, some assessment can be made of the sensitivity of the discrete choice model to changes in the functional form of the representative component of utility. Such information can usefully inform any assessment of the predictions of the discrete choice model in the context of changes in policy variables, and can facilitate comparison of the fit of alternative types of discrete choice models.

Table 10.9 shows how the results of such a search procedure are typically portrayed. In this case, the maximized log-likelihood value is greatest when $\xi = 1.5$ and $\bar{\mu} = 0.001$ but the difference between this value and that for the usual linear-in-parameters, $\xi = 1$, form is small and it is unlikely that a likelihood ratio test would reject the null hypothesis that $\xi = 1$.

Table 10.9
Values of maximized log–likelihood functions at various ξ and $\bar{\mu}$ combinations.

		$\bar{\mu}$			
		0.001	10	20	30	40
	0	−961.539	−957.853	−957.379	−957.126	−956.938
	0.5	−949.320	−947.669	−947.472	−947.382	−947.322
ξ	1.0	−943.821	−943.821	−943.821	−943.821	−943.821
	1.5	−943.804	−944.994	−945.152	−945.206	−945.228
	2.0	−947.146	−949.142	−949.466	−949.577	−949.622

Note: Since, for $\bar{\mu}$ equal to zero, the Box–Tukey transformation is not defined when the explanatory variable is zero, the arbitrary value 0.001 is used rather than 0.

10.5.2 Sample design and parameter estimation

Just as the majority of applications of discrete choice models assume a simple linear-in-parameters additive form for the representative component of utility, so the majority of applications also assume that parameter estimation will be conducted using standard maximum likelihood techniques on sample survey data collected from simple or exogenously stratified random sample designs. However, just as there have been attempts to broaden the range of functional forms used in the application of discrete choice models, so there has been increased sensitivity to the implications and potential of alternative sampling designs.

The most basic sample design assumption is simple random sampling. In large samples this will yield reliable estimates of the market shares of the choice alternatives and of the mean levels of the attributes of the choice alternatives and the socio-economic characteristics of the decision-makers. In addition, such a sample design is conventionally assumed in the derivation of the standardly-applied maximum likelihood parameter estimation techniques. In most practical research contexts, however, simple random sampling can prove unnecessarily inefficient and costly. Given minimal *a priori* knowledge concerning attribute distributions within the population, considerable savings of resources and increases in efficiency can be achieved by appropriate stratification. Consequently, many applications of discrete choice modelling have emphasized sample survey procedures which embody stratification procedures of either of two types. The majority have used stratification criteria *exogenous* to the selected choice alternatives (e.g. the use of sex, age, area, occupation criteria to stratify a home interview sample survey pertaining to travel mode choice on the journey to work). In this procedure, the population is first classified into subsets on the basis of one or more exogenous variables and a random sample is then drawn from each group. However, the search for more refined and cost-effective sampling procedures in discrete choice modelling has frequently led to the adoption of an *endogenous* (choice-based) approach to sample stratification. In this procedure, the classification of the population into the subsets which will then be randomly sampled is based upon choices already made. That is to say, a random sample is drawn of individuals who are already committed to one particular choice alternative (e.g. in a study of travel mode choice, we may select 500 individuals using each travel mode; bus, car, train, etc., and this is referred to as 'on-board' choice-based sampling). In other words, in exogenously stratified sampling the analyst selects decision-makers and observes their choices, whilst in endogenously stratified samples the analyst selects choice alternatives and observes decision-makers choosing them.

Under certain standard regularity conditions, the maximum likelihood estimation procedures normally adopted in discrete choice modelling produce parameter estimators that are consistent, asymptotically normal and asymptotically efficient using either simple random or exogenously stratified random sampling procedures (see Section 2.4.2). However, endogenous (choice-based) stratified sampling presents additional estimation difficulties and, in such circumstances, the conventional maximum likelihood estimators will be inconsistent and asymptotically biased. These difficulties were neither fully acknowledged nor overcome prior to the work of Manski and Lerman (1977), Lerman and Manski (1979), Manski and McFadden (1981b) and Coslett (1981).

In a recent summary of the state-of-the-art of endogenous stratification, Manski (1981, 77–80) describes three alternative procedures which circumvent the associated likelihood estimation problems. The first of these assumes that the attribute density function, while not known, can *a priori* be restricted to a particular parametric family of functions prior to the maximization of the likelihood. The problem with this procedure is that the parametric restriction on the attribute density function may or may not be realistic and, in addition, computation is often costly. The second alternative does not require an assumption about the attribute density function. This is the so-called 'weighted exogenous sampling maximum likelihood' estimation procedure in which the estimators are obtained on the basis that the proportions of the population selecting each choice alternative (the aggregate market shares) are known. The third alternative is due to Coslett (1981) and involves the use of joint maximum likelihood estimation of the choice model parameters and the attribute density function.

Although these estimation procedures can present additional computational burdens for the researcher, they do produce parameter estimators with the desirable properties of consistency and asymptotic efficiency. This is important, for once suitable estimators are available a properly designed, endogenously stratified (choice-based) sample has the capacity in some circumstances to provide more precise estimates than a simple or exogenously stratified sample of the same total size. Likewise, if estimates are required to meet some prespecified level of precision, use of a properly designed endogenously stratified (choice-based) sample can often help to reduce the size and cost of the sample required. This is particularly true where certain choice alternatives are selected only very infrequently. In these circumstances a very large simple random sample may be needed to provide useful information on the infrequently chosen alternatives, and it may not be possible by stratifying on exogenous variables to find individuals with a high probability of selecting the infrequently chosen alternatives. For this reason, it has often been found to be a valuable practice in discrete choice modelling to supplement and enrich large exogenously stratified household samples with small choice-based surveys of the alternatives which occur infrequently but are of interest in the analysis.

The work of Manski, Lerman, McFadden and Cosslett serves to stress the intrinsic and centrally important relationship between sample survey design and statistical analysis, and research continues on this same theme (e.g. Daganzo 1980 has recently considered optimal sampling strategies for discrete choice models). However, there has as yet been surprisingly little linkage with the broader, and recently rapidly developing, literature on categorical data and complex (clustered, multistage, etc.) sample designs (e.g. Altham 1979; Brier 1980; Cohen 1976; Fellegi 1980; Holt, Scott and Ewings 1980), which was introduced in the context of Section 9.6.4.

10.5.3   Panel data and dynamic modelling

As we have seen in the preceding section, discrete choice models are normally estimated using cross-sectional sample survey data. However, in certain circumstances the researcher will have access to longitudinal survey data. It is well known in the standard statistics/econometrics literature that the use of a pooled time series of cross-sectional samples (i.e. panel data) is often much more efficient, in both statistical and behavioural terms, than the estimation of separate relationships for each cross-sectional sample. It is not unreasonable, therefore, to suggest that the same is likely to be true for discrete choice models, and that specifications analogous to those adopted in standard linear modelling of a time series of cross-sectional samples will be useful.

Tardiff (1980) adopts this approach and suggests that a useful replacement for the normal utility function [10.3] in such circumstances is

$$U_{rit} = x'_{rit}\boldsymbol{\beta} + \sum_s \beta^*_{rs} C_{si(t-1)} + \tilde{\varepsilon}_{ri} + \varepsilon^*_{rit} \qquad [10.46]$$

where U_{rit} = the utility of choice alternative r to individual i at time period t,

$C_{si(t-1)}$ = 1 if individual i chooses alternative s in the previous period $(t-1)$ and 0 otherwise,

$\tilde{\varepsilon}_{ri}$ = an error component that varies among individuals but not time periods, and

ε_{rit}^{*} = an error component that varies among both individuals and time periods.

In addition to permitting the use of data from a time series of cross-sectional samples, the revised specification of the utility function introduces two additional elements. First, through the use of the $\sum_{n} C_{si(t-1)}$ term, it allows choice in one period $(t-1)$ to influence choice in the following period. If the estimate of the associated parameter β_{rs}^{*} is positive (negative), it indicates an increased (decreased) choice probability in the subsequent time period. Second, the use of a component structure for the error term (i.e. the error term ε_{ri} in equation [10.3] is decomposed into the two components $\tilde{\varepsilon}_{ri}$ and ε_{rit}^{*}) allows for the fact (in $\tilde{\varepsilon}_{ri}$) that some unobserved attributes may remain constant across time periods for particular individuals, but that there will also be pure random disturbance (i.e. ε_{rit}^{*}).

By setting various elements in [10.46] to zero, Tardiff (1980) is able to consider three special cases of the general specification. Case 1 is where the researcher assumes that $\beta_{rs}^{*} = 0$ and $\tilde{\varepsilon}_{ri} = 0$ for all s, r and i. If these assumptions are valid the observations for a given individual over time can be treated as independent and the usual static discrete choice models can be applied directly to the dynamic problem. Case 2 is where the researcher assumes that $\tilde{\varepsilon}_{ri} = 0$ for all r and i. In this case previous choices are explicitly considered, but error terms are treated as completely random and independent across time periods and thus the usual type of discrete choice models can be applied directly. Case 3 is where the researcher assumes that $\beta_{rs}^{*} = 0$ for all r and s. In this case previous choices are not considered but both error components are considered (i.e. some unobserved attributes are now assumed to remain constant across time periods for particular individuals). As a result, error terms are assumed to be correlated over time and normal estimation procedures are no longer valid. In these circumstances, Tardiff (1980) and Heckman (1981) suggest two alternative procedures: (i) a 'fixed effects' approach in which the $\tilde{\varepsilon}_{ri}$ terms are explicitly identified using ASCs and then normal discrete choice models are applied directly; (ii) a 'random-effects' approach in which the more complex error variance structure is dealt with directly, in a manner analogous to the MNP model work on correlated random components among choice alternatives. If a Case 3 specification is erroneously estimated as a Case 1 type, or the 'full' specification [10.46] is erroneously estimated as a Case 2 type, further research by Tardiff (1979) suggests that exclusion of the $\tilde{\varepsilon}_{ri}$ component can result in two sources of bias in the parameter estimates: bias resulting from possible correlation between the error component and the other explanatory variables, and bias resulting from changes in the distribution of the overall error term of the utility function.

In behavioural terms, the use of panel data in discrete choice modelling and the associated utility function specification [10.46] serves to focus attention on the intertemporal nature of many choice processes. Effects of experience, time discounted preferences, the learning process, habit persistence, and so on, become centrally important issues. In other words many aspects of complex choice behaviour ignored in standard cross-sectional discrete modelling, and for which omission discrete choice modelling has been criticized in the past are now being incorporated (at least to a limited extent) in the theoretical and empirical studies of Heckman (1981), Tardiff (1980), Daganzo and Sheffi (1979; 1982) and Johnson and Hensher (1982). Recently panel data surveys have become more important in both transportation science and human geography, and there is evidence to suggest that the next few years will see considerable developments in the form, estimation and application of discrete choice models for panel data.[†]

10.5.4 Specification analysis: improper exclusion or inclusion of explanatory variables

An implict assumption common to all the previous sections of this chapter is that our discrete choice models are correctly specified in the sense of including all relevant explanatory variables. If the assumption is not true, two types of problems can arise: (i) from the exclusion of relevant explanatory variables from the model; (ii) from the inclusion of explanatory variables which are not part of the choice process.

Tardiff (1979) has provided what is currently the most detailed analysis of these problems. In this study he partitions the set of explanatory variables into two groups x_{ri}^1 the relevant and included explanatory variables, and x_{ri}^2 the excluded or improperly included (superfluous) variables. The conventional expression [10.3] for the utility of choice alternative r is then re-expressed as:[‡]

$$U_{ri} = x_{ri}^{1\prime} \, \boldsymbol{\beta}^1 + x_{ri}^{2\prime} \, \boldsymbol{\beta}^2 + \varepsilon_{ri} \qquad\qquad [10.47]$$

Using [10.47], Tardiff shows that the problem of inclusion of explanatory variables which are not part of the choice process is relatively straightforward. Standard consistency arguments (Manski and Lerman 1977) can be used to show that estimates of $\boldsymbol{\beta}^1$ converge in probability to the true values and that estimates of $\boldsymbol{\beta}^2$ converge in probability to zero. As in the case of conventional linear regression models, it appears that the inclusion of superfluous variables does not affect the consistency of the parameter estimates of the relevant variables but makes them inefficient.

The problem of excluded explanatory variables (underspecification) is somewhat more complex. In the special and relatively simple case where excluded explanatory variables are assumed to be correlated only with the included explanatory variables for the *same* choice alternative, i.e. where

$$x_{ri}^{2\prime} = x_{ri}^{1\prime} \, \boldsymbol{\beta}_r^3 + \bar{\bar{\boldsymbol{\varepsilon}}}_{ri}' \qquad\qquad [10.48]$$

Tardiff (1979) notes that the utility of choice alternative r can be

[†] Since completion of this text some important papers on this topic have appeared in the geographical literature by authors such as Davies, Pickles and Crouchley. See Dunn and Wrigley (1985) for references to this work and for a discussion of a class of logistic models for panel data.

[‡] In this expression the $\boldsymbol{\beta}$ vectors do not have a subscript r to indicate the choice alternative, i.e. all explanatory variables are assumed to be generic. However the reader should note that the assumption is relaxed in equation [10.49], i.e. explanatory variables are allowed to be alternative specific.

expressed (by substituting [10.48] into [10.47] as:

$$U_{ri} = x_{ri}^{1\prime} \, (\boldsymbol{\beta}^1 + \boldsymbol{\beta}_r^3 \boldsymbol{\beta}^2) + (\bar{\bar{\boldsymbol{\varepsilon}}}_{ri}' \boldsymbol{\beta}^2 + \varepsilon_{ri}) \qquad\qquad [10.49]$$

It can immediately be seen that the error structure of [10.49] differs from that of [10.47] and [10.3]. This implies that, depending on the distribution of $\bar{\bar{\boldsymbol{\varepsilon}}}_{ri}' \, \boldsymbol{\beta}^2$, [10.49] may produce choice models of either the same general family or completely different families than those produced by [10.47] and [10.3]. For example, if both ε_{ri} and $\bar{\bar{\boldsymbol{\varepsilon}}}_{ri}' \, \boldsymbol{\beta}^2$ are multivariate normally distributed, then both [10.49] and [10.47]–[10.3] yield MNP models, though the variance–covariance matrices of the resulting models will obviously differ. However, if ε_{ri} are independent and identical type-I extreme value distributed components then [10.47] and [10.3] yield a logit model but [10.49] yields selection probabilities from a S_B distribution (Westin 1974).

Because underspecification of a MNP model does not result in another type of choice model when the $\bar{\bar{\boldsymbol{\varepsilon}}}_{ri}$ of equation [10.48] are multivariate normally distributed, Tardiff (1979) focusses on this case as a means of determining the likely effects of underspecification on the parameter estimates of discrete choice models. To do this he first notes that [10.49] must be slightly modified to the form

$$U_{ri}^* = U_{ri}/k = x_{ri}^{1\prime}(\boldsymbol{\beta}^1 + \boldsymbol{\beta}_r^3 \boldsymbol{\beta}^2)/k + (\bar{\bar{\boldsymbol{\varepsilon}}}_{ri}' \boldsymbol{\beta}^2 + \varepsilon_{ri})/k \qquad [10.50]$$

The reason for this is that in the estimation of any choice model the magnitude of the parameter estimates are relative to the standard deviations of the error terms and, since all utilities can be multiplied by a constant and still convey the same information, it is necessary to arbitrarily fix one of the standard deviations (e.g. that of the first error component). When this is done in [10.47] the standard deviation of the corresponding error component in [10.49] is larger by a constant k. Since the interpretation of the parameter estimates from alternative model specifications requires the same normalization procedure, this implies that [10.49] should be multiplied by $1/k$ to give [10.50].

Equation [10.50] shows that the underspecification produces two sources of bias in the parameter estimates of the included variables: the $\boldsymbol{\beta}_r^3 \boldsymbol{\beta}^2$ term and the k term. As an example of the possible directions of bias, the simple case where both x_{ri}^1 and x_{ri}^2 consist of single variables (i.e. x_{ri}^1 and x_{ri}^2) and the associated β^1 and β^2 are both positive can be considered. In this case it can readily be seen that:

(a) if x_{ri}^2 is negatively correlated with x_{ri}^1 ($\beta_r^3 < 0$) both sources of bias contribute to the parameter estimate of x_{ri}^1 being biased down from β^1;

(b) if x_{ri}^1 and x_{ri}^2 are uncorrelated ($\beta_r^3 = 0$) then k will still result in the parameter estimate of x_{ri}^1 being biased downward from β^1;

(c) if x_{ri}^2 is positively correlated with x_{ri}^1 ($\beta_r^3 > 0$) the bias of the parameter estimate of x_{ri}^1 is ambiguous as the $\beta_r^3 \beta^2$ term contributes positively and the k term contributes negatively.

Although these conclusions regarding the effects of underspecification are confined to the MNP model, Tardiff believes that it is reasonable to expect the general findings on the sources and direction of bias to also apply to the MNL model. The reason for this is that the MNP results hold for the special case of the identity probit model (in which the variance–covariance matrix takes the form $\Sigma^{\mathbf{I}} = (\pi^2/6)\mathbf{I}$) and we have already noted (in Section 10.4.2) how the identity probit model and the MNL are virtually equivalent.

10.5.5   *Wider themes of empirical application*

This chapter has been concerned primarily with statistical models. As such, it has made no attempt to discuss the many wider issues which must be confronted in the empirical application of discrete choice models. For example, issues such as definition and representation of individual choice sets; the incorporation of important features of complex and constrained behaviour; the possibility of using choice experiment data from repeated experimentally controlled trials to complement conventional sample survey data; the possibility of transferring 'well defined' models to other geographical locations and to other time periods; and so on. These and other issues are discussed by Wrigley and Longley (1984), and it has been suggested that there is potential for a considerable enrichment of discrete choice models in both spatial and behavioural terms (see Burnett and Hanson 1982; Pipkin 1981). Significant research along these lines is beginning to emerge and it has major implications for behavioural geography (Thrift 1981; Rushton 1982) and for the potential integration of choice/preference-oriented and constraints/allocation-oriented approaches in urban geography.

PART 5

Towards integration

CHAPTER 11

An alternative framework

Categorical data analysis continues to be a major focus for advances in statistical methodology, whilst in geography and environmental science categorical data methods are becoming more widely adopted and are providing the core of a broader movement towards the more effective analysis of qualitative spatial data (Nijkamp *et al.* 1984). As such this final chapter can offer no 'conclusions'. The pace of development is still rapid and, despite its length, this book is essentially only a springboard from which the reader can explore the more specialized statistical literature, and from which he will be able to monitor and appreciate the many developments in theory and application which will occur in the next decade. In this spirit this final chapter will offer no summing up but will merely outline an alternative perspective from which the reader can view current methods and likely developments in the field.

11.1 The central classification scheme reconsidered

Of central importance to Part 1 of this book was the classificatory framework shown in Table 1.1. This framework provides an important pedagogic tool as it links together the conventional linear models discussed in existing statistics textbooks for geographers and environmental scientists and the new approaches to the analysis of categorical data. However, despite the pedagogic value of this framework there is nothing sacrosanct about it. For example, it is perhaps a valid general criticism of Table 1.1 that it rests upon a distinction between response and explanatory variables (i.e. a dependency distinction) which may be difficult to draw in some exploratory studies and that, in placing cell (g) at what might seem a loosely connected extremity, it fails to highlight the central importance of the Poisson distribution and the general log-linear model in the analysis of categorical data. For example, R.L. Plackett (1974), using the term 'factors' to refer to explanatory variables, has argued that: 'although the concepts of response and factor have a logical appeal and are widely used, they are often difficult to distinguish in practice . . . the basic (categorical) statistical model is one where the frequencies are assigned independent Poisson distributions. Under this model, constraints on frequencies are equivalent to the definition of factors . . . consequently a failure to distinguish fully between responses and factors serves only to modify the presentation of the results.' In addition, Table 1.1 suggests an approach to categorical data analysis in which a series of models, each with its own form, tradition and notation, are presented and linked together by a series of mathematical equivalences which occur at certain transition points. Although this has the advantage of introducing the reader to these various traditions and to the wide literature which now exists on categorical data analysis, and although it also brings with it benefits which flow from a 'horses for courses' approach to statistical modelling, it has the deficiencies which inevitably result from a separate treatment of each of the models and modelling traditions in categorical data analysis. Clearly, there are significant economies of derivation, notation, estimation and inferential procedure to be derived from a generalized framework which is capable of accommodating all categorical data models and linking these in a unified manner with conventional continuous data linear models. In this final chapter we will briefly consider such a framework: that provided by Nelder and Wedderburn's (1972) family of 'generalized linear models' (GLMs) and implemented by the GLIM computer package (see also Section 5.6.2 and Ch. 7).

11.2   The GLM framework

At the centre of the generalized linear models (GLMs) approach lies the concept that many apparently separate statistical models and modelling traditions can be reconciled in a single system with a common notation and unified estimation procedure. The potential benefits of this perspective are threefold. First, the algebraic differences which have characterized the categorical data models of Chapters 1 to 10 become largely redundant. Second, the consistent notation of GLMs allows users to more readily perceive the interrelatedness of models and encourages a more extensive experimentation in data analysis. Third, the GLM perspective not only simplifies the various categorical data methods to a single family of models, but also reconciles the methodologies of categorical data analysis with those of conventional continuous data analysis. Two perceptual gaps affecting the assimilation of categorical data methods in geography and environmental science are thus closed: the gap between the various categorical data approaches which has resulted in users limiting their attention to just one or two model types, and the gap between the categorical and continuous data literatures which has tended to inhibit the appreciation and use of categorical data methods by geographers and environmental scientists trained solely in the use of conventional continuous data linear models.

Generalized linear models as derived from the work of Nelder and Wedderburn (1972) may be expressed algebraically as

$$Y_i = \mu_i + \varepsilon_i \qquad\qquad\qquad\qquad\qquad\qquad\qquad\qquad \textbf{[11.1]}$$

where

Y_i is a response variable which is assumed to come from the exponential family of probability distributions

μ_i is the expected value (mean) of Y_i, i.e., $\mu_i = E(Y_i)$

ε_i is a randomly distributed error

A model of the variability of Y_i requires (i) information on the characteristics of other variables thought to influence the variation in Y_i; (ii) information on the linkage which relates the expected mean of Y_i to these other variables; and (iii) a more precise definition of the theoretical probability distribution assumed for Y_i.

The effect of other variables on the variability of Y_i is expressed as a 'linear predictor' (η_i) of the following form

$$\eta_i = \sum_k \beta_k X_{ik} \qquad [11.2]$$

where the

X_{ik} are known 'explanatory' variables, and the

β_k are parameters which are either known in advance or have to be estimated from the data

This linear predictor is related to the expected value of Y_i by a 'link function' (g)

$$\eta_i = g(\mu_i) \qquad [11.3]$$

which may also be written as

$$\mu_i = g^{-1}(\eta_i) \qquad [11.4]$$

where g^{-1} is the inverse of the link function. As the link (and its inverse) is normally identical for each observation, Y_i, there is no need to distinguish individual links by subscripts. The linear predictor and inverse link function may be combined to form a generalized linear model

$$Y_i = g^{-1}(\eta_i) + \varepsilon_i \qquad [11.5]$$

which may also be written as

$$Y_i \simeq F\left(g^{-1}\left[\sum_k \beta_k X_{ik}\right]\right) \qquad [11.6]$$

where F represents the assumed exponential family of probability distributions. These three components – linear predictor, link function and 'error distribution' – need to be stated explicitly in order to specify the model of variability in Y_i. Some further details of each component are now given. (Further details and examples are provided in McCullagh and Nelder 1983; Nelder 1974, 1983; Baker and Nelder 1978; Arminger 1983; O'Brien 1983; O'Brien and Wrigley 1984).

11.2.1   *The linear predictor*

In any GLM the expected mean of Y_i is related to the linear sum of a set of explanatory variables in the model. These explanatory variables may be continuous or categorical or a mixture of both, and the linear sum is termed the 'linear predictor'. Represented in matrix terms the linear predictor is given by $\mathbf{X}\boldsymbol{\beta}$, where the $\boldsymbol{\beta}$ parameters are either known in advance or require estimation from the data. Ultimately, the exact form of the linear predictor and the interpretation to be given the β depends on the processes of data generation and the experimental or observational design of the analysis.

11.2.2   The link function

The relationship of the expected value of Y_i to the linear predictor is specified by the link function which must be a monotonic differentiable function. A variety of link functions are available in practice. For categorical regression models appropriate links are provided by the logit and probit transformations. Similarly, general log-linear models (cell g problems) can be obtained using a linear link on the logarithmic scale. Continuous data models, which may also be accommodated in GLMs (e.g., analysis of variance, multivariate regression), usually require an identity or unitary link specified in the same measurement scale as the original observations. However, these are merely the most probable links a user might require in an analysis of data; other variants are possible, and may be specified within GLMs.

In addition, 'composite link functions' can be specified (Thompson and Baker 1981). Instead of a one-to-one relationship between the expected values μ_i and the linear predictors η_i, composite link functions allow each μ_i to be a function of several η s. In practice, this is done by defining each μ to be a linear combination of some intermediate quantities (say γ_i) which are themselves functions of the corresponding η_i.

11.2.3   The error distribution

The link function and linear predictor provide the mechanism for relating the expected value of Y_i to explanatory information in the model. However, such information is insufficient to assess the worth of the model. An important characteristic of statistical modelling is the ability to test hypothetical relationships in data for significance and strength. This possible only if the model can be associated with a tenable theoretical probability distribution for Y_i. The normal distribution provides the necessary mechanism for continuously distributed data. The Poisson distribution however is central in categorical data analysis, underlying, at least initially, all analyses involving such data. Other important distributions in this context are the binomial and the multinomial, and the relationship of these to the Poisson was considered in Section 1.4.

In the class of generalized linear models it is assumed that the response variable Y_i comes from the exponential family of probability distributions. As Figure 11.1 shows, this family includes the important distributions for categorical data analysis (Poisson, binomial, multinomial) and many other distributions which have played important roles in various areas of geographical and environmental science research. In practice, the exponential family will often be sufficiently broad to handle a majority of the statistical analysis requirements of many researchers. In addition, the exponential family has several attractive theoretical properties, not least those which enable parameters of

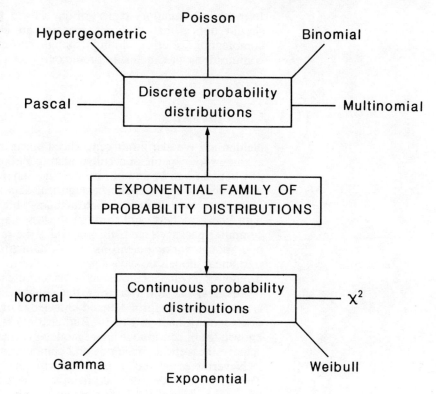

Fig. 11.1
Some members of the exponential family of distributions

exponential distribution models to be estimated with ease using straightforward computational methods.

11.2.4 *Some examples of GLMs*

Table 11.1 lists some examples of generalized linear models which can be derived from equations [11.5] and [11.6] along with their appropriate error distributions and link functions. The models presented correspond

Table 11.1
Examples of generalized linear models

Model	Error distribution	Link function
Linear regression	Normal	Identity
ANOVA (fixed effects)	Normal	Identity
ANOVA (random effects)	Gamma	Identity
Logistic/logit regression	Binomial or Multinomial	Logit
Binary probit regression	Binomial	Probit
Linear logit model for cell f problems	Binomial or Multinomial	Logit
General log-linear model for cell g problems	Poisson	Logarithmic
Log-linear model for cell f problems	Poisson	Logarithmic

to modelling situations typical of the cells of Table 1.1, and the reader should notice that different models can be specified by identical combinations of error and link (though such models will have different compositions of the linear predictor).

11.3   Conclusion

Nelder and Wedderburn's generalized linear models approach provides a framework capable of accommodating all the categorical data models of Chapters 1 to 10 and of revealing the inherent methodological unity of the recent approaches to categorical data analysis. Furthermore, it links these models in a unified manner with conventional continuous data linear models, and through the associated GLIM (and PRISM) computer package (see Ch. 7) provides the researcher with a powerful operational working environment for data analysis. However, generalized linear models do have a specialized terminology and notation, and it requires a resonable investment of the researcher's time to overcome these initial barriers. Moreover, the generalized linear models approach is not a substitute for a detailed knowledge of the models and methods discussed in Chapters 1 to 10. Rather, such knowledge (and equivalent knowledge of continuous linear models) is assumed and is built upon to achieve theoretical parsimony and computational parsimony in GLIM.

Nevertheless, if taken together with the pedagogic framework of Table 1.1 and with a detailed reading of at least Chapters 1 to 6 of this book, the generalized linear models approach provides an important additional perspective on categorical data analysis, and on the linkage of categorical and continuous data methodologies. In addition, the excellent specifications of the GLIM package and its now worldwide implementation, provides a powerful and flexible statistical modelling environment in which readers of this book can gain first-hand experience of using the models and methods discussed in Chapters 1 to 10.

REFERENCES

Aitchison, J. and **Bennett, J.A.** (1970) Polychotomous quantal response by maximum indicant, *Biometrika*, **57**, 253–62.

Aitkin, M. (1978) The analysis of unbalanced cross-classifications (with discussion), *Journal of the Royal Statistical Society*, Series A, **141**, 195–223.

Aitkin, M. (1979) A simultaneous test procedure for contingency table models, *Applied Statistics*, **28**, 233–42.

Aitkin, M. (1980) A note on the selection of log-linear models, *Biometrics*, **36**, 173–8.

Albright, Lerman, R. and **Manski, C.F.** (1977) *Report on the development of an estimation program for the multinomial probit model*, final report prepared by Cambridge Systematics Inc. (Boston) for US Federal Highway Administration.

Altham, P.M.E. (1976) Discrete variable analysis for individuals grouped into families, *Biometrika*, **63**, 263–9.

Altham, P.M.E. (1979) Detecting relationships between categorical variables observed over time: a problem of deflating a chi-squared statistic, *Applied Statistics*, **28**, 115–25.

Amemiya, T. (1975) Qualitative response models, *Annals of Economic and Social Measurement*, **4**, 263–72.

Amemiya, T. and **Nold, F.** (1975) A modified logit model, *Review of Economics and Statistics*, 255–7.

Anderson, R.C. (1955) Pebble lithology of the Marseilles Till Sheet in northeastern Illinois, *Journal of Geology*, **63**, 214–27.

Anscombe, F.J. (1953) Discussion of H. Hotelling's paper, *Journal of the Royal Statistical Society*, Series B, **15**, 229–30.

Arminger, G. (1984) Analysis of qualitative data and of latent class models with generalized linear models, Ch. 4 in Nijkamp, P., Leitner, H. and Wrigley, N. (Eds), *Measuring the Unmeasurable: Analysis of Qualitative Spatial Data*, Martinus Nijhoff, The Hague.

Asher, H.B. (1976) *Causal Modelling*, Sage, Beverly Hills, California.

Atherton, T.J. and **Ben-Akiva, M.E.** (1976) Transferability and updating of disaggregate travel demand models, *Transportation Research Record*, **610**, 12–18.

Baker, R.J. and **Nelder, J.A.** (1978) *The GLIM System: Release 3*, Numerical Algorithms Group, Oxford.

Beaver, R.J. (1977) Weighted least squares analysis of several univariate Bradley-Terry models, *Journal of the American Statistical Association*, **72**, 629–34.

Belsley, D.A., **Kuh, E.** and **Welsh, R.E.** (1980) *Regression Diagnostics: Identifying Influential Data and Sources of Collinearity*, Wiley, New York.

Ben-Akiva, M.E. (1973) Structure of passenger travel demand models. Department of Civil Enginering, MIT, Ph.D. dissertation.

Ben-Akiva, M.E. (1977) *Choice models with simple choice set generating processes* Working Paper, Centre for Transportation Studies. M.I.T., Massachusetts, U.S.A.

Ben-Akiva, M.E. and **Lerman, S.R.** (1979) Disaggregate travel and mobility choice models and measures of accessibility, pp. 654–79 in Hensher, D.A. and Stopher, P.R. (Eds), *Behavioural travel modelling*, Croom Helm, London.

Benedetti, J.K. and **Brown, M.B.** (1978) Strategies for the selection of log-linear models, *Biometrics* **34**, 680–6.

Bennett, R.J. (1979) *Spatial Time Series: Analysis, Forecasting and Control*, Pion, London.

Berkman, J. and **Brownstone, D.** (1979) *QUAIL 4.0 Users and Programmers Manuals*, Dept. of Economics, Princeton University.

Berkson, J. (1953) A statistically precise and relatively simple method of estimating the bioassay with quantal response, based on the logistic function, *Journal of the American Statistical Association*, **48**, 565–99.

Berkson, J. (1955) Maximum likelihood and minimum χ^2 estimates of the logistic function, *Journal of the American Statistical Association,* **50**, 130–62.

Besag, J.E. (1974) Spatial interaction and the statistical analysis of lattice systems, *Journal of the Royal Statistical Society*, Series B, **36**, 192–236.

Besag, J.E. (1975) The statistical analysis of non-lattice data, *The Statistician*, **24**, 179–95.

Bhapkar, V.P. (1979) On tests of marginal symmetry and quasi-symmetry in two and three-dimensional contingency tables, *Biometrics*, **35**, 417–26.

Bibby, J. (1977) The general linear model – a cautionary tale, pp.35–79 in Muircheartaigh, C.A. and Payne, C. (Eds) *The Analysis of Survey Data. Vol. 2: Model Fitting*, John Wiley, London.

Bielawski, E. and **Waters, N.M.** (1983) The use of probability surface mapping in predicting cultural affiliation from site locations in Aston Bay, Somerset Island, Northwest Territories, Mimeo, Dept, of Geography, University of Calgary.

Birch, M.W. (1963) Maximum likelihood in three-way contingency tables, *Journal of the Royal Statistical Society*, Series B, **25**, 220–33.

Birnbaum, I. (1982) The causal analysis of contingency tables: a return to first principles, *Quality and Quantity*, **16**, 217–41.

Bishop, Y.M.M. (1967) Multidimensional contingency tables: cell estimates, Department of Statistics, Harvard University, Ph.D. Dissertation.

Bishop, Y.M.M. Fienberg, S.E. and **Holland, P.W.** (1975) *Discrete Multivariate Analysis: Theory and Practice*, MIT Press, Cambridge, Massachusetts.

Blalock, H.M. (1964) *Causal Inferences in Non-Experimental Research*, University of North Carolina Press, Chapel Hill, North Carolina.

Blyth, C.R. (1972) On Simpson's paradox and the sure-thing principle, *Journal of the American Statistical Association*, **67**, 364–66.

Bock, R.D. (1975) *Multivariate Statistical Methods in Behavioural Research*, McGraw-Hill, New York.

Bock, R.D. and **Yates, G.** (1973) *MULTIQUAL: Log-linear Analysis of Nominal and Ordinal Qualitative Data by the Method of Maximum Likelihood*, International Educational Services, Chicago, Illinois.

Bowlby, S. and **Silk, J.** (1982) Analysis of qualitative data using GLIM: two examples based on shopping survey data, *Professional Geographer*, **34**, 80–90.

Bradley, R.A. and **Terry, M.B.** (1952) Rank analysis of incomplete block designs: 1 the method of paired comparisons, *Biometrika*, **39**, 324–45.

Brier, S.S. (1980) Analysis of contingency tables under cluster sampling, *Biometrika*, **67**, 591–96.

Brown, M.B. (1976) Screening effects in multidimensional contingency tables, *Applied Statistics*, **25**, 37–46.

Brown, M.B. (1981) BMDP routine P4F: two-way and multiway frequency tables – measures of association and the log-linear model (complete and incomplete tables), in Dixon, W.J. (Ed.) *BMDP Statistical Software 1981*, University of California Press, Los Angeles.

Bulmer, M.G. (1967) *Principles of Statistics*, Second Edition, Oliver and Boyd, Edinburgh.

Burnett, P. and **Hanson, S.** (1982) The analysis of travel as an example of complex human behaviour in spatially-constrained situations: definitions and measurement issues, *Transportation Research A*, **16A**, 87–102.

Chapman, G.P. (1981) Q-analysis, pp.235–47 in Wrigley, N. and Bennett, R.J. (Eds) *Quantitative Geography: A British View*, Routledge and Kegan Paul, London.

Chatterjee, S. and **Price, B.** (1977) *Regression Analysis by Example*, John Wiley, New York.

Chung, C.F. (1978) *Computer Program for the Logistic Model to Estimate the Probability of Occurence of Discrete Events*, Geological Survey of Canada, Paper 78–11, Ottawa.

Clark, C.E. (1961) The greatest of a finite set of random variables, *Operations Research*, **9**, 145–62.

Clarke, S.H. and **Koch, G.G.** (1976) The influence of income and other factors on whether criminal defendants go to prison, *Law and Society Review*, **11**, 57–92.

Clayton, D.G. (1974) Some odds ratio statistics for the analysis of ordered categorical data, *Biometrika*, **61**, 525–31.

Cliff, A.D. (1977) Quantitative methods: time series methods for modelling and forecasting, *Progress in Human Geography*, **1**, 492–502.

Cliff, A.D. and **Ord, J.K.** (1972) Testing for spatial autocorrelation among regression residuals, *Geographical Analysis*, **3**, 51–62.

Cliff, A.D. and **Ord, J.K.** (1973) *Spatial Autocorrelation*, Pion, London.

Cliff, A.D. and **Ord, J.K.** (1981) *Spatial Processes: Models and Applications*, Pion, London.

Cochran, W.G. (1954) Some methods for strengthening the common chi-square test, *Biometrics*, **10**, 417–51.

Cohen, J.E. (1976) The distribution of the chi-squared statistic under clustered sampling from contingency tables, *Journal of the American Statistical Association*, **71**, 665–670.

Coleman, J.S. (1964) *Introduction to Mathematical Sociology*, Free Press, New York.

Coslett, S.R. (1981) Efficient estimation of discrete choice models, pp.51–111 in Manski, C.F. and McFadden, D. (Eds) *Structural Analysis of Discrete Data with Econometric Applications*, MIT Press, Cambridge, Massachusetts.

Costanzo, C.M., Halperin, W.C., Gale, N.D. and **Richardson, G.D.** (1982) An alternative method for assessing goodness-of-fit for logit models, *Environment and Planning A*, **14**, 963–71.

Cox, D.R. (1970) *The Analysis of Binary Data*, Methuen, London.

Cragg, J.G. and **Lisco, T.E.** (1968) *PROLO-User Manual*, Dept. of Economics, University of Chicago.

Crittle, F.J. and **Johnson, L.W.** (1980) *Basic Logit (BLOGIT) - Technical Manual*, Australian Road Research Board, Technical Manual ATM No.9, Vermount South, Victoria, Australia.

Daganzo, C.F. (1979) *Multinomial probit: the theory and its application to demand forecasting*, Academic Press, New York.

Daganzo, C.F. (1980) Optimal sampling strategies for statistical models with discrete dependent variables, *Transportation Science*, **14**, 324–45.

Daganzo, C.F., Bouthelier, F and **Sheffi, Y.** (1977) Multinomial probit and qualitative choice: a computationally efficient algorithm, *Transportation Science*, **11**, 338–58.

Daganzo, C.F. and **Schonfeld, L.** (1978) *CHOMP User's Manual*, ITS Research Report UCB-ITS-RR-78–7, Institute of Transportation Studies, University of California, Berkeley.

Daganzo, C.F. and **Sheffi, Y.** (1979) Estimation of choice models from panel data, paper presented at 1979 Meeting of the Regional Science Association, Los Angeles, California.

Daganzo, C.F. and **Sheffi, Y.** (1982) Multinomial probit with time-series data: unifying state dependence and serial correlation models, *Environment and Planning A*, **14**, 1377–88.

Daly, A.J. and **Zachary, S.** (1978) Improved multiple choice models, pp.335–57 in Hensher, D.A. and Dalvi, M.Q. (Eds) *Determinants of Travel Choice*, Teakfield, Farnborough, Hampshire.

Darroch, A.G. and **Ornstein, M.D.** (1980) Ethnicity and occupational structure in Canada in 1871: the vertical mosaic in historical perspective, *Canadian Historical Review*, **61**, 305–33.

Davis, J.C. (1973) *Statistics and Data Analysis in Geology*, Wiley, New York.

Dear, M.J. and **Wittman, I.** (1980) Conflict over the location of mental health facilities, pp.345–62 in Herbert, D.T. and Johnston, R.J. (Eds) *Geography and the Urban Environment, Vol.3*, John Wiley, Chichester.

Debreu, G. (1960) Review of Luce, R.D. *Individual Choice Behaviour*, *American Economic Review*, **50**, 186–8.

Deming, W.E. and **Stephan, F.F.** (1940a) On a least squares adjustment of a sampled frequency table when the expected marginal totals are known, *Annals of Mathematical Statistics*, **11**, 427–44.

Deming, W.E. and **Stephan, F.F.** (1940b) The sampling procedure of the 1940 population census, *Journal of the American Statistical Association*, **35**, 615–30.

Diggle, P.J. (1979) Statistical methods for spatial point patterns in ecology, in Cormack, R.M. and Ord, J.K. (Eds) *Spatial and Temporal Analysis in Ecology*, International Cooperative Publishing House, Maryland, U.S.A.

Domencich, T.A. and **McFadden, D.** (1975) *Urban Travel Demand: A Behavioural Analysis*. North-Holland, Amsterdam.

Doveton, J.H. (1973) Numerical analysis relating location of hydrocarbon traps to structure and stratigraphy of the Mississippian 'B' of Stafford County, South-Central Kansas, *Technical Report, KOX Project*, Geological Research Section, Kansas Geological Survey.

Duncan, O.D. (1966) Path analysis: sociological examples, *American Journal of Sociology*, **72**, 1–16.

Duncan, O.D. (1974) Footnotes, *Proceedings of the American Sociological Association*.

Duncan, O.D. (1975) *Introduction to Structural Equation Models*, Academic Press, New York.

Duncan, O.D. (1979) How destination depends on origin in the occupational mobility table, *American Journal of Sociology*, **84**, 793–803.

Dunn, R. and **Wrigley, N.** (1985) Beta-logistic models of urban shopping centre choice, *Geographical Analysis*, **17** (2).

Ebdon, D. (1977) *Statistics in Geography: A Practical Approach*, Blackwell, Oxford.

Edwards, A.W.F. (1972) *Likelihood*, Cambridge University Press, Cambridge.

Engleman, L. (1979) BMDP routine PLR: stepwise logistic regression, in Dixon, W.J. (Ed) *BMDP-79 Biomedical Computer Programs P-Series*, University of California Press, Los Angeles.

Everitt, B.S. (1977) *The Analysis of Contingency Tables*, Chapman and Hall, London.

Evers, M. and **Namboodri, N.K.** (1978) On the design matrix strategy in the analysis of categorical data, pp.86–111 in Schuessler, K.F. (Ed.) *Sociological Methodology 1979*, Jossey-Bass, San Francisco.

Fay, R.E. and **Goodman, L.A.** (1975) *The ECTA Program: Description for Users*, Dept. of Statistics, University of Chicago.

Fellegi, P. (1980) Approximate tests of independence and goodness of fit based on stratified multistage samples, *Journal of the American Statistical Association*, **75**, 261–75.

Ferguson, R.I. (1977) *Linear Regression in Geography*, Concepts and Techniques in Modern Geography, **15**, Geo Abstracts, Norwich.

Fienberg, S.E. (1968) The estimation of cell probabilities in two-way contingency tables, Department of Statistics, Harvard University, Ph.D. dissertation.

Fienberg, S.E. (1977) *The Analysis of Cross-Classified Categorical Data*, MIT Press, Cambridge, Massachusetts.

Fienberg, S.E. (1978) A note on fitting and interpreting parameters in models for categorical data, pp.112–18 in Schuessler, K.F. (Ed.) *Sociological Methodology 1979*, Jossey-Bass, San Francisco.

Fienberg, S.E. (1979) The use of chi-squared statistics for categorical data problems, *Journal of the Royal Statistical Society, Series B*, **41**, 54–64.

Fienberg, S.E. (1980) *The Analysis of Cross-Classified Categorical Data*, Second Edition, MIT Press, Cambridge, Massachusetts.

Fienberg, S.E. and **Holland, P.W.** (1970) Methods for eliminating zero counts in contingency tables, pp.233–60 in Patil, G.P. (Ed.) *Random Counts in Models and Structures*, Pennsylvania State University Press, University Park, Pennsylvania.

Fienberg, S.E. and **Holland, P.W.** (1973) Simultaneous estimation of multinomial cell probabilities, *Journal of the American Statistical Association*, **68**, 638–91.

Fienberg, S.E. and **Mason, W.M.** (1978) Identification and estimation of age-period-cohort models in the analysis of discrete archival data, pp.1–67 in Schuessler, K.F. (Ed.) *Sociological Methodology 1979*, Jossey-Bass, San Francisco.

Fingleton, B. (1981) Log-linear modelling of geographical contingency tables, *Environment and Planning A*, **13**, 1539–51.

Fingleton, B. (1983) Log-linear models with dependent spatial data, *Environment and Planning A*, **15**, 801–13.

Finney, D.J. (1971) *Probit Analysis*, Third Edition, Cambridge University Press, Cambridge.

Fleiss, J.L. and **Everitt, B.S.** (1971) Comparing the marginal totals of square contingency tables, *British Journal of Mathematical and Statistical Psychology*, **24**, 117–23.

Forthofer, R.N. and **Koch, G.G.** (1973) An analysis of compounded functions of categorical data, *Biometrics*, **29**, 143–57.

Forthofer, R.N., Starmer, C.F. and **Grizzle, J.E.** (1971) A program for the analysis of categorical data by linear models, *Journal of Biomedical Systems*, **2**, 3–48.

Galbraith, R.A. and **Hensher, D.A.** (1982) Intra-metropolitan transferability of mode choice models, *Journal of Transport Economics and Policy*, **16**, 7–29.

Gart, J.J. and **Zweifel, J.R.** (1967) On the bias of various estimators of the logit and its variance with application to quantal bioassay, *Biometrika*, **54**, 181–7.

Gatrell, A.C. (1981a) Multidimensional scaling, pp.151–63 in Wrigley, N. and Bennett, R.J. (Eds) *Quantitative Geography: A British View*, Routledge and Kegan Paul, London.

Gatrell, A.C. (1981b) On the structure of urban social areas: explorations using Q-analysis, *Transactions of the Institute of British Geographers*, New Series, **6**, 228–45.

Gaudry, M.J.I. (1980) Dogit and logit models of travel mode choice in Montreal, *Canadian Journal of Economics*, **13**, 268–79.

Gaudry, M.J.I. and **Dagenais, M.G.** (1979) The dogit model, *Transportation Research B*, **13B**, 105–11.

Gaudry, M.J.I. and **Wills, M.J.** (1978) Estimating the functional form of travel demand models, *Transportation Research*, **12**, 257–89.

Glick, B.J. (1979) Tests for space-time clustering used in cancer research, *Geographical Analysis*, **11**, 202–8.

Goldfeld, S.M. and **Quandt, R.F.** (1972) *Nonlinear Methods in Econometrics*, North-Holland, Amsterdam.

Golledge, R.G., Hubert, L.J. and **Richardson, G.D.** (1981) The comparison of related data sets: examples from multidimensional scaling and cluster analysis, *Papers of the Regional Science Association*, **48**, 57–66.

Goodman, L.A. (1964) Simultaneous confidence limits for cross-product ratios in contingency tables, *Journal of the Royal Statistical Society*, Series B, **26**, 86–102.

Goodman, L.A. (1968) The analysis of cross-classified data: independence, quasi-independence and interaction in contingency tables with or without missing cells, *Journal of the American Statistical Association*, **63**, 1091–1131.

Goodman, L.A. (1969) How to ransack social mobility tables and other kinds of cross-classification tables, *American Journal of Sociology*, **75**, 1–40.

Goodman, L.A. (1970) The multivariate analysis of qualitative data: interactions among multiple classifications, *Journal of the American Statistical Association*, **65**, 226–56.

Goodman, L.A. (1971a) The analysis of multidimensional tables: stepwise procedures and direct estimation methods for building models for multiple classifications, *Technometrics*, **13**, 33–61.

Goodman, L.A. (1971b) Partitioning of chi-square analysis of marginal contingency tables and estimation of expected frequencies in multidimensional contingency tables, *Journal of the American Statistical Association*, **66**, 339–44.

Goodman, L.A. (1972a) A general model for the analysis of surveys, *American Journal of Sociology*, **77**, 1035–86.

Goodman, L.A. (1972b) A modified multiple regression approach to the analysis of dichotomous variables, *American Sociological Review*, **37**, 28–46.

Goodman, L.A. (1972c) Some multiplicative models for the analysis of cross-classified data, pp.649–96 in Le Carn, L. (Ed.) *Proceedings of the Sixth Berkeley Symposium on Mathematical Statistics and Probability Vol. 1*, University of California Press, Berkeley, California.

Goodman, L.A. (1973a) Causal analysis of data from panel studies and other kinds of surveys, *American Journal of Sociology*, **78**, 1135–91.

Goodman, L.A. (1973b) The analysis of multidimensional contingency tables when some variables are posterior to others: a modified path analysis approach, *Biometrika*, **60**, 179–92.

Goodman, L.A. (1979a) Multiplicative models for the analysis of occupational mobility tables and other kinds of cross-classification tables, *American Journal of Sociology*, **84**, 804–19.

Goodman, L.A. (1979b) A brief guide to the causal analysis of data from surveys, *American Journal of Sociology*, **84**, 1078–95.

Gould, P.R. (1975) *People in Information Space: The Mental Maps and Information Surfaces of Sweden*, Lund Studies in Geography B42, Dept. of Geography, University of Lund, Sweden.

Gould, P.R. (1980) Q-analysis, or a language of structure: an introduction for social scientists, geographers and planners, *International Journal of Man-Machine Studies*, **12**, 169–99.

Gregory, S. (1978) *Statistical Methods and the Geographer*, Fourth Edition, Longman, London.

Griffin, L.R. (1984) The utility of log-linear models in the analysis of the geography of mental health, Dept. of Geography, University of Bristol, Ph.D. dissertation.

Griffith, D.A. (1980) Towards a theory of spatial statistics, *Geographical Analysis*, **12**, 325–39.

Grizzle, J.E., Starmer, C.F. and **Koch, G.G.** (1969) Analysis of categorical data by linear models, *Biometrics*, **25**, 489–504.

Gurland, J., Lee, I. and **Dahm, P.A.** (1960) Polychotomous quantal response in biological assay, *Biometrics*, **16**, 382–98.

Haberman, S.J. (1970) The general log-linear model. Department of Statistics, University of Chicago, Ph.D. dissertation.

Haberman, S.J. (1972) Log-linear fit for contingency tables, *Applied Statistics*, **21**, 218–25.

Haberman, S.J. (1973a) The analysis of residuals in cross-classified tables, *Biometrics*, **29**, 205–220.

Haberman, S.J. (1973b) *C-TAB: Analysis of Multidimensional Contingency Tables by Log-Linear Models: User's Guide*, International Educational Services, Chicago, Illinois.

Haberman, S.J. (1974a) *The Analysis of Frequency Data*, University of Chicago Press, Chicago.

Haberman, S.J. (1974b) Log-linear models for frequency tables with ordered classifications, *Biometrics*, **30**, 589–600.

Haberman, S.J. (1978) *Analysis of Qualitative Data. Volume 1, Introductory Topics*, Academic Press, New York.

Haberman, S.J. (1979) *Analysis of Qualitative Data. Volume 2, New Developments*, Academic Press, New York.

Haggett, P. (1981) The edges of space, pp.51–70 in R.J. Bennett (Ed.) *European Progress in Spatial Analysis*, Pion, London.

Haggett, P., Cliff, A.D. and **Frey, A.** (1977) *Locational Analysis in Human Geography*, Second Edition, Edward Arnold, London

Haining, R.P. (1980) Spatial autocorrelation problems, pp.1–44 in Herbert, D.T. and Johnston, R.J.(Eds) *Geography and the Urban Environment Vol. 3*, John Wiley, Chichester.

Haining, R.P. (1982) Interaction models and spatial diffusion processes, *Geographical Analysis*, **14**, 95–108.

Halperin, W.C., Richardson, G.D., Gale, N.D. and **Costanzo, C.M.** (1984) A generalized procedure for comparing models of spatial choice, *Environment and Planning A*, **16**, 1289–1301.

Hammond, R. and **McCullagh, P.S.** (1978) *Quantitative Techniques in Geography: An Introduction*, Oxford University Press, Oxford.

Hanushek, E.A. and **Jackson, J.E.** (1977) *Statistical Methods for Social Scientists*, Academic Press, New York.

Harbaugh, J.W., Doveton, J.H. and **Davis, J.C.** (1977) *Probability Methods in Oil Exploration*, John Wiley, New York.

Hauser, R.M. (1978) A structural model of the mobility table, *Social Forces*, **56**, 919–53.

Hausman, J. and **Wise, D.A.** (1978) A conditional probit model for qualitative choice: discrete decisions recognizing interdependence and heterogenous preferences, *Econometrica*, **46**, 403–26.

Heckman, J.J. (1977) Simultaneous equations models with both continuous and discrete endogenous variables with and without structural shift in the equation, in Goldfeld, S.M. and Quandt, R.E. (Eds) *Studies in Nonlinear Estimation*, Ballinger, Cambridge, Massachusetts.

Heckman, J.J. (1978) Dummy endogenous variables in a simultaneous equation system. *Econometrica*, **46**, 931–59.

Heckman, J.J. (1981) Statistical models for discrete panel data, pp.114–78 in Manski, C.F. and McFadden, D. (Eds) *Structural Analysis of Discrete Data: with Econometric Applications*, MIT Press, Cambridge, Massachusetts.

Hensher, D.A. (1979) Individual choice modelling with discrete commodities: theory and application to the Tasman Bridge re-opening, *Economic Record*, **50**, 243–61.

Hensher, D.A. (1980) The demand for location and accommodation – a qualitative choice approach, in *Housing Economics*, Australian Government Publishing Services, Canberra.

Hensher, D.A. (1981) A practical concern about the relevance of alternative-specific constants for new alternatives in simple logit models, *Transportation Research B*, **15B**, 407–410.

Hensher, D.A. (1984) BLOGIT—An abridged users guide, with example program inputs and outputs and interpretative guidelines, School of Economic and Financial Studies, Macquarie University, NSW 2113 Australia.

Hensher, D.A. and **Johnson, L.W.** (1981a) *Applied Discrete Choice Modelling*, Croom Helm, London.

Hensher, D.A. and **Johnson, L.W.** (1981b) Behavioural response and the form of the representative component of the indirect utility function in travel choice models. *Regional Science and Urban Economics*, **11**, 559–72.

Hensher, D.A. and **Louviere, J.J.** (1981) An integration of probabilistic discrete choice models and experimental design data: an application in cultural economics, School of Economic and Financial Studies, Macquarie University, NSW 2113, Australia.

Hensher, D.A. and **McLeod, P.B.** (1977) Towards an integrated approach to the identification and evaluation of the transport determinants of travel choice, *Transportation Research*, **11**, 77–93.

Hensher, D.A. and **Manefield, T.** (1981) A structured logit model of automobile acquisition and type choice: some preliminary evidence, School of Economics and Financial Studies, Macquarie University, NSW, Australia.

Hepple, L.W. (1974) The impact of stochastic process theory upon spatial analysis in human geography, pp.89–142 in Board, C., Chorley, R.J., Haggett, P. and Stoddart, D.R. (Eds), *Progress in Geography 6*, Edward Arnold, London.

Higgins, J.E. and **Koch G.G.** (1977) Variable selection and generalized chi-square analysis of categorical data applied to a large cross-sectional occupational health survey, *International Statistical Review*, **45**, 51–62.

Holt, D. (1979) Log-linear models for contingency table analysis: on the interpretation of parameters, *Sociological Methods and Research* **7**, 330–6.

Holt, D., Scott, A.J. and **Ewings, P.D.** (1980) Chi-squared tests with survey data, *Journal of the Royal Statistical Society, Series A*, **143**, 303–20.

Horowitz, J. (1981a) Indentification and diagnosis of specification errors in the multinomial logit model, *Transportation Research B*, **15B**, 345–60.

Horowitz, J. (1981b) Testing the multinomial logit against the multinomial probit without estimating the probit parameters, *Transportation Science*, **15**, 153–63.

Horowitz, J. (1982) Specification tests for probabilistic choice models, *Transportation Research A*, **16A**, 383–94.

Huang, D.S. (1970) *Regression and Econometric Methods*, Wiley, New York.

Hubert, L.J. and **Golledge, R.G.** (1981) A heuristic method for the comparison of related structures, *Journal of Mathematical Psychology*, **23**, 214–26.

Hubert, L. J. and **Golledge, R.G.** (1982) Comparing rectangular data matrices, *Environment and Planning A*, **14**, 1087–95.

Hubert, L. J., Golledge, R.G. and **Costanzo, C. M.** (1981) Generalized procedures for evaluating spatial autocorrelation, *Geographical Analysis*, **13**, 224–33.

Imrey, P. B., Johnson, W. D. and **Koch, G. G.** (1976) An incomplete contingency table approach to paired-comparison experiments, *Journal of the American Statistical Association*, **71**, 614–23.

Ireland, C. T., Ku, H. H. and **Kullback, S.** (1969) Symmetry and marginal homogeneity of a $r \times r$ contingency table, *Journal of the American Statistical Association*, **64**, 1323–41.

Johnson, L. W. and **Hensher, D. A.** (1982) Application of multinomial probit to a two-period panel data set, *Transportation Research A*, **16A**, 457–64.

Johnson, J. and **Wanmali, S.** (1981) A Q-analysis of periodic market sy[...] *Geographical Analysis,* **13**, 262–75.

Johnston, J. (1972) *Econometric Methods* Second Edition, McGraw-Hill, Ne[...] York.

Johnston, R. J. (1978) *Multivariate Statistical Analysis in Geography*, Longman, London.

Jones, R. H. (1975) Probability estimation using a multinomial logistic function, *Journal of Statistical Computer Simulation,* **3**, 315–29.

King. L. J. (1969) *Statistical Analysis in Geography*, Prentice Hall, Englewood Cliffs, New Jersey.

Kirkland, J. S. (1969) Housing filtration in Kingston 1953–1968, Department of Geography, Queens University, Canada, M. A. dissertation.

Kmenta, J. (1971) *Elements of Econometrics*, Macmillan, New York.

Knoke, D. and **Burke, P. J.** (1980) *Log-Linear Models*, Sage, Beverly Hills, California.

Koch, G. G. (1978) The interface between statistical methodology and statistical practice, *Proceedings of the ASA Social Statistics Section 1977*, 205–14, American Statistical Association, Washington, D.C.

Koch, G. G., Freeman, D. H. and **Freeman, J. L.** (1975) Strategies in the multivariate analysis of data from complex surveys, *International Statistical Review*, **43**, 59–78.

Koch, G. G., Freeman, J. L. and **Lehnen, R. G.** (1976) A general methodology for the analysis of ranked policy preference data, *International Statistical Review*, **44**, 1–28.

Koch, G. G., Imrey, P. B. and **Reinfurt, D.W.** (1972) Linear model analysis of categorical data with incomplete response vectors, *Biometrics*, **28**, 663–92.

Koch, G. G., Landis, J. R., Freeman, J. L., Freeman, D. H. and **Lehnen, R.G.** (1977) A general methodology for the analysis of experiments with repeated measurement of categorical data, *Biometrics*, **33**, 133–58.

Koch, G. G. and **Reinfurt, D. W.** (1971) The analysis of categorical data from mixed models, *Biometrics*, **27**, 157–73.

Koch, G. G. and **Reinfurt, D. W.** (1974) An analysis of the relationship between driver injury and vehicle age for automobiles involved in North Carolina accidents during 1966–70, *Accident Analysis and Prevention*, **6**, 1–18.

Koppelman, F. S. (1981) Non-linear functions in models of travel choice behaviour, *Transportation*, **10**, 127–46.

Kritzer, H. M. (1978) An introduction to multivariate contingency table analysis, *American Journal of Political Science*, **22**, 187–226.

Kritzer, H. M. (1981) *NONMET II. A Program for the Analysis of Contingency Tables and Other Types of Non-Metric Data by Weighted Least Squares*, Version 6.11, Dept. of Political Science, University of Wisconsin, Madison.

Krumbein, W. C. and **Graybill, F. A.** (1965) *An Introduction to Statistical Models in Geology*, McGraw-Hill, New York.

Kullback, S. (1971) Marginal homogeneity of multidimensional contingency tables, *Annals of Mathematical Statistics*, **42**, 594–606.

Landis, J. R., Cooper, M.M., Kennedy, T. and **Koch, G. G.** (1978) PARCAT: A computer program for testing average partial association in three-way contingency tables, *Proceedings of the ASA Statistical Computing Section 1977*, 288–92, American Statistical Association, Washington, D.C.

Landis, J. R. and **Koch, G. G.** (1977) The measurement of observer agreement for categorical data, *Biometrics*, **33**, 159–74.

Landis, J. R., Stanish, W. M., Freeman, J. L. and **Koch, G. G.** (1976) A computer program for the generalized chi-square analysis of categorical data using weighted least squares (GENCAT), *Computer Programs in Biomedicine*, **6**, 196–231.

Law, C. M. and **Warnes, A.M.** (1980) The characteristics of retired migrants, pp.175–222 in Herbert, D.T. and Johnston, R.J. (Eds) *Geography and the Urban Environment, Vol. 3*, John Wiley, Chichester.

Lee, S. K. (1977) On the asymptotic variances of û terms in loglinear models of multidimensional contingency tables, *Journal of the American Statistical Association*, **72**, 412–9.

Lee, S. K. (1978) An example for teaching some basic concepts in multi-dimensional contingency table analysis, *The American Statistician*, **32**, 69–70.

Lehnen, R. G. and **Koch, G. G.** (1974a) A general linear approach to the analysis of nonmetric data: applications for political science, *American Journal of Political Science*, **18**, 283–313.

Lehnen, R. G. and **Koch G. G.** (1974b) The analysis of categorical data from repeated measurement research designs, *Political Methodology*, **1**, 103–23.

Lerman, S. R. and **Manski, C. F.** (1979) Sample design for discrete choice: state of the art, *Transportation Research*, **13A**, 29–44.

Lewis, P. (1977) *Maps and Statistics*, Methuen, London.

Li, M. M. (1977) A logit model of home ownership, *Econometrica*, **45**, 1081–97.

Longley, P. A. and **Wrigley, N.** (1984) Scaling residential preferences: a methodological note, *Tijdschrift voor Economische en Sociale Geographie*, **75**, 292–99.

Louviere, J. J. (1981a) On the identification of the functional form of the utility expression and its relationship to discrete choice, Appendix B, pp.385–415 in Hensher, D.A. and Johnson, L.W. (Eds) *Applied Discrete Choice Modelling*, Croom Helm, London.

Louviere, J. J. (1981b) A conceptual and analytical framework for understanding spatial and travel choices, *Economic Geography*, **57**, 304–14.

Luce, R. D. (1959) *Individual Choice Behaviour: A Theoretical Analysis*, John Wiley, New York.

Luce, R. D. and **Suppes, P.** (1965) Preference, utility and subjective probability. In Luce, R.D., Bush, R. and Galanter, E. C. (Eds) *Handbook of Mathematical Psychology Vol.3*, John Wiley, New York.

McCarthy, P. S. (1980) A study of the importance of generalized attributes in shopping choice behaviour, *Environment and Planning A*, **12**, 1269–86.

McCarthy, P. S. (1982) Further evidence of the temporal stability of disaggregate travel demand models, *Transportation Research B*, **16B**, 263–78.

McCullagh, P. (1980) Regression models for ordinal data (with discussion), *Journal of the Royal Statisticial Society*, Series B, **42**, 109–42.

McCullagh, P. and **Nelder, J. A.** (1983) *Generalized Linear Models*, Chapman and Hall, London.

McFadden, D. (1968) The revealed preferences of a government bureaucracy, University of California, Berkeley: *Institute of International Studies, Technical Report*, W-17.

McFadden, D. (1974) Conditional logit analysis of qualitative choice behaviour, pp.105–42 in Zarembka, P. (Ed.) *Frontiers in Econometrics*, Academic Press, New York.

McFadden, D. (1979) Quantitative methods for analysing travel behaviour of individuals: some recent developments, pp.279–318 in Hensher, D.A. and Stopher, P.R. (Eds) *Behavioural Travel Modelling*, Croom Helm, London.

McFadden, D. (1981) Econometric models of probabilistic choice, pp. 198–272 in Manski, C.F. and McFadden, D. (Eds) *Structural Analysis of Discrete Data with Econometric Applications*, MIT Press, Cambridge, Massachusetts.

McFadden, D., Train, K. and **Tye, W.** (1976) Diagnostic tests for the independence from irrelevant alternatives property of the multinomial logit model, Working Paper No. 7616, Urban Travel Demand Forecasting Project, University of California, Berkeley.

McFadden, D., Train, K. and **Tye, W.** (1977) An application of diagnostic tests for the independence from irrelevant alternatives property of the multinomial logit model, *Transportation Research Record*, **637**, 39–45.

Macdonald, K. I. (1977) Path analysis, pp.81–104 in O'Muircheartaigh, C.A. and Payne, C. (Eds) *The Analysis of Survey Data, Vol.2 Model Fitting*, John Wiley, London.

Magidson, J., Swan, J. H. and Berk, R. A. (1981) Estimating non-hierarchical and nested log-linear models, *Sociological Methods and Research*, **10**, 3–49.

Manski, C. F. (1974) The conditional/polytomous logit program: instructions for use. Working Paper, Carnegie-Mellon University.

Manski, C. F. (1981) Structural models for discrete data: the analysis of discrete choice, pp. 58–109 in Leinhardt, S. (Ed.) *Sociological Methodology 1981*, Jossey-Bass, San Francisco.

Manski, C. F. and Lerman, S. R. (1977) The estimation of choice probabilities from choice-based samples, *Econometrica*, **45**, 1977–88.

Manski, C. F. and McFadden, D. (Eds) (1981a) *Structural Analysis of Discrete Data with Econometric Applications*, MIT Press, Cambridge, Massachusetts.

Manski, C. F. and McFadden, D. (1981b) Alternative estimators and sample design for discrete choice analysis, pp.2–50 in Manski, C. F. and McFadden, D. (Eds) *Structural Analysis of Discrete Data with Econometric Applications*, MIT Press, Cambridge, Massachusetts.

Mantel, N. (1966) Models for complex contingency tables and polychotomous dosage response curves, *Biometrics*, **22**, 83–95.

Mantel, N. (1967) The detection of disease clustering and a generalized regression approach, *Cancer Research*, **27**, 209–20.

Mantel, N. (1979) Multidimensional contingency table analysis (Letter to the Editor). *The American Statistician*, **33**, 93.

Mantel, N. and Brown, C. (1973) A logistic re-analysis of Ashford and Sowden's data on respiratory symptoms in British coal miners, *Biometrics*, **29**, 649–65.

Mantel, N. and Byar, D.P. (1978) Marginal homogeneity, symmetry and independence, *Communications in Statistics: Part A - Theory and Methods*, **7**, 953–76.

Mantel, N. and Haenszel, W. (1959) Statistical aspects of the analysis of data from retrospective studies of disease, *Journal of the National Cancer Institute*, **22**, 719–48.

Mather, P. M. (1976) *Computational Methods of Multivariate Analysis in Physical Geography*, John Wiley, Chichester.

Matthews, J. A. (1981) *Quantitative and Statistical Approaches to Geography: A Practical Manual*, Pergamon, Oxford.

Meyer, R. J. and Eagle, T. C. (1981) A parsimonious multinomial choice model recognising alternative interdependence and context-dependent utility functions, Carnegie-Mellon University, Graduate School of Industrial Administration, Working Paper 26–80–81.

Meyer, R.J. and Eagle, T. C. (1982) Context-induced parameter instability in a disaggregate stochastic model of store choice, *Journal of Marketing Research*, **19**, 62–71.

Miller, E. J. and Lerman, S. R. (1981) Disaggregate modelling and decisions of retail firms: a case study of clothing retailers, *Environment and Planning A*, **13**, 729–46.

Mosteller, F. (1968) Association and estimation in contingency tables, *Journal of the American Statistical Association*, **63**, 1–28.

Nelder, J. A. (1974) Log-linear models for contingency tables: a generalization of classical least squares, *Applied Statistics*, **23**, 323–9.

Nelder, J. A. (1975) *GLIM Manual: Release 2*, Numerical Algorithms Group, Oxford.

Nelder, J. A. (1976) Hypothesis testing in linear models (Letter to the Editor) *The American Statistician*, **30**, 101.

Nelder, J. A. (1984) Statistical models for qualitative data, Chap. 2 in Nijkamp,

P., Leitner, H.and Wrigley, N. (Eds) *Measuring the Unmeasurable: Analysis of Qualitative Spatial Data,* Martinus Nijhoff, The Hague.

Nelder, J. A. and **Wedderburn, R. W. M.** (1972) Generalized linear models, *Journal of the Royal Statistical Society,* Series A, **135**, 370–84.

Nerlove, M. and **Press, S. J.** (1973) *Univariate and Multivariate Log-Linear and Logistic Models,* RAND Corporation Report R-1306-EDA/NIH, Santa Monica, California.

Nijkamp, P. (1982) Soft econometric models: an analysis of regional income determinants, *Regional Studies,* **16**, 121–8.

Nijkamp, P., Leitner, H. and **Wrigley, N.** (Eds) (1984) *Measuring the Unmeasurable: Analysis of Qualitative Spatial Data,* Martinus Nijhoff, The Hague.

Norcliffe, G. B. (1977) *Inferential Statistics for Geographers,* Hutchinson, London.

O'Brien, L.G. (1982) Categorical data analysis for geographical research: with applications to public sector residential mobility, Dept. of Geography, University of Bristol, Ph.D. dissertation.

O'Brien, L. G. (1983) Generalized linear modelling using the GLIM system, *Area,* **15**, 327–36.

O'Brien, L. G. and **Wrigley, N.** (1980) Computer programs for the analysis of categorical data, *Area,* **12**, 263–8.

O'Brien, L. G. and **Wrigley, N.** (1984) A generalized linear models approach to categorical data analysis: theory and applications in geography and regional science, pp.231–52 in Bahrenberg, G., Fischer, M. and Nijkamp, P. (Eds) *Recent Developments in Spatial Analysis: Methodology, Measurement, Models,* Gower, Aldershot.

Odland, J. and **Balzer, B.** (1979) Localized externalities, contagious processes and the deterioration in urban housing: an empirical analysis, *Socio-Economic Planning Sciences,* **13**, 87–93.

Odland, J. and **Barff, R.** (1982) A statistical model for the development of spatial patterns: applications to the spread of housing deterioration, *Geographical Analysis,* **14**, 327–39.

Olsen, R. J. (1978) Comment on 'The effect of unions on earnings and earnings on unions: a mixed logit approach', *International Economic Review,* **19**, 259–61.

Orum, A. and **McCranie, E. W.** (1970) Class, tradition and partisan alignments in a southern urban electorate, *Journal of Politics,* **32**, 156–76.

Ostrowski, A. M. (1966) *Solutions of Equations and Systems of Equations,* Academic Press, New York.

Parks, R. W. (1980) On the estimation of multinomial logit models from relative frequency data, *Journal of Econometrics,* **13**, 293–303.

Payne, C. (1977) The log-linear model for contingency tables, pp.105–44 in C.A. O'Muircheartaigh and C.Payne (Eds) *The Analysis of Survey Data: Vol.2, Model Fitting,* John Wiley, London.

Pearson, K. (1900) On a criterion that a given system of deviations from the probable in the case of a correlated system of variables is such that it can be reasonably supposed to have arisen from random sampling, *Philosophical Magazine,* **50**, 157–175.

Pearson, K. (1904) On the theory of contingency and its relation to association and normal correlation. *Drapers' Co. Research Memoirs, Biometric Series No. 1,* London.

Pindyck, R. S. and **Rubinfeld, D. L.** (1976) *Econometric Models and Economic Forecasts,* McGraw-Hill, New York.

Pipkin, J. S. (1981) Cognitive behavioural geography and repetitive travel, pp.145–181 in Cox, K.R. and Golledge, R. G. (Eds) *Behavioural Problems in Geography Revisited,* Methuen, London.

Plackett, R. L. (1974) *The Analysis of Categorical Data*, Griffin, London.

Pregibon, D. (1980) Discussion of P. McCullagh's paper, *Journal of the Royal Statistical Society*, Series B, **42**, 138–9.

Pregibon, D. (1981) Logistic regression diagnostics, *The Annals of Statistics*, **9**, 705–24.

Pregibon, D. (1982) Resistant fits of some commonly used logistic models with medical applications, *Biometrics*, **38**, 485–98.

Pringle, D. G. (1980) *Causal Modelling: The Simon-Blalock Approach*, Concepts and Techniques in Modern Geography 27, Geo Abstracts, Norwich.

Quigley, J. M. (1976) Housing demand in the short run – an analysis of polytomous choice, *Explorations in Economic Research*, **3**, 76–102.

Rao, P. V. and **Kupper, L. L.** (1967) Ties in paired-comparison experiments: a generalization of the Bradley-Terry model, *Journal of the American Statistical Association*, **62**, 194–204.

Reynolds, H. T. (1977) *The Analysis of Cross-Classifications*, Free Press, New York.

Richards, M. G. and **Ben-Akiva, M. E.** (1975) *A Disaggregate Travel Demand Model*, Saxon House, Farnborough.

Robson, B. T. (1975) *Urban Social Areas*, Oxford University Press, Oxford.

Rogers, A. (1971) *Matrix Methods in Urban and Regional Analysis*, Holden-Day, San Francisco.

Rushton, G. (1982) Review of Gold, J. R. *An Introduction to Behavioural Geography*, *Environment and Planning A*, **14**, 275–6.

Schmidt, P. (1978) Estimation of a simultaneous equations model with jointly dependent continuous and qualitative variables: the union-earnings question revisited, *International Economic Review*, **19**, 453–65.

Schmidt, P. and **Strauss, R. P.** (1975a) The prediction of occupation using multiple logit models, *International Economic Review*, **16**, 471–86.

Schmidt, P. and **Strauss, R. P.** (1975b) Estimation of models with jointly dependent qualitative variables: a simultaneous logit approach, *Econometrica*, **43**, 745–55.

Schmidt, P. and **Strauss, R. P.** (1976) The effect of unions on earnings and earnings on unions: a mixed logit approach, *International Economic Review*, **17**, 204–12.

Sheffi, Y., Hall, R. and **Daganzo, C. F.** (1982) On the estimation of the multinomial probit model, *Transportation Research A*, **16A**, 447–56.

Silk, J. (1979) *Statistical Concepts in Geography*, George Allen and Unwin, London.

Simon, G. (1974) Alternative analyses for the singly-ordered contingency table, *Journal of the American Statistical Association*, **69**, 971–6.

Simpson, E. H. (1951) The interpretation of interaction in contingency tables, *Journal of the Royal Statistical Society*, Series B, **13**, 238–41.

Smith, C. J. (1980) Neighbourhood effects on mental health, pp. 363–415 in Herbert, D. T. and Johnston, R. J. (Eds) *Geography and the Urban Environment Vol. 3*, John Wiley, Chichester.

Smith, T. R. and **Slater, P. B.** (1981) A family of spatial interaction models incorporating information flows and choice set constraints applied to U. S. interstate labour flows, *International Regional Science Review*, **6**, 15–31.

Sobel, K. (1980) Travel demand forecasting with the nested multinomial logit model, *Transportation Research Record*, **775**, 48–55.

Sokal, R. R. and **Rohlf, F. J.** (1969) *Biometry: The Principles and Practice of Statistics in Biological Research*, W.H. Freeman, San Francisco.

Southworth, F. (1981) Calibration of multinomial logit models of mode and destination choice, *Transportation Research A*, **15A**, 315–25.

Sparman, J. and **Daganzo, C. F.** (1980) *TROMP: User's Manual*, ITS Research Report UCB-ITS-RR-80-11, Institute of Transportation Studies, University of California, Berkeley.

Stone, R. (1954) Linear expenditure systems and demand analysis: an application to the pattern of British demand, *Economic Journal*, **64**, 511–27.

Stopher, P. R. and **Meyburg, A. H.** (1979) *Survey Sampling and Multivariate Analysis for Social Scientists and Engineers*, D. C. Heath, Lexington, Massachusetts.

Stouffer, S. A., Suchmann, E. A., De Vinney, L.C., Star, S. A. and **Williams, R. M.** (1949) *The American Soldier: Adjustment During Army Life. Studies in Social Psychology in World War II, Volume 1*. Princeton University Press, Princeton, New Jersey.

Talvitie, A. and **Kirshner, D.** (1978) Specification, transferability and the effect of data outliers in modelling the choice of mode in urban travel, *Transportation*, **7**, 311–31.

Tardiff, T. J. (1976) A note on goodness-of-fit statistics for probit and logit models, *Transportation*, **5**, 377–88.

Tardiff, T. J. (1979) Specification analysis for quantal choice models, *Transportation Science*, **13**, 179–90.

Tardiff, T. J. (1980) Definition of alternatives and representation of dynamic behaviour in spatial choice models, *Transportation Research Record*, **723**, 25–30.

Taylor, P. J. (1977) *Quantitative Methods in Geography*, Houghton Mifflin, Boston.

Theil, H. (1969) A multinomial extension of the linear logit model. *International Economic Review*, **10**, 251–9.

Theil, H. (1970) On the estimation of relationships involving qualitative variables, *American Journal of Sociology*, **76**, 103–54.

Theil, H. (1971) *Principles of Econometrics*, John Wiley, New York.

Thompson, R. and **Baker, R. J.** (1981) Composite link functions in generalized linear models, *Applied Statistics*, **30**, 125–31.

Thrift, J. (1981) Behavioural geography, pp. 352–365 in Wrigley, N. and Bennett, R. J. (Eds) *Quantitative Geography: A British View*, Routledge and Kegan Paul, London.

Till, R. (1974) *Statistical Methods for the Earth Scientist*, Macmillan, London.

Timmermans, H. J. P. (1984) Decision models for predicting preferences among multiattribute choice alternatives, pp.337–54 in Bahrenberg, G., Fischer, M. M. and Nijkamp, P. (Eds) *Recent Developments in Spatial Data Analysis: Methodology, Measurement, Models*, Gower, Aldershot.

Tversky, A. (1972a) Elimination-by-aspects: a theory of choice, *Psychological Review*, **79**, 281–99.

Tversky, A. (1972b) Choice-by-elimination, *Journal of Mathematical Psychology*, **9**, 341–67.

Tversky, A. and **Sattath, S.** (1979) Preference trees, *Psychological Review*, **86**, 542–73.

Unwin, D. J. (1975) *An Introduction to Trend Surface Analysis*, Concepts and Techniques in Modern Geography 5, Geo Abstracts, Norwich.

Upton, G. J. G. (1977) A memory model for voting transitions in British elections, *Journal of the Royal Statistical Society*, Series A, **140**, 86–94.

Upton, G. J. G. (1978a) *The Analysis of Cross-Tabulated Data*, John Wiley, Chichester.

Upton, G. J. G. (1978b) Factors and responses in multidimensional contingency tables, *The Statistician*, **27**, 43–8.

Upton, G. J. (1980) More basic concepts in multidimensional contingency table analysis, Mimeo, Dept. of Mathematics, University of Essex.

Upton, G. J. G. (1981) Log-linear models, screening, and regional industrial surveys, *Regional Studies*, **15**, 33–45.

Upton, G. J. G. and **Fingleton, B.** (1979) Log-linear models in geography, *Transactions of the Institute of British Geographers*, New Series 4, 103–15.

Upton, G. J. G. and **Sarlvik, B.** (1981) A loyalty-distance model for voting change, *Journal of the Royal Statistical Society, Series A*, **144**, 247–59.

Veblen, T. T. and **Stewart, G. H.** (1982) On the conifer regeneration gap in New Zealand: the dynamics of *Libocedrus Bidwillii* stands on South Island, *Journal of Ecology*, **70**, 413–36.

Wald, A. (1943) Tests of statistical hypotheses concerning several parameters when the number of observations is large, *Transactions American Mathematical Society*, **54**, 426–82.

Westin, R. B. (1974) Predictions from binary choice models, *Journal of Econometrics*, **2**, 1–16.

Westin, R. B. and **Gillen, D. W.** (1978) Parking location and transit demand – a case study of endogenous attributes in disaggregate mode choice models, *Journal of Econometrics*, **8**, 75–101.

Whittaker, J and **Aitkin, M.** (1978) A flexible strategy for fitting complex log-linear contingency table models, *Biometrics*, **34**, 487–95.

Whittemore, A. S. (1978) Collapsibility of multi-dimensional contingency tables, *Journal of the Royal Statistical Society, Series B*, **40**, 328–40.

Wilkinson, G. N. and **Rogers, C. E.** (1973) Symbolic description of factorial models for analysis of variance, *Applied Statistics*, **22**, 392–9.

Williams, K. (1976) Analysis of multidimensional contingency tables, *The Statistician*, **25**, 51–5.

Williams, H. W. C. L. (1977) On the formation of travel demand models and economic evaluation measures of user benefit. *Environment and Planning A*, **9**, 285–344.

Williams, H. W. C. L. (1981) Random utility theory and probabilistic choice models, pp. 46–84 in Wilson, A. G, Coelho, J. D., Macgill, S. M. and Williams, H. W.C. L. *Optimization in Locational and Transport Analysis,* John Wiley, Chichester.

Williams, H. W. C. L. and **Ortuzar, J. D.** (1982) Behavioural theories of dispersion and the mis-specification of travel demand models, *Transportation Research B*, **16B**, 167–219.

Williams, O. D. and **Grizzle, J. E.** (1972) Analysis of contingency tables having ordered response categories, *Journal of the American Statistical Association*, **67**, 55–63.

Wilson, A. G. and **Kirkby, M.** (1980) *Mathematics for Geographers and Planners*, Second Edition, Oxford University Press, Oxford.

Wonnacott, R. J. and **Wonnacott, T. H.** (1979) *Econometrics,* Second Edition, John Wiley, New York.

Wrigley, N. (1975) Analysing multiple alternative dependent variables, *Geographical Analysis*, **7**, 187–95.

Wrigley, N. (1976) *An Introduction to the Use of Logit Models in Geography,* Concepts and Techniques in Modern Geography 10, Norwich, Geo Abstracts.

Wrigley, N. (1977a) Probability surface mapping: a new approach to trend surface mapping, *Transactions of the Institute of British Geographers,* New Series 2, 129–40.

Wrigley, N. (1977b) *Probability Surface Mapping. An Introduction with Examples and Fortran Programs*, Concepts and Techniques in Modern Geography 16, Geo Abstracts, Norwich.

Wrigley, N. (1979a) Developments in the statistical analysis of categorical data, *Progress in Human Geography*, **3**, 315–55.

Wrigley, N. (Ed.) (1979b) *Statistical Applications in the Spatial Sciences*, Pion, London.

Wrigley, N. (1980a) Paired-comparison experiments and logit models: a review and illustration of some recent developments, *Environment and Planning A*, **12**, 21–40.

Wrigley, N. (1980b) Categorical data, repeated measurement research designs, and regional industrial surveys, *Regional Studies*, **14**, 455–71.

Wrigley, N. (1980c) Sons of the green giant, *Progress in Human Geography*, **4**, 611–16.

Wrigley, N. (1981a) Categorical data analysis, pp.111–22 in Wrigley, N. and Bennett, R. J. (Eds) *Quantitative Geography: A British View*, Routledge and Kegan Paul, London.

Wrigley, N. (1981b) Quantitative methods: a view on the wider scene, *Progress in Human Geography*, **5**, 548–61.

Wrigley, N. (1982) Quantitative methods: developments in discrete choice modelling, *Progress in Human Geography*, **6**, 547–62.

Wrigley, N. (1984) Quantitative methods: diagnostics revisited, *Progress in Human Geography*, **8**, 525–35.

Wrigley, N. and **Bennett R. J.** (Eds) (1981) *Quantitative Geography: A British View*, Routledge and Kegan Paul, London.

Wrigley, N. and **Dunn, R.** (1984) Diagnostics and resistant fits in logit choice models, pp.44–66 in Pitfield, D. E. (Ed.) *London Papers in Regional Science, Vol. 14, Discrete Choice Models in Regional Science*, Pion, London.

Wrigley, N. and **Longley, P. A.** (1984) Discrete choice modelling in urban analysis, pp. 45–94 in Herbert, D. T. and Johnston, R. J. (Eds.) *Geography and the Urban Environment, Vol. 6*, John Wiley, Chichester.

Yeates, M. (1974) *An Introduction to Quantitative Analysis in Human Geography*, McGraw-Hill, New York.

Author index

Example index

Subject index